Peninsula Watershed Historical Ecology Study

PREPARED FOR
THE SAN FRANCISCO PUBLIC UTILITIES COMMISSION

PREPARED BY
SAN FRANCISCO ESTUARY INSTITUTE

Authors

Sean Baumgarten

Steven Hagerty

Emily Clark

Robin Grossinger

Erica Spotswood

Erin Beller

Micha Salomon

Lauren Stoneburner

Lydia Vaughn

Design, Production, and Cartography

Ruth Askevold

Lauren Stoneburner

Ellen Plane

Amy Richey

FUNDED BY

San Francisco Public Utilities Commission

AUGUST 2021

SAN FRANCISCO ESTUARY INSTITUTE PUBLICATION #1029

SUGGESTED CITATION

Baumgarten S, Hagerty S, Clark E, Grossinger R, Spotswood E, Beller E, Salomon M, Stoneburner L, Vaughn L. *Peninsula Watershed Historical Ecology Study.* 2021. Prepared for the San Francisco Public Utilities Commission. A Report of SFEI-ASC's Resilient Landscapes Program, Publication #1029, San Francisco Estuary Institute-Aquatic Science Center, Richmond, CA.

Version 1.1, August 2021.

REPORT AVAILABILITY

Report and GIS layers are available on the SFEI-ASC website at https://www.sfei.org/projects/peninsula-watershed-historical-ecology-study.

COVER CREDITS

(top) Photographs courtesy SFPUC; (bottom) Detail from 1868 San Mateo County map. *(Easton 1868, courtesy San Francisco History Center, San Francisco Pubic Library)*

CONTENTS

Conifer Forest. *(Photo courtesy SFPUC)*

Acknowledgments

The Peninsula Watershed Historical Ecology Study was funded by the San Francisco Public Utilities Commission (SFPUC). We are grateful to the many current and former SFPUC staff who contributed to the project through scientific support, data transfer, technical review, field tours, and interviews, including Tim Ramirez, Carla Schultheis, Ellen Natesan, Jessica Appel, Scott Simono, Mia Ingolia, John Fournet, Jeremy Lukins, Carin Apperson, Aaron Brinkerhoff, Jim Avant, Josh Milstein, and others.

We thank the technical advisory committee (TAC) and additional advisors for their thoughtful guidance and technical review throughout the course of the project. TAC members included James Bartolome (UC Berkeley), Tom Parker (San Francisco State University), Alison Forrestel (Golden Gate National Recreation Area), Peter Alagona (UC Santa Barbara), Toni Corelli (botanist), David Freyberg (Stanford University), and David Ackerly (UC Berkeley). Heath Bartosh (Nomad Ecology) assisted with interpretation of historical botanical records and analysis of serpentine grasslands, and provided technical review of draft text. M.V. Eitzel provided advice regarding statistical analyses and data visualization, and reviewed draft text regarding quantitative analyses of landscape change over time. Jonathan Cordero (Association of Ramaytush Ohlone) reviewed draft text regarding indigenous presence, native land management, and fire history.

SFEI staff members Samuel Shaw and Gloria Desanker assisted with report development. Additional assistance with data collection, literature review, rephotography, and data analysis was provided by two interns from the Bill Lane Center for the American West at Stanford University, Nicholas Mascarello and Miranda Vogt. Two temporary GIS Technicians, Eddie Fitzsimmons (San Francisco State University) and Melissa Jaffe (UC Berkeley), assisted with vegetation/land cover classification in the aerial imagery point analysis (page 27). Thank you to Michael Svanevik for helpful advice regarding historical sources.

We are indebted to all of the staff and volunteers at archives and institutions that we visited or contacted during the course of the project (page 21). In particular we would like to thank Mike Housh, Leslie Fisher, Robin Scheswohl, Jonathan Chow, and William Xie at SFPUC; Carol Peterson at the San Mateo County History Museum; Richard Vallejos, Dave Holbrook, and Joe LaClair at the San Mateo County Planning and Building Department; Liliya Latman and Julie Casagrande at San Mateo County Public Works; Susan Goldstein and Thomas Carey at the San Francisco Public Library; and Paul Vaughn at the San Mateo Public Library.

Aerial photo of Upper and Lower Crystal Springs reservoirs, April 1960.
(Photo by Marshall Moxom, X-5552_4-11-1960, courtesy SFPUC)

CHAPTER 1

Introduction
and Summary of Findings

Background and Study Goals

Nestled in the rugged coastal mountains between San Francisco and Silicon Valley lies one of the ecological treasures of the San Francisco Bay Area: the Peninsula Watershed. Home to mountain lions, marbled murrelets, towering old-growth Douglas-firs, and an immense diversity of other plants and animals, the Peninsula Watershed is a unique and wild expanse of open space just minutes from one of the most urbanized parts of California.

The Peninsula Watershed has been integral to the story of San Francisco's growth ever since the Gold Rush. The rapid influx of settlers to San Francisco during the Gold Rush spurred a sudden demand for a reliable water source, which led to the formation of the Spring Valley Water Works (later purchased by the Spring Valley Water Company [SVWC]) in 1858 (Hanson 2005). Over the subsequent 70 years, SVWC bought up large swaths of land on the Peninsula, and constructed a complex system of dams, tunnels, and pipes to capture and transport water to San Francisco. Within the Peninsula Watershed, this system includes the Crystal Springs and San Andreas reservoirs, located in the San Andreas Creek, Laguna Creek, and Upper San Mateo Creek basins along the San Andreas Fault; and Pilarcitos Reservoir the west, located in the Pilarcitos Creek Basin along the Pilarcitos Fault.

The City of San Francisco purchased SVWC in 1930, and today the Peninsula Watershed, managed by the San Francisco Public Utilities Commission (SFPUC), continues to be a key source of water for San Francisco and for other communities in the South and East Bay. Despite the past 150 years of reservoir construction and other hydrologic modifications, the construction of transportation and utility corridors, and the large-scale suburban development that has occurred to the east, the Peninsula Watershed has remained largely undeveloped and is managed to protect water quality, water supply, wildlife habitat, and a range of other natural and cultural resources. The watershed supports some of the largest intact remnants of contiguous habitat in the region, including extensive oak woodlands, old-growth Douglas-fir forests, serpentine grasslands, chaparral, and coastal scrub.

Over the past 250 years since Spanish explorers first set foot on the watershed, however, changes in disturbance regimes and other large-scale anthropogenic modifications, including fire suppression, homesteading, livestock grazing, agriculture, tree planting, introduction of plant pathogens, spread of invasive species, and climate change, have altered vegetation dynamics and changed the distribution and structure of vegetation communities throughout the watershed. The changes have raised many questions about the historical ecology of the watershed: What was the extent, distribution, and composition of terrestrial, riparian, and wetland habitats prior to Euro-American modification? How have vegetation distributions changed over the past two centuries, and what are the implications of those changes for species support? Are there remnant patches of relatively unmodified habitat present in the watershed, or areas that are currently in a state of recovery? Where are current habitat characteristics most similar to or different from historically documented conditions? How have key natural and anthropogenic disturbance regimes and processes changed over time?

The *Peninsula Watershed Historical Ecology Study* aims to advance understanding of landscape conditions of the Peninsula Watershed prior to major Euro-American modification, and to provide insights into the nature and drivers of vegetation change since the first Spanish explorers set foot in the watershed 250 years ago. The primary goal of the research was to examine the historical extent, distribution, and composition of terrestrial vegetation types and their trajectories of change within the watershed. To the extent possible, research also addressed historical riparian, wetland, and estuarine habitats; hydrology and sediment dynamics; wildlife support; land use history; and a range of other topics. Findings from the study will inform a variety of watershed management activities and will support SFPUC and other land managers in the region in identifying appropriate restoration targets and priorities.

Serpentine grassland flowers in the Peninsula Watershed. *(Photo courtesy SFPUC)*

Historical Ecology

Historical ecology is an interdisciplinary field of research that focuses on examining and reconstructing past landscape conditions and processes. Researchers draw on a range of data sources and analytical approaches, including archival documents (e.g., maps, photographs, and textual documents), ethnographic data, physical and paleoecological evidence (e.g., stratigraphy, pollen cores, tree rings), and statistical models (e.g., species distribution models) to reveal and synthesize information about the historical landscape (Egan and Howell 2005). This research can enhance understanding of historical vegetation distribution and composition, wildlife presence and abundance, hydrologic and geomorphic patterns and processes, disturbance regimes, and a range of other topics.

Research into past landscape patterns and processes provides baseline information that can help contextualize current conditions and inform a range of contemporary management and restoration activities. An understanding of past habitat configuration, species composition, and ecological function can help in quantifying ecosystem change, identifying appropriate restoration targets, and recognizing remnant habitats (Jackson and Hobbs 2009, Suding et al. 2015). Many of the underlying physical controls and processes (e.g., topography, geology) that determined past landscape conditions still operate today, and thus understanding the role of these drivers in the past can reveal their continuing influence and help in identifying contemporary restoration opportunities, strategies, and constraints (Hayward et al. 2012). Understanding how landscapes responded to past environmental variability and disturbance, including human activities, can help in anticipating how they will respond to climate change and other environmental changes in the future (Swetnam et al. 1999).

While historical ecology can provide critical guidance for land management and restoration, it is not a prescriptive tool or restoration panacea. Contemporary ecosystems face many novel constraints and stressors, and the idea of "returning" to what once was is often impractical, particularly in the face of climate change. However, maintaining or re-integrating key ecological functions provided by historical landscapes is, in many cases, not only practical but critical to building future ecological resilience. Those characteristics that enabled ecosystems to persist and respond to dynamic conditions in the past—for example, native vegetation communities adapted to the local setting, intact physical processes and disturbance regimes, connectivity between habitat patches, and a diversity of habitat and landscape features—will be the same characteristics that will confer ecological resilience in the future (Safford et al. 2012b, Beller et al. 2015).

Environmental Setting

The Peninsula Watershed is located on the San Francisco Peninsula, at the northern end of the Santa Cruz Mountains, approximately six miles south of San Francisco, in San Mateo County. The term "Peninsula Watershed" actually refers to an administrative entity, managed by SFPUC, which includes portions of several hydrologic watersheds: 17,140 ac in the San Mateo Creek watershed (which drains eastward to San Francisco Bay), 4,590 ac in Pilarcitos Creek watershed (which drains westward to the Pacific Ocean), and small portions of several surrounding watersheds. The study area for this project includes the entire Peninsula Watershed, as well as the additional downstream portions of the San Mateo Creek watershed (the Lower San Mateo Creek Basin), comprising a total of 25,950 ac (Fig. 1.1). Eighty-eight percent of the study area is owned/managed by SFPUC. Seventy-seven percent of the study area falls within unincorporated San Mateo County, while smaller portions of the study area fall within cities such as Hillsborough, San Mateo, and Pacifica.

The San Francisco Peninsula falls within the Mediterranean climate zone, with cool, dry summers; cool, wet winters; and summer fog along the coast (Kauffman 2003). Annual precipitation within the study area has averaged 32 in (mean for 1981-2010), ranging from 21 in near the mouth of San Mateo Creek to 39 in in the higher elevation portions of the watershed (Fig. 1.2). More than 90% of precipitation falls between the months of October and April (USDA 2012). Fog drip also supplies a significant amount of moisture to certain parts of the watershed (Oberlander 1956), with the mean daily duration of summertime fog and low cloud cover generally increasing to the north (Torregrosa et al. 2016; Fig. 1.3). The average annual maximum temperature within the study area was 68 °F for the years 1981-2010, while the average annual minimum temperature over the same period was 47 °F (USDA 2012).

The San Andreas Valley runs northwest-southeast through the eastern portion of the watershed, and is occupied by three reservoirs: San Andreas, Lower Crystal Springs, and Upper Crystal Springs.[1] The valley is a rift zone formed by the San Andreas Fault, a major right lateral strike-slip fault that extends for 750 miles along the boundary of the Pacific and North American plates (Brabb et al. 1998). On the San Francisco Peninsula, the average late Holocene slip rate along the San Andreas Fault is estimated at approximately 17 mm/yr (Hall et al. 1999). In addition to the 1906 San Francisco earthquake (magnitude 7.9), evidence suggests that large earthquakes occurred along this section of the San Andreas Fault in 1865, 1838, and in the late 1500s to mid-1600s. During the 1906 San Francisco earthquake, displacement along the fault within the Peninsula Watershed was between 2.4-2.7 m (Fig. 1.4); no measurable displacement has occurred since then (Hall et al. 1999). Another inactive strike-slip fault, the Pilarcitos Fault, is located west of the San Andreas Fault (Hall et al. 1999).

---

1   The southern portion of San Andreas Valley is referred to in this report as the Laguna Creek Basin, named for the creek that flows south-to-north into Upper Crystal Springs Reservoir. It is also frequently referred to as "Cañada de Raymundo" in historical sources, a name taken from the Mexican land grant of that name.

Figure 1.1. The 25,950 acre study area for this project (area enclosed by the solid white line) encompasses the SFPUC Peninsula Watershed lands as well as additional downstream portions of the San Mateo Creek watershed and the Filoli Estate (indicated by the dashed white line and hash marks). *(imagery courtesy ESRI)*

San Bruno

San Francisco Bay

Millbrae

Sweeney Ridge

San Pedro Valley

Portola Ridge

San Andreas Reservoir

San Andreas Valley

Burlingame

Fifield Ridge

Sawyer Ridge

San Andreas Creek

Buri Buri Ridge

Whiting Ridge

Spring Valley Ridge

Montara Mountain

Pilarcitos Reservoir

Upper San Mateo Creek

Lower San Mateo Creek

San Mateo

Cahill Ridge

Pilarcitos Creek (and Canyon)

Scarper Peak

Ox Hill

Stone Dam

Sherwood Point

Lower Crystal Springs Reservoir

N

0   2
Miles

Upper Crystal Springs Reservoir

Adobe Gulch

Kings Mountain Ridge

Laguna Creek Basin

Pulgas Ridge

Pise Hill

Filoli Estate

Laguna Creek

Kings Mountain

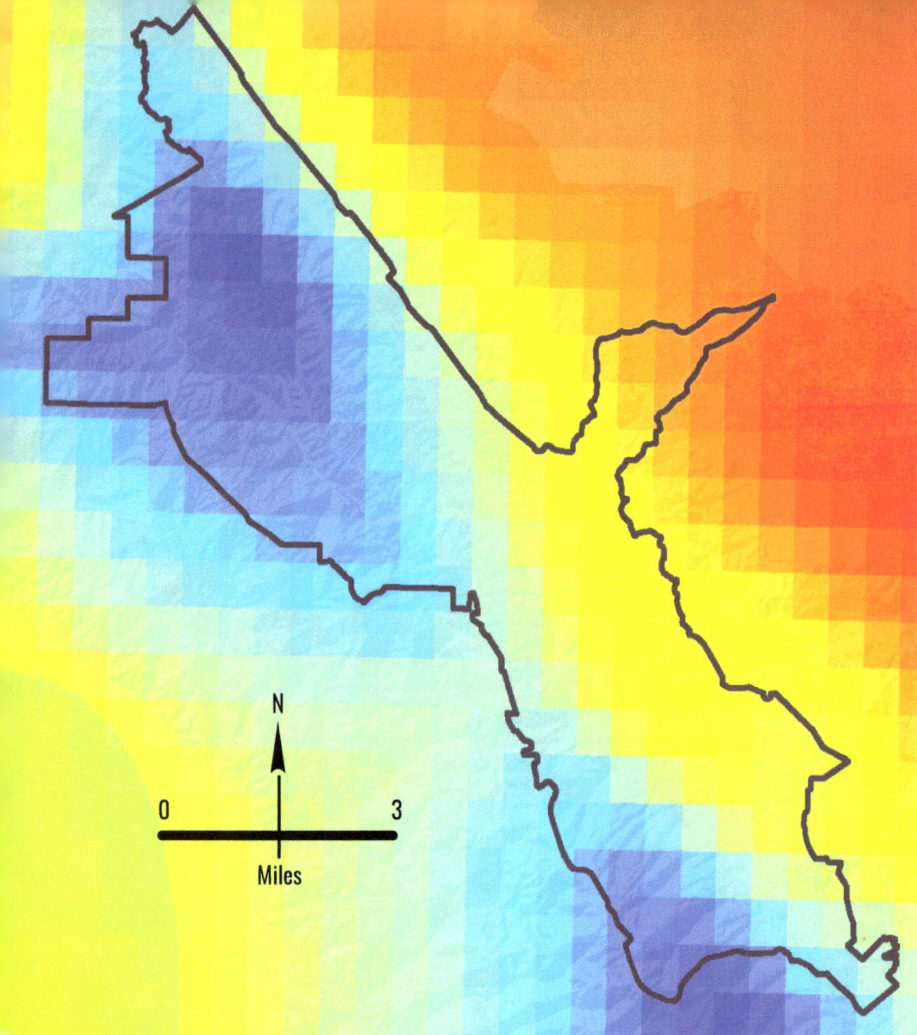

Annual Precipitation

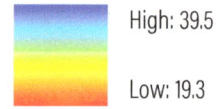

High: 39.5

Low: 19.3

Figure 1.2. Average annual precipitation (1981-2010), in/yr (USDA 2012).

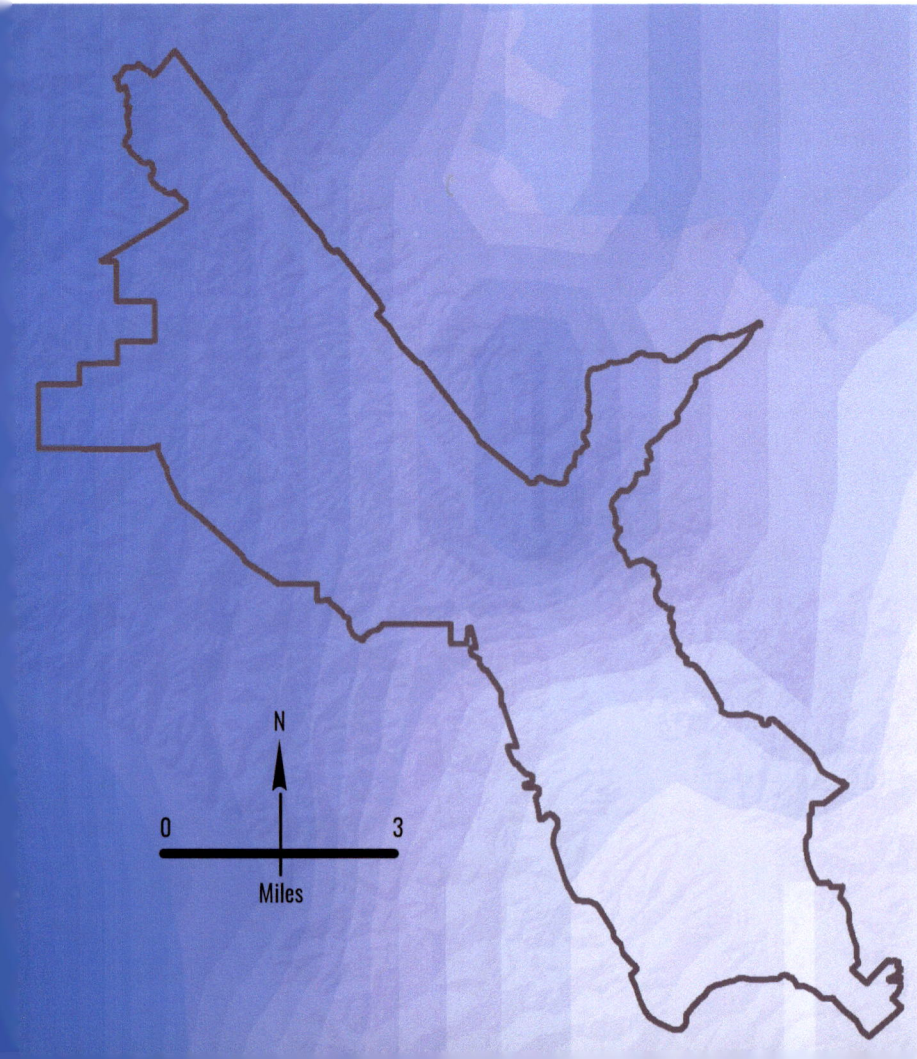

Fog Belt Zones

High: 14.5

Low: 5

Figure 1.3. Average daily summertime fog and low cloud cover (1999-2009), hrs/day (Torregrosa et al. 2016).

Figure 1.4. This photograph, taken shortly after the 1906 earthquake, shows the offset in a fence between San Andreas and Lower Crystal Springs reservoirs resulting from movement of the San Andreas Fault. *(BANC PIC 1957.007 Ser. 2 :087, courtesy The Bancroft Library, UC Berkeley)*

A series of ridges and intervening valleys, also oriented in a generally northwest-southeast direction, cross the watershed: Sweeney Ridge in the north; Portola, Sawyer, Fifield, Spring Valley, and Whiting ridges and Montara Mountain in the northwest; Buri Buri and Pulgas ridges on the east; Cahill Ridge in the center; and Kings Mountain Ridge in the south. The upper slopes of Montara Mountain are comprised of granitic rocks and decomposed granite, which extend as far east as Pilarcitos Canyon, while the lower slopes northwest of Pilarcitos Lake are made up of Paleocene sandstone, shale, and conglomerate. Ridges and valleys in the northern part of the watershed, between Montara Mountain and the San Andreas Valley, are comprised primarily of rocks in the Franciscan Complex, including sandstone interbedded with siltstone and shale and outcroppings of greenstone (basaltic rocks). The slopes of Kings Mountain Ridge, south of Highway 92, consist of sandstones and siltstones in the Eocene Whiskey Hill Formation. Large areas of Buri Buri and Pulgas ridges are made up of a mélange of Franciscan Complex greywacke, siltstone, and shale. This area also includes substantial outcroppings of serpentinite; smaller outcroppings of serpentinite are found along Fifield Ridge (Brabb et al. 1998).

Previous Research

Previous studies have summarized and revealed a large amount of information about the cultural and ecological history of the Peninsula Watershed and surrounding areas, and served as invaluable resources for this historical ecology investigation. George Oberlander's 1953 dissertation on *The Taxonomy and Ecology of the Flora of the San Francisco Watershed Reserve* examines the distribution of vegetation types throughout the watershed, summarizes information from key historical sources such as the late 18th century Spanish expeditions, and discusses changes in vegetation distribution prior to the mid-20th century (Oberlander 1953). A 2014 dissertation by John Dingman, entitled *Modeling the Impacts of Climate Change on the Douglas-fir Forest within the San Francisco Peninsula Watershed using GIS and LiDAR*, documents changes in Douglas-fir distribution over the 20th century and models future changes in the extent of Douglas-fir forest under climate change (Dingman 2014).

Several volumes provided in-depth summaries of regional and local land use history. A series of books by historian Frank Stanger, including *South from San Francisco*, *Sawmills in the Redwoods*, and *A California Rancho under Three Flags: A History of Rancho Buri Buri in San Mateo County*, were invaluable resources for understanding the land use history of the Peninsula (Stanger 1938, 1963, 1967). Marianne Babal's 1990 report entitled *The Top of the Peninsula: A History of Sweeney Ridge and the San Francisco Watershed Lands, San Mateo County, California*, summarizes key aspects of the land use history of the Peninsula Watershed and provides an inventory of cultural resources in the watershed (Babal 1990). *A History of the Municipal Water Department and Hetch Hetchy System* (Hanson 2005) summarizes the history of the Spring Valley Water Company and San Francisco's water supply system. The 2010 *Historic Resource Study for Golden Gate National Recreation Area in San Mateo County*, produced by the National Park Service and Mitch Postel (San Mateo County Historical Association), documents the history of GGNRA sites on the Peninsula, including Sweeney Ridge and the Phleger Estate adjacent to the Peninsula Watershed (Postel 2010).

Anthropological information about the indigenous inhabitants of the Peninsula is summarized in *Ohlone/Costanoan Indians of the San Francisco Peninsula and their Neighbors, Yesterday and Today*, by Milliken et al. (2009). Archaeological and historical ecological research conducted in the Quiroste Valley area (Cuthrell 2013a,b; Lightfoot et al. 2013a,b; Evett and Cuthrell 2013), other locations in the Santa Cruz Mountains (Stephens and Fry 2005, Striplen 2014), and the Monterey Bay area (Greenlee and Langenheim 1990) provide essential resources for understanding historical fire regimes, traditional burning practices, and other indigenous land management practices in the region.

Coast live oak (*Quercus agrifolia*) woodland near Upper Crystal Spring Reservoir. *(Photo courtesy SFPUC)*

Report Organization

Chapter 1 provides an overview of project goals, background about the field of historical ecology, an overview of the environmental setting of the study, and a summary of key findings. Chapter 2 summarizes the methods, data sources, and key assumptions underpinning the research. Chapter 3 discusses the major drivers—both physical and anthropogenic—of landscape change within the watershed over the past 250 years. Chapter 4 summarizes major findings from several quantitative analyses and presents a framework for synthesis of other qualitative data sources. Chapters 5-8 provide detailed discussion and evidence of historical vegetation patterns and changes in extent and distribution with respect to each of the four major terrestrial vegetation types: grasslands, shrublands, hardwood forests, and conifer forests. Chapter 9 discusses the historical ecology of wetlands, streams, and riparian habitats within the watershed. Chapter 10 discusses overall conclusions from the research, including management implications and proposed future direction directions.

Several technical appendices provide additional detail about methods, data sources, and terminology. Appendix A includes crosswalks between the simplified vegetation types used throughout the report (see page 22) and the Wieslander VTM mapping (Kelly et al. 2005) and contemporary vegetation mapping (MRLCC 2001, Schirokauer et al. 2003). Appendix B provides an index map to parcels referenced in the 1907-14 SVWC court case.

Summary of Findings

HISTORICAL LANDSCAPE PATTERNS PRIOR TO MAJOR EURO-AMERICAN MODIFICATION

The watershed supported extensive areas of grassland, shrubland, and hardwood forest

Grasslands dominated the eastern side of San Andreas Valley, and many of the ridges throughout the watershed, including portions of Sawyer, Fifield, Spring Valley, and Kings Mountain ridges. Shrublands, including a variety of chaparral and coastal scrub communities, dominated the northwestern portion of the watershed and the upper eastern slope of Kings Mountain Ridge. Grasslands and shrublands likely formed a heterogeneous mosaic, in some areas controlled in large part by patterns of indigenous use of fire as a vegetation management tool. Hardwood forest, in many areas dominated by oaks (*Quercus* spp.) and California bay (*Umbellularia californica*), occupied canyons and lower hillslopes in San Andreas Valley, the Laguna Creek Basin, Lower San Mateo Creek Canyon, and Upper San Mateo Creek Canyon.

Several groves of conifer forest existed along the western side of the watershed

Douglas-fir (*Pseudotsuga menziesii*)-dominated forests occupied hillslopes on the western side of Pilarcitos Canyon, comprising the most extensive patch of conifer forest within the watershed historically. Coast redwood (*Sequoia sempervirens*)-dominated forests occupied a small portion of the watershed along Kings Mountain Ridge.

Wetlands occupied large portions of San Andreas Valley

A range of wetland and aquatic habitats, including sag ponds, freshwater emergent wetlands, and willow thickets occupied the valley floor within San Andreas Valley. A large sag pond known as Laguna Grande was located along Laguna Creek near the northern end of present-day Upper Crystal Springs Reservoir, while a series of smaller sag ponds existed along San Andreas Creek further north. Springs were a common hydrological feature. Tidal marsh and mudflats existed at the outlet of San Mateo Creek.

Streams supported a diversity of riparian habitats

A variety of riparian habitats existed within the watershed, including mixed riparian forest along Pilarcitos Creek, willow- and oak-dominated forest and scrub along San Andreas and Laguna creeks, and oak and bay riparian forest along Lower San Mateo Creek.

LANDSCAPE CHANGES BETWEEN THE LATE 18TH AND MID-20TH CENTURIES

Agricultural development and reservoir construction in the mid- to late 19th century resulted in substantial loss of wetland, riparian forest, grassland, and hardwood forest habitats within San Andreas Valley

A number of farms and homesteads were established in San Andreas Valley and the Laguna Creek Basin during the 1850s-60s, and in many cases native vegetation was cleared for cultivation. Construction of San Andreas Dam (1868), Upper Crystal Springs Dam (1877), and Lower Crystal Springs Dam (1888) inundated large areas of San Andreas Valley and the Laguna Creek Basin, submerging both farms and much of the remaining wetland, riparian, grassland, and hardwood forest habitat. As a result, seeps, other wetland and riparian habitats, and serpentine grasslands are today largely confined to relatively narrow areas on the margins of the reservoirs and adjacent urban development.

Conifer forest appears to have expanded substantially

Between the mid-19th and mid-20th centuries, conifer forest extent within the watershed appears to have increased substantially, likely driven by a combination of factors including natural successional processes, absence of indigenous burning, active fire suppression, climatic variability, and deliberate planting of Monterey pine (*Pinus radiata*), Monterey cypress (*Hesperocyparis macrocarpa*), and Douglas-fir. For the most part, the commercial logging operations that targeted the extensive redwood forests on the Peninsula during the mid-19th century were focused to the south of the Peninsula Watershed, though some commercial logging did occur in the southwestern corner of the watershed and may have temporarily reduced conifer cover within the watershed overall.

Grassland extent appears to have increased somewhat due to livestock grazing, clearing of woody vegetation, and large wildfires, while shrubland extent likely decreased somewhat

The cessation of indigenous burning in the late 18th century may have resulted in an initial expansion of shrublands (coastal scrub) into former grassland areas that would have been maintained in part by frequent fires. However the introduction of grazing and the clearing of land for cultivation in the San Andreas Valley and other parts of the watershed in the mid-19th century, in addition to several large fires that occurred during the late 19th century, likely contributed to a modest increase in grassland extent and a reduction in shrubland extent (though much of this cleared land was subsequently inundated by reservoir construction). Forest encroachment (resulting from fire suppression) may have also contributed to the decrease in shrubland extent over this time period.

LANDSCAPE CHANGES
SINCE THE MID-20TH CENTURY

Large areas of grassland, and to a lesser extent hardwood forest, have been lost due to residential development and other direct modifications

Large areas formerly occupied by grasslands, primarily on the eastern side of San Andreas Valley, have been lost due to residential development associated with the cities of Hillsborough and San Mateo, transportation corridors (including Highway 280, Highway 92, Highway 35, Cañada Road and SFPUC maintenance roads), and recreational facilities. Of the grasslands present in 1928-32, approximately 24% (1,526 ac) were lost due to development by 1995-2001. In addition, hardwood forests have been lost along Lower San Mateo Creek, and to a lesser extent south of Upper Crystal Springs Reservoir, due to residential development associated with the city of Hillsborough and estate inholdings. Development pressures, combined with deliberate tree planting, have also facilitated the establishment of non-indigenous invasive trees throughout former grassland areas.

Forest cover has increased throughout the watershed since the mid-20th century, displacing grasslands and shrublands

Between 1928-32 and 1995-2001, grassland cover decreased by an estimated 70% (4,392 ac) and shrubland cover decreased by 3% (395 ac). Over the same time period, hardwood forest cover increased by an estimated 2-7% (119-408 ac) and conifer forest cover increased by 78-91% (1,728-2,016 ac); planted, non-indigenous tree species contributed substantially to this increase in forest cover. The most notable vegetation shifts have been conversion from grassland to hardwood forest; conversion from shrubland to hardwood and conifer forest; and conversion of hardwood forest to conifer forest. Throughout the watershed, this period has been characterized by extensive tree planting programs, along with a lack of disturbances such as fire and grazing, enabling the establishment of trees in formerly unforested areas.

Serpentine grasslands have been more persistent than non-serpentine grasslands

Between 1946-48 and 2015-17, approximately 71% of non-serpentine grasslands and 52% of serpentine grasslands converted to other land cover types. Twenty-four percent of all serpentine grasslands and 47% of non-serpentine grasslands were lost due to tree or shrub encroachment, while an additional 28% of serpentine grasslands and 24% of non-serpentine grasslands were displaced by development, transportation corridors, or water.

Diverse chaparral and coastal scrub communities west of the San Andreas Fault have been lost due to afforestation and tree planting, but low diversity shrublands have been gained east of the San Andreas Fault as a result of coyote brush encroachment

Gradual loss of shrublands since the 1930s-40s appears to have been highest in areas dominated by diverse chaparral and coastal scrub communities west of San Andreas Fault. Hardwood forest has displaced shrublands along portions of the eastern and southwestern slopes of Sawyer Ridge, while conifer forest (including planted stands of non-indigenous conifers such as Monterey cypress) has displaced shrublands around Upper San Mateo Creek, Pilarcitos Lake, and Kings Mountain Ridge. Shrub gain has occurred where coyote brush (*Baccharis pilularis*) has expanded into grassland (e.g., along the western slope of Pulgas Ridge) or into the understory of oak savannas and woodlands.

Construction and mitigation projects, as well as the spread of plant pathogens, have continued to alter vegetation communities in the watershed

In recent decades, SFPUC has conducted mitigation activities through its Bioregional Habitat Restoration (BHR) Program, which has successfully restored wetlands, grasslands, and other native habitats in numerous locations throughout the watershed. The spread of invasive species, including novel plant pathogens, has had widespread impacts; Sudden Oak Death, for instance, has resulted in substantial tree mortality within the watershed's oak woodlands.

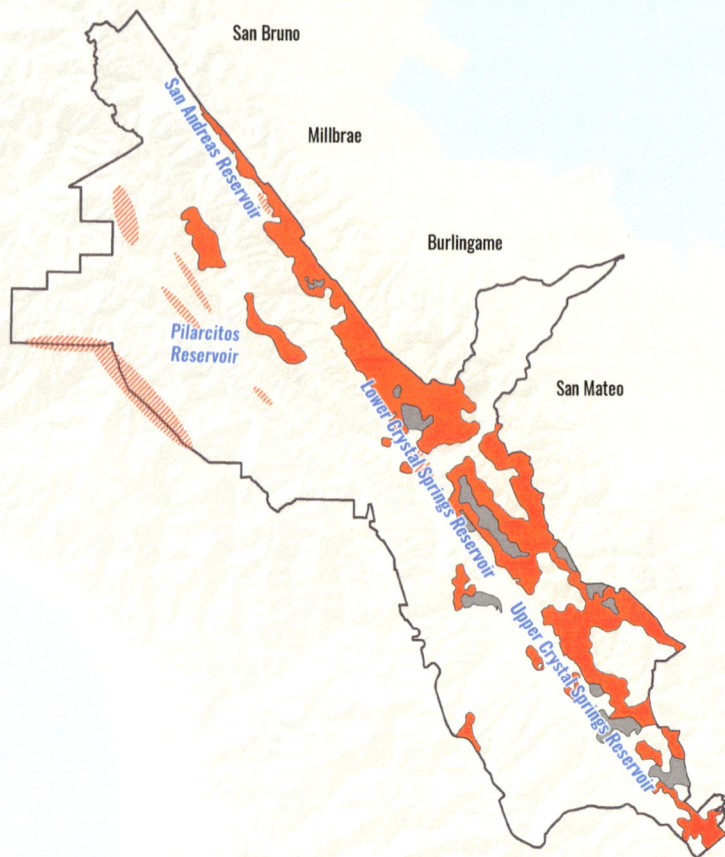

Grassland

- Large areas of grassland on the eastern side of the watershed have been lost due to agricultural and urban development, reservoir construction, tree plantings, and woody vegetation encroachment.

- Serpentine grasslands have been more persistent than non-serpentine grasslands.

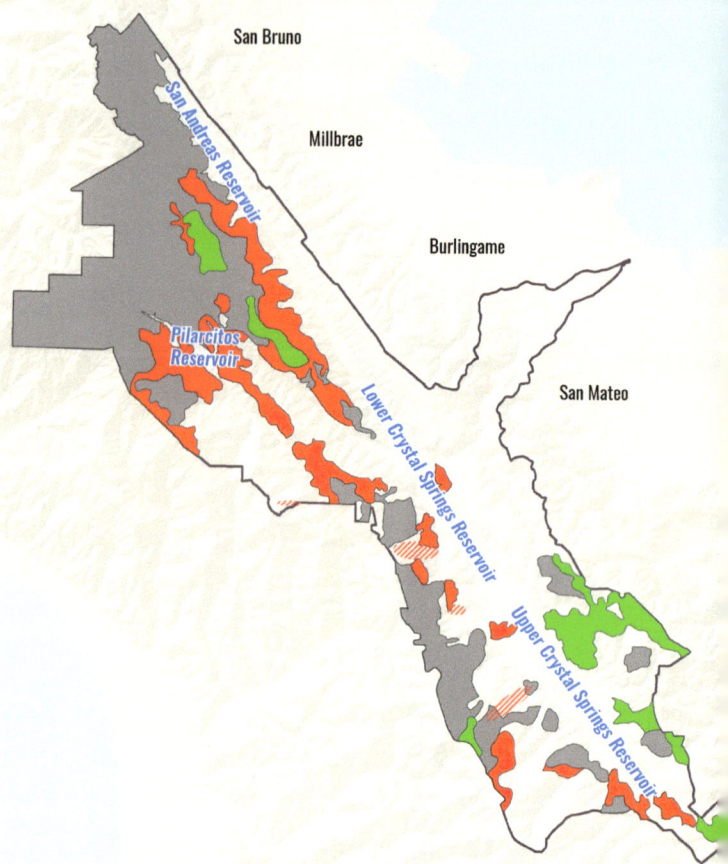

Shrubland

- Shrublands have persisted across much of the northwestern portion of the watershed and along Kings Mountain Ridge.

- Diverse chaparral and coastal scrub communities have been lost due to afforestation and tree planting, while low diversity shrublands have been gained elsewhere as a result of coyote brush encroachment.

Hotspots of vegetation change and persistence. These maps show generalized "hotspots" of vegetation change (loss and gain) and persistence across the watershed since at least the 1940s for four major terrestrial vegetation types: grassland, shrubland, hardwood forest, and conifer forest. Delineation of the hotspots was based primarily on the results of the aerial imagery point analysis (pages 86-93), but was also informed by the vegetation mapping comparison (pages 94-101), General Land Office survey analysis (pages 102-107), and other historical evidence (see page 35-36 for more information on methodology). The earliest documented evidence varies between hotspots: in some cases, there is no direct evidence about vegetation cover prior to the 1930s, while in other cases there is much earlier evidence. Separate polygons were created to indicate vegetation loss that occurred prior to the 1940s. The hotspots are generalized representations meant to highlight dominant trends of vegetation change at the watershed scale, but they simplify more complex changes that have occurred at finer scales, and should not be interpreted as precise vegetation change maps.

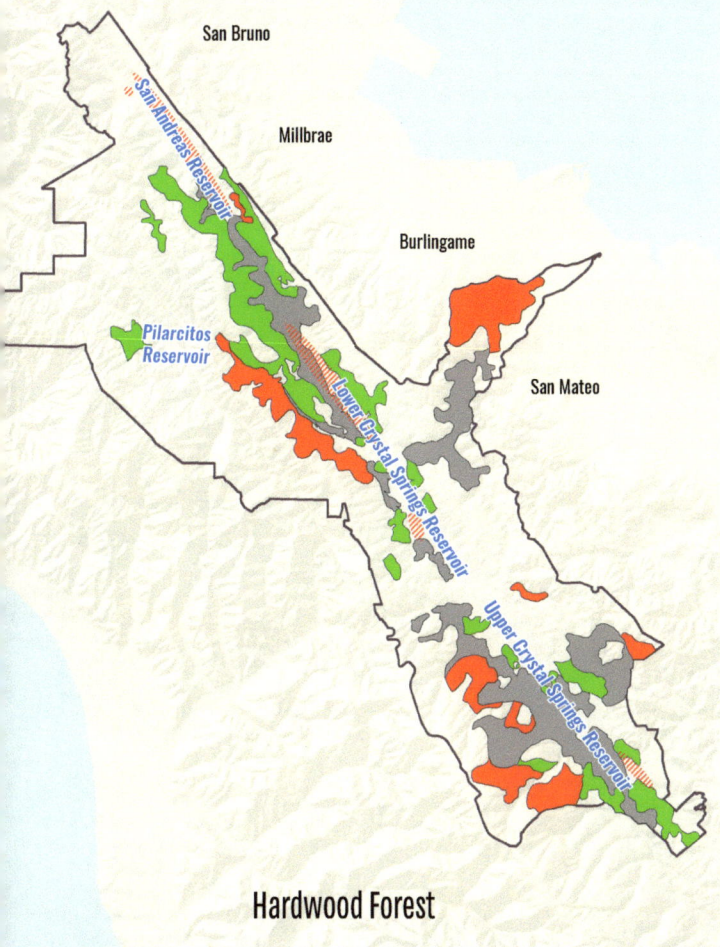

Hardwood Forest

- Reservoir construction, agricultural and urban development, and conifer encroachment have displaced hardwood forest in many parts of the watershed; Sudden Oak Death has resulted in recent tree mortality.

- Hardwood forest gain has occurred due to tree planting and afforestation of grasslands and shrublands.

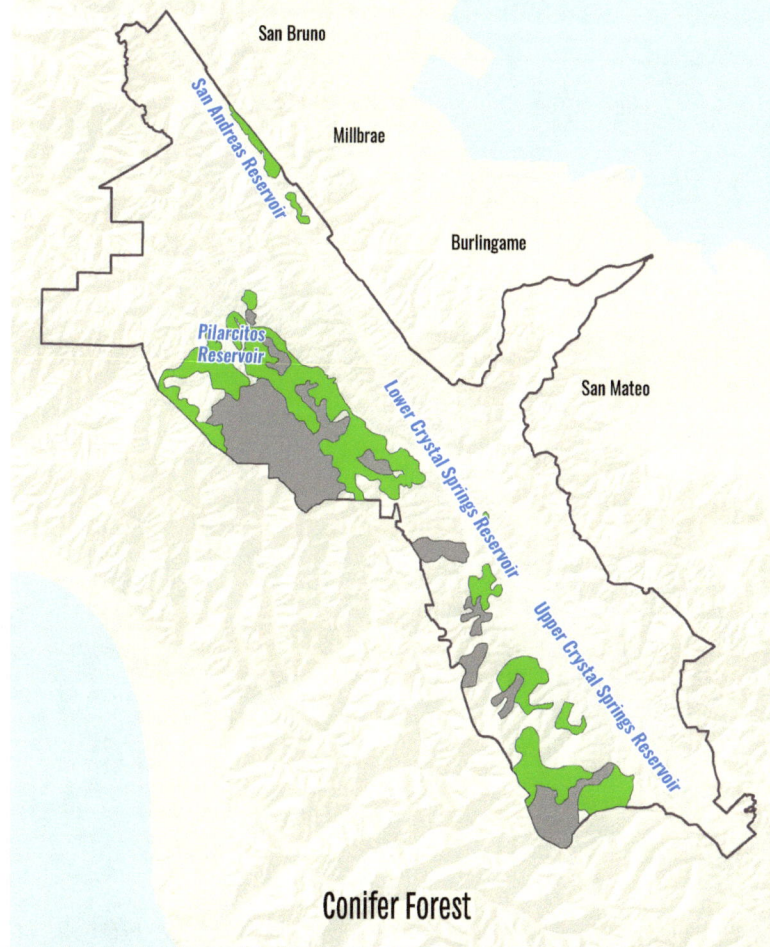

Conifer Forest

- Historical nodes of conifer forest on the western side of the watershed have largely persisted.

- Substantial expansion of conifer forest has occurred likely due to a combination of natural successional processes, fire suppression, climatic variability, and active planting of species such as Monterey cypress, Monterey pine, and Douglas-fir.

Vegetation Change and Persistence

- Loss before 1940
- Gain
- Loss
- Persistence

Conducting repeat photography, or "rephotography," of historical photographs in the hills above Pilarcitos Reservoir. *(Photo by SFEI-ASC)*

CHAPTER 2

Methods

Overview

This chapter summarizes the approaches taken to gather and evaluate historical data for the Peninsula Watershed study area. In prior historical ecology studies, SFEI-ASC has typically developed a digital map representing average landscape conditions prior to major Euro-American development, which serves to synthesize and illustrate historical ecological patterns. For several reasons, it was decided that a different approach would be more practical in this study. First, compared to lowland and coastal habitats, there is often less spatial data available for upland terrestrial habitats in the historical record. The lack of navigable waterways for exploration and commerce, and the lack of broad alluvial valleys for settlement and agriculture, usually meant that upland areas were less meticulously documented by early sources, and thus are more difficult to reconstruct spatially. Second, unlike wetlands and other

lowland habitats, whose locations on the landscape are often largely determined by persistent physical controls such as topography, soil type, and hydrology, the boundaries between upland habitats are often more fluid and less predictable, constantly fluctuating due to disturbance regimes and other processes. Third, because the main objective of this study was to document trajectories of terrestrial vegetation change over time, a simple comparison between the "historical" (pre-Euro-American modification) and "modern" landscapes was deemed insufficient. Rather, greater emphasis was placed on reconstructing trends in vegetation extent and distribution through time, in relation to land use modifications and other drivers of landscape change.

In lieu of digital maps of the historical landscape, several alternative approaches have been taken to analyze and synthesize the historical data to develop an overall understanding of historical landscape patterns and vegetation trajectories. First, a review of scientific literature on vegetation succession and response to disturbance in coastal California, along with a review of the Peninsula Watershed's land use history, were used to develop a series of hypotheses about expected vegetation shifts. Second, a series of three quantitative analyses, focusing on relatively comprehensive and spatially explicit datasets, was used to estimate the extent and distribution of terrestrial vegetation communities at several time periods dating back to the mid-19th century, and to reconstruct vegetation trajectories over time. Finally, the results of these analyses served as a framework for synthesizing more qualitative historical sources, such as landscape photographs and narrative descriptions (some dating back to the Spanish expeditions of the late 18th century); the overall findings from this synthesis process are summarized in a series of generalized vegetation change "hotspot" maps illustrating the dominant vegetation trends in different parts of the watershed.

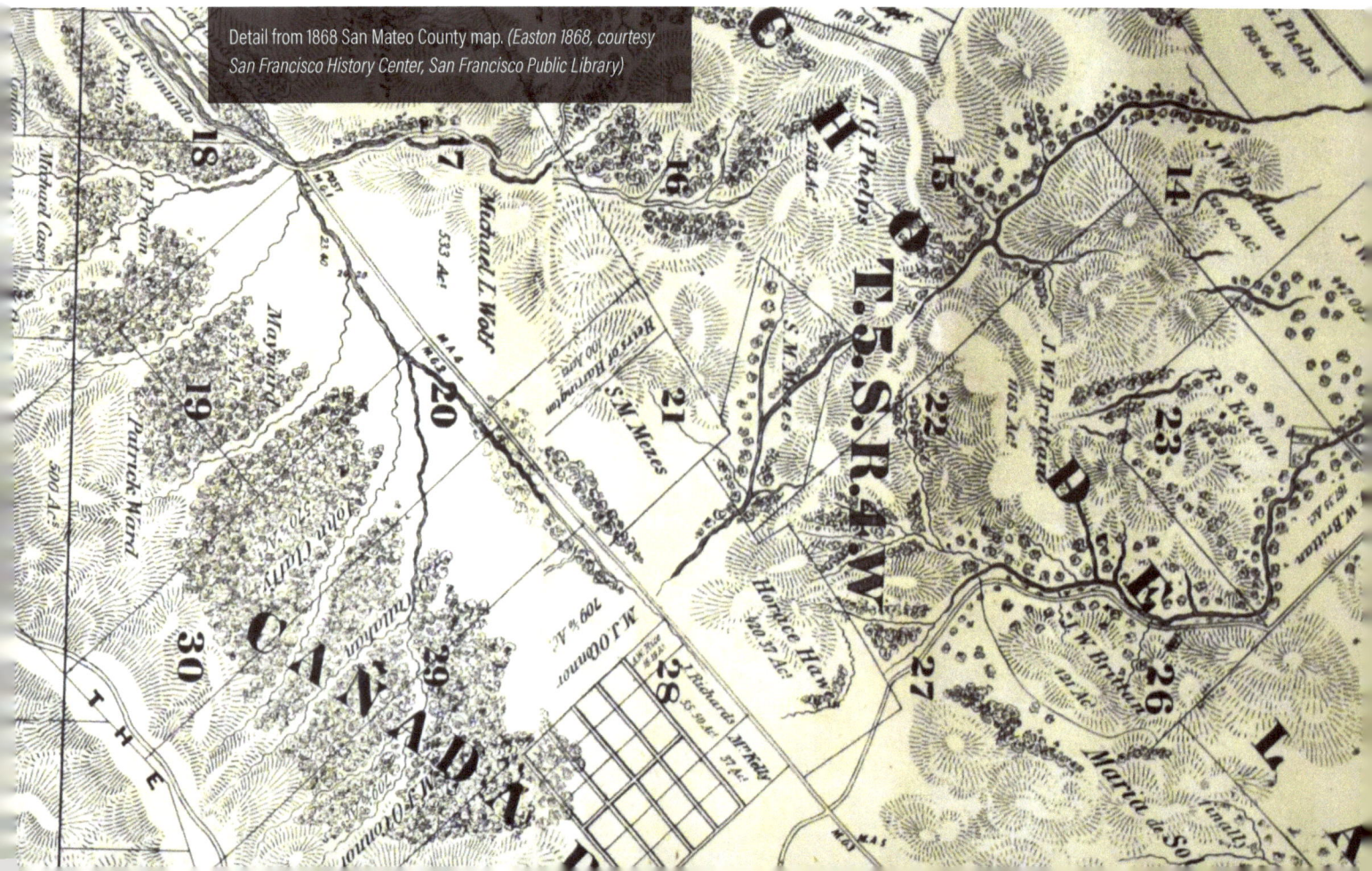

Detail from 1868 San Mateo County map. (Easton 1868, courtesy San Francisco History Center, San Francisco Public Library)

Data Collection and Compilation

The data collected represent a large body of information on the historical (late 18th through early 20th century) ecological and physical conditions of the Peninsula Watershed. Data were collected from 18 local, regional, state, and federal archives (Table 2.1), and approximately 50 online databases such as the Online Archive of California, the David Rumsey Map Collection, and the California Digital Newspaper Collection. The assembled dataset is composed of a variety of historical data types, including maps, photographs, drawings, and textual documents. Altogether, the dataset includes approximately 3,000 landscape photographs and drawings, 475 maps, and 190 individual textual sources.

The earliest written accounts of the Peninsula Watershed are contained in the diaries of Spanish explorers from the Portolá (1769), Rivera (1774), and Anza (1776) expeditions. Land grant court case files (including diseños, or sketch maps) from the early 19th century also provide early glimpses of vegetation cover and land use. General Land Office (GLO) surveys, conducted in the mid-19th century throughout much of the watershed, produced some of the earliest spatially accurate and detailed descriptions/depictions of the study area. Beginning in the mid-to-late 19th century, many other sources document ecological and physical conditions in the watershed, including newspaper accounts, travelogues, landscape photographs, botanical specimen records, SVWC records, and a wide range of maps.

Early 20th century sources provide increasingly detailed and spatially explicit information about watershed conditions. While these sources post-date major hydrologic modifications and altered disturbance regimes, they provide the first comprehensive information about vegetation cover and are essential in helping with the interpretation of earlier sources. Historical aerial photographs dating from 1946-48 cover the full watershed and show fine scale vegetation patterns and land use modifications. Aerial photographs were acquired from NETR Online/Historic Aerials. Wieslander Vegetation Type Mapping (VTM), produced from surveys conducted in the 1930s, represents the earliest comprehensive vegetation mapping within the watershed. Though the spatial accuracy of the Wieslander VTM mapping does not meet modern standards, at coarse scales it captures patterns of vegetation distribution that can be refined through intercalibration with other sources.

Because of the unique history of land use and private land ownership in the Peninsula Watershed, the archival materials documenting the study area differ notably compared with other northern California watersheds. SVWC gradually acquired lands within the watershed between the mid-1800s and early 1900s, and as a result SVWC records represent a massive repository of information about the watershed over this time period. Key SVWC records include a number of early survey maps, thousands of landscape photographs, San Francisco Water books, and court case documents. One particularly valuable dataset is a set of documents from an early 20th century SVWC court case (SVWC vs. San Francisco 1916), which includes detailed descriptions of vegetation cover in many parcels throughout the watershed, along with accompanying maps and photographs. There are numerous references to parcels described in this 1907-14 SVWC court case in chapters 5-8; Appendix B provides an index map showing the locations of these parcels. While SVWC ownership resulted in a large number of company records, it also means that there are relatively few parcel and subdivision maps, oral histories, or other archival sources typically associated with more developed parts of the Bay Area.

The data were compiled in a number of ways to make them more useable and comparable. Of the approximately 475 maps collected, 56 were georeferenced in a GIS database. Approximately 100 pages of relevant textual excerpts were transcribed, and spatially explicit textual data (including GLO survey notes from 277 points) were geolocated and added to the GIS database. Seventy-nine landscape photos (dating from ca. 1860-1944) were geolocated in GIS.

To further examine and illustrate patterns of vegetation change in particular parts of the watershed, a subset of 24 geolocated historical landscape photographs were "rephotographed" in the field. Rephotographs, or repeat photographs, were taken in August, 2018. The location and direction of the rephotographs was matched as closely as possible to the historical photographs. A selection of paired historical photographs and rephotographs are presented in chapters 5-8.

In addition to the historical data collection and compilation, an extensive literature review was conducted to synthesize research findings pertaining to vegetation succession and response to disturbance in coastal California. A discussion and summary of this literature is included in Chapter 3.

Table 2.1. Archives visited during the historical data collection process.

Institution	
Bay Area regional archives	**Location**
The Bancroft Library, UC Berkeley	Berkeley
UC Berkeley Earth Science and Map Library	Berkeley
UC Berkeley Bioscience and Natural Resources Library	Berkeley
California Historical Society	San Francisco
Stanford University, Special Collections	Palo Alto
Stanford University, Branner Library	Palo Alto
San Francisco Public Library	San Francisco
Sacramento state archives	
California State Archives	Sacramento
California State Library	Sacramento
California State Railroad Museum	Sacramento
Local and county archives	
SFPUC Historical and Engineering archives	San Francisco/Millbrae
San Mateo Public Library	San Mateo
San Mateo County Historical Association	San Mateo
San Mateo County Recorder	San Mateo
San Mateo County Public Works	San Mateo
San Mateo County Planning and Building Department	San Mateo
Other	
Water Resources Collections & Archives (WRCA), UC Riverside	Riverside
Bureau of Land Management	Sacramento
Online databases (many)	N/A

Vegetation Classification

Throughout the report, a simplified set of vegetation types has been used to represent vegetation cover within the study area. Because different data sources describe vegetation cover in different ways (with varying levels of accuracy and specificity), simplifying vegetation types into these broad categories was necessary to ensure that diverse historical and contemporary data sources were comparable at multiple timesteps. The broad vegetation types represent the highest level of detail that could be consistently identified and mapped across the study area over time. However, these simplified vegetation types (or lifeform types) are generalizations that lump together multiple vegetation communities, or alliances, with different species composition. There is significant heterogeneity within each vegetation type, and where possible more information is provided about the historical distribution and the trajectory over time of more specific vegetation communities.

Terrestrial vegetation communities are classified as one of four broad vegetation types: grassland, shrubland, hardwood forest, and conifer forest. These four categories are used consistently in quantitative analyses of vegetation change (see Chapter 4), and form the basis for the discussions of historical ecology and change over time in chapters 5-8. In some datasets (e.g., the National Land Cover Dataset used to classify land cover in a small portion of the modern vegetation map), it was not possible to consistently differentiate hardwood and conifer forest, and thus a combined forest (hardwood or conifer) class was assigned to a subset of features. A separate class for riparian/wetland habitat types was distinguished where possible (e.g., in the modern vegetation map), though it was not possible to consistently differentiate riparian/wetland features in other datasets (e.g., Wieslander VTM mapping or aerial photographs). Brief descriptions and characteristic species of each of the four broad terrestrial vegetation types, as well as the riparian/wetland class, are provided on page 25 (Table 2.2). Two additional classes were used to represent water and developed/disturbed areas.

Figure 2.1 (opposite page) shows the modern vegetation and land cover types within the study area, based on data from Schirokauer et al. (2003) and MRLCC (2001). For more information about data sources used to develop the modern vegetation map, see page 30 and Appendix A.

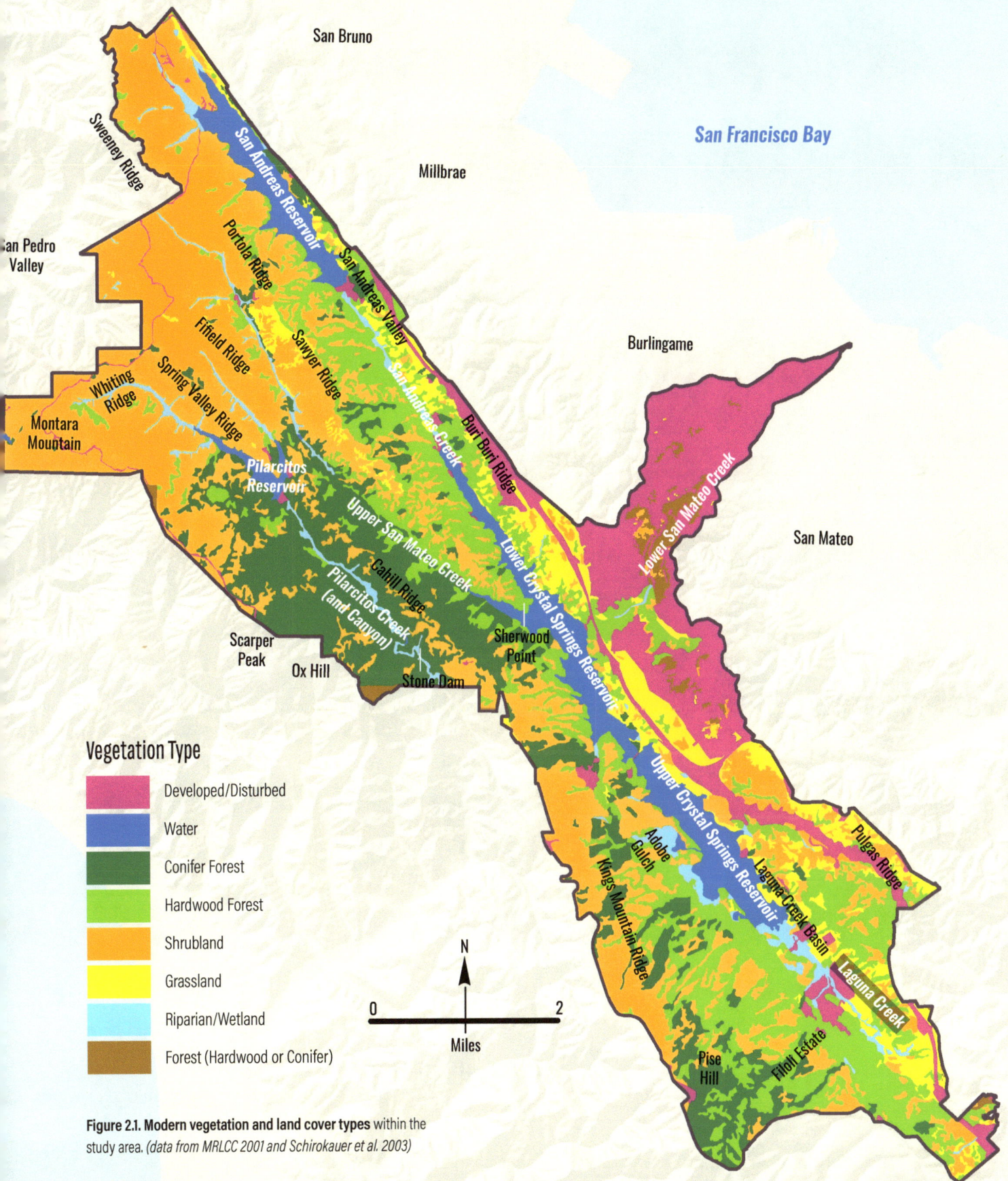

Vegetation Type

- Developed/Disturbed
- Water
- Conifer Forest
- Hardwood Forest
- Shrubland
- Grassland
- Riparian/Wetland
- Forest (Hardwood or Conifer)

N

0   2
Miles

Figure 2.1. Modern vegetation and land cover types within the study area. *(data from MRLCC 2001 and Schirokauer et al. 2003)*

Map labels:
San Francisco Bay
San Bruno
Millbrae
Burlingame
San Mateo
Sweeney Ridge
San Pedro Valley
Whiting Ridge
Montara Mountain
Spring Valley Ridge
Fifield Ridge
Portola Ridge
San Andreas Reservoir
San Andreas Valley
San Andreas Creek
Sawyer Ridge
Buri Buri Ridge
Pilarcitos Reservoir
Upper San Mateo Creek
Cahill Ridge
Pilarcitos Creek (and Canyon)
Scarper Peak
Ox Hill
Stone Dam
Sherwood Point
Lower Crystal Springs Reservoir
Lower San Mateo Creek
Upper Crystal Springs Reservoir
Adobe Gulch
Kings Mountain Ridge
Laguna Creek Basin
Pulgas Ridge
Laguna Creek
Pise Hill
Filoli Estate

Conifer forest

Hardwood forest

(Photos this page and facing page courtesy SFPUC)

Table 2.2. Descriptions and characteristic species of the simplified vegetation types used throughout the report to facilitate comparison of disparate data sources from multiple time periods. (Sources: Jackson and Bartolome 2002, Schirokauer et al. 2003, ABI et al. 2003, Ford and Hayes 2007, Bartolome et al. 2007, H. Bartosh pers. comm., S. Simono pers. comm.).

CONIFER FOREST

Conifer forests are characterized by cone-bearing gymnosperm species that typically produce evergreen, needle-shaped leaves. Within the Peninsula Watershed, conifer forests are dominated by Douglas-fir (*Pseudotsuga menziesii*) or Coast redwood (*Sequoia sempervirens*), as well as planted stands of Monterey pine (*Pinus radiata*), Monterey cypress (*Hesperocyparis macrocarpa*), and fir (*Abies* spp.).

HARDWOOD FOREST

Hardwood forests are composed of evergreen or deciduous flowering trees (angiosperms) which produce seeds enclosed by ovaries. Within the Peninsula Watershed, hardwood forests are dominated by oak woodlands. Characteristic species include coast live oak (*Quercus agrifolia*), California bay (*Umbellularia californica*), California buckeye (*Aesculus californica*), and Pacific madrone (*Arbutus menziesii*).

GRASSLAND

Grasslands are herbaceous plant communities dominated by annual and perennial grasses and forbs. Characteristic species in the Peninsula Watershed include native species such as purple needle grass (*Stipa pulchra*), foothill needle grass (*Stipa lepida*), and California oatgrass (*Danthonia californica*), as well as non-native species such as oats (*Avena* spp.) and bromes (*Bromus* spp.). A substantial portion of the grasslands that have persisted on the watershed are serpentine, and support a distinctive and specialized flora.

SHRUBLAND

Shrublands encompasses vegetation communities dominated by woody vegetation generally shorter that ~3-6 ft (~1-2 m). These communities can be broadly divided into coastal scrub, which occurs in cool coastal areas and is dominated by a mix of drought-deciduous and evergreen shrubs, and chaparral, dominated by evergreen sclerophyllous shrubs (see page 164). Species characteristic of coastal scrub include coyote brush (*Baccharis pilularis*), California coffeeberry (*Frangula californica)*, and poison oak (*Toxicodendron diversilobum*), while species characteristic of chaparral include chamise (*Adenostoma fasciculatum*), manzanita (*Arctostaphylos* spp.), and California lilac (*Ceanothus* spp.).

RIPARIAN/WETLAND

The Riparian/Wetland category includes species associated with waterways and wet environments. The soils in riparian and wetland areas are subject to intermittent flooding and fluctuating water tables. Characteristic species within the watershed include arroyo willow (*Salix lasiolepis*), Pacific willow (*Salix lasiandra*), Oregon ash (*Fraxinus latifolia*), American dogwood (*Cornus sericea*), California bulrush (*Schoenoplectus californicus*), and broadleaf cattail (*Typha latifolia*).

Grassland

Shrubland

Riparian/Wetland

Quantitative Analyses

In order to quantify vegetation dynamics in the study area and provide a framework for synthesizing other qualitative data sources, several key analyses were performed to understand trajectories of change. These analyses focused on datasets that provided relatively comprehensive and standardized information about vegetation cover in the watershed. The first analysis classified land cover in historical (1946-48) and contemporary (2015-17) aerial imagery.[1] The second analysis compared vegetation mapping from Wieslander Vegetation Type Mapping conducted in the 1930s and contemporary vegetation and land cover mapping (1995-2001). The third analysis examined vegetation cover documented in GLO surveys (conducted 1852-1864), and assessed changes in vegetation cover at the same locations using historical and contemporary aerial imagery. The following sections describe the methods used in each of these three analyses; analysis results are presented in Chapter 4.

---

1 Historical aerial photos from 1946 cover the majority of the watershed; photos from 1948 cover a small portion of the southern-most part of the watershed (imagery from historicaerials.com). Modern aerial photos from 2015 (Pictometry, Inc. 2015) cover the majority of the watershed; photos from 2017 (Sanborn Map Co, Inc. 2017) cover the Lower San Mateo Creek Basin.

Oaks and grassland. *(Photo courtesy SFPUC)*

AERIAL IMAGERY POINT ANALYSIS (1946-48 TO 2015-17)

The aerial imagery point analysis assessed landscape change in the study area between two timesteps, broadly representing historical and modern conditions. Aerial photographic surveys conducted in 1946-48 (Fig. 2.2) and 2015-17 were used to classify vegetation throughout the watershed at 5,000 randomly distributed points. These points were created in ArcGIS (ArcMap 10.7), buffered with a 7.5 m radius, and given attribute fields to record vegetation type, certainty of classification, forest type (this was done separately because differentiating conifer and hardwood forest was often more difficult than differentiating forest from non-forest), and certainty of forest type classification. Four independent mappers (including a GIS Specialist at SFEI-ASC, an Environmental Analyst at SFEI-ASC, and two GIS technicians) each classified 1,150 randomly determined unique points, as well as 400 randomly determined duplicate points that were classified by all mappers in order to assess classification uncertainty. Of the 400 duplicate points, 200 were classified prior to review by the GIS Specialist, who then provided feedback and helped to calibrate and standardize vegetation identification among the other three mappers. The second 200 duplicate points were then classified following this review and calibration. Where classification of the duplicate points differed among the four mappers, the final classification was determined by the GIS Specialist.

Mappers classified each buffered point at standardized scales—1:600 to 1:3,000 for classification and 1:3,000 to 1:10,000 for assessing landscape context. Guidelines for identifying characteristics of each vegetation type were provided. Texture, shadow, and color were used to guide classification and recognize distinct features of the various vegetation classes. The classification was determined by majority cover of the buffered area around each point. The 7.5 m radius buffer was chosen to approximate the average diameter (15 m) of a tree crown so there would be less conflict in determining the majority cover for each point. It was not possible to differentiate riparian or wetland vegetation from upland habitat types in the aerial imagery point analysis; woody riparian vegetation is thus generally lumped together with forest or shrubland vegetation types, while herbaceous wetland vegetation is generally lumped with grassland.

Caution should be exercised in interpreting the results of the aerial imagery point analysis, as inferring percent cover from a random point drop may potentially be unreliable. Clustering of points is possible based on chance alone, and the random point drop may not have stratified evenly across cover types. Thus, the estimates for each vegetation type derived from the aerial imagery point analysis should technically be interpreted as

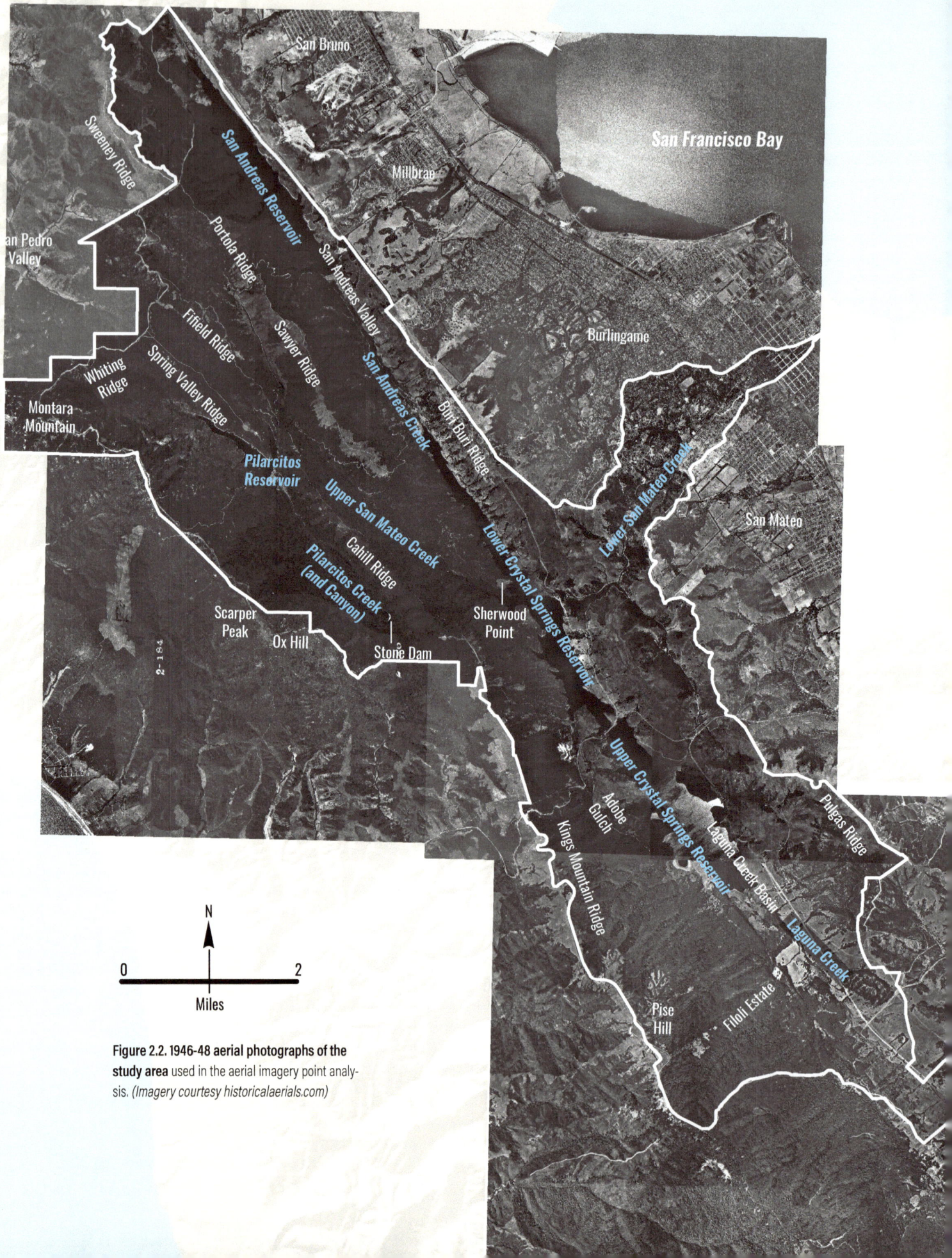

Figure 2.2. 1946-48 aerial photographs of the study area used in the aerial imagery point analysis. *(Imagery courtesy historicalaerials.com)*

Labels on map:

San Bruno
San Francisco Bay
Sweeney Ridge
Millbrae
San Andreas Reservoir
San Pedro Valley
Portola Ridge
San Andreas Valley
Burlingame
Fifield Ridge
Sawyer Ridge
San Andreas Creek
Whiting Ridge
Spring Valley Ridge
Bluff Buri Ridge
Montara Mountain
Pilarcitos Reservoir
Upper San Mateo Creek
Lower San Mateo Creek
San Mateo
Cahill Ridge
Pilarcitos Creek (and Canyon)
Lower Crystal Springs Reservoir
Scarper Peak
Ox Hill
Sherwood Point
Stone Dam
2-184
Upper Crystal Springs Reservoir
Adobe Gulch
Pulgas Ridge
Kings Mountain Ridge
Laguna Creek Basin
Laguna Creek
Pise Hill
Filoli Estate

N

0   2
Miles

frequencies, rather than proportions, though comparison with other data sources such as historical and modern vegetation mapping provides some confidence in generalizations of percent cover from these frequencies (M.V. Eitzel pers. comm.).

Uncertainty in the frequency of each vegetation type based on the aerial imagery point analysis, as well as the overall reliability of the analysis for the historical (1946-48) and modern (2015-17) time periods, were estimated in two ways. First, the range of the frequencies generated by all four mappers provides a measure of the classification uncertainty of each vegetation type. The range was derived from the second set of 200 points classified by all four mappers after review and calibration, and is provided for both historical and modern time periods. Because these ranges are only calculated using the subset of 200 duplicate points, rather than the full set of 5,000 points, they are only an estimate of what the true range would be for the full 5,000 points (and indeed, in several cases the ranges do not include the frequency of each vegetation type estimated from the 5,000 points). These ranges are provided in Chapter 4 (page 86) and shown visually using gray boxes on the trajectory diagrams for each vegetation type.

A second method, involving calculation of Krippendorff's Alpha statistic, was used to assess uncertainty across all of the vegetation/land cover types (grassland, shrubland, hardwood forest, conifer forest, water, developed/disturbed) for both the historical and modern time periods. Krippendorff's Alpha is a measure of inter-rater reliability—in other words, the amount of agreement in the classification, or "ratings," assigned by multiple independent mappers, or "raters." A value of 1 indicates complete reliability, a value of 0 indicates a lack of reliability, and a value less than 0 indicates systematic disagreement among raters. Values greater than 0.8 are generally considered reliable, while values between 0.667 and 0.8 can be used to draw tentative conclusions (Krippendorff 2004). Krippendorff's Alpha was calculated from all four mappers' classifications of the 200 duplicate, calibrated points (it does not privilege the final classification from the GIS Specialist which is displayed in the graphs); calculated values for Krippendorff's Alpha are provided on page 86.

VEGETATION MAPPING COMPARISON (1928-32 TO 1995-2001)

The vegetation mapping comparison analyzed changes in the extent and distribution of vegetation types using historical (Wieslander VTM) and modern vegetation mapping (MRLCC 2001, Schirokauer et al. 2003) for the study area.

Albert Everett Wieslander was a forester tasked with mapping vegetation through California in the early 20th century. Surveyors, often situated along ridges or peaks, identified dominant plant species in different portions of the surrounding landscape and recorded observations directly on topographic maps; the resulting products were called Vegetation Type Maps (VTM). Surveyors also established vegetation plots in certain areas and recorded detailed notes about plant species composition, percent cover, and a range of other information about plot condition. Wieslander VTM mapping in the Peninsula Watershed, which was conducted during the period from approximately 1928-1932, represents the first comprehensive vegetation mapping within the watershed.

A vector dataset of the Wieslander VTM mapping for California, digitized by researchers at UC Berkeley and UC Davis, was downloaded and crosswalked to the broad vegetation classes described on page 25 (Kelly et al. 2005; Fig. 2.3). The local georeferencing of the vector polygons was modified slightly to align better with contemporary landscape features within the Peninsula Watershed. Modern vegetation mapping covering most of the Peninsula Watershed was produced in 1995 (and updated in 2001) by the National Park Service Golden Gate National Recreation Area, based on aerial photographs and field validation (Schirokauer et al. 2003; see Fig. 2.1). Data from Lower San Mateo Creek were missing from the NPS vegetation mapping, so for this area, 2001 National Land Cover Dataset (NLCD) data from USGS were used (MRLCC 2001). Classification crosswalks were developed to reclassify vegetation classes used in the Wieslander VTM, NPS vegetation mapping, and NLCD data to the simplified vegetation/land cover classifications described on page 25, and are provided in Appendix A.

To compare aggregate change over time, these layers were projected into NAD 1983 (2011) California (Teale) Albers (Meters), and then each vegetation/land cover class was summarized by area. To analyze pathways of change over time, the union tool was applied in ArcGIS to the modern vegetation and Wieslander VTM data. Pivot tables in Microsoft Excel were then generated to characterize change across time periods by alliance.

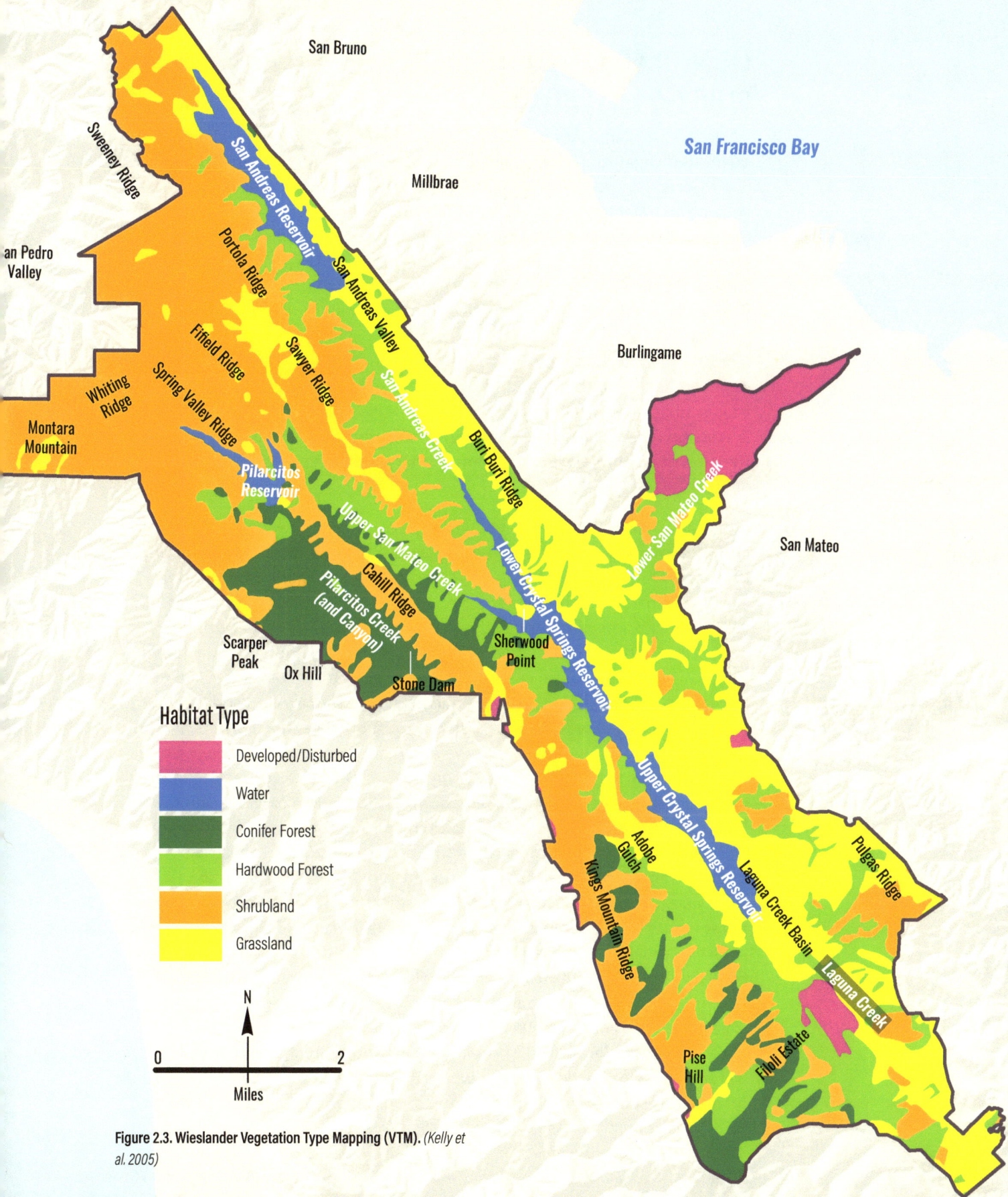

Figure 2.3. Wieslander Vegetation Type Mapping (VTM). *(Kelly et al. 2005)*

San Francisco Bay

San Bruno

Sweeney Ridge

Millbrae

an Pedro Valley

San Andreas Reservoir

Portola Ridge

San Andreas Valley

San Andreas Creek

Burlingame

Fifield Ridge

Sawyer Ridge

Buri Buri Ridge

Whiting Ridge

Spring Valley Ridge

Lower San Mateo Creek

Montara Mountain

Pilarcitos Reservoir

Upper San Mateo Creek

Lower Crystal Springs Reservoir

San Mateo

Cahill Ridge

Pilarcitos Creek (and Canyon)

Sherwood Point

Scarper Peak

Ox Hill

Stone Dam

Upper Crystal Springs Reservoir

Adobe Gulch

Laguna Creek Basin

Pulgas Ridge

Kings Mountain Ridge

Pise Hill

Filoli Estate

Laguna Creek

Habitat Type

- Developed/Disturbed
- Water
- Conifer Forest
- Hardwood Forest
- Shrubland
- Grassland

N

0 — 2

Miles

GENERAL LAND OFFICE SURVEY ANALYSIS (1852-64 TO 2015-17)

While the aerial imagery point analysis and vegetation mapping comparison quantified vegetation changes over the past 70-90 years, a third analysis using General Land Office (GLO) survey data extended the temporal scope of vegetation trajectories back to the mid-19th century. Established in 1812, the GLO was tasked with surveying and overseeing the sale of public lands in the western United States. In areas not claimed through the land grant system, the U.S. Public Land Survey divided the land into a grid of 1x1 mile squares (known as "sections"). Surveyors systematically walked section boundaries, keeping detailed field notes on the natural and cultural features encountered along the way. Separate surveys were conducted to delineate the boundaries of private land grants. Notes and plat maps from these surveys are useful for their ecological information and have been extensively utilized in historical landscape reconstruction and land cover change research (Buordo 1956, Radeloff et al. 1999, Collins and Montgomery 2001, Brown 2005, Whipple et al. 2011).

Information from 15 surveys, including 9 public land surveys and 6 land grant surveys conducted between 1852 and 1864, was compiled in the GIS database. Out of a total of 277 geolocated points, sufficient information existed to characterize vegetation cover during the mid-19th century at 175 individual locations throughout the watershed (Fig. 2.4). The information available at each location varied, but included a combination of qualitative descriptions of vegetation cover (e.g., "dense chaparral," "oak grove," "barren rolling hills") as well as records of "bearing trees," which were used as reference points in order to establish the location of section corners and half mile points ("quarter sections"). Surveyors recorded up to four bearing trees at each section corner and two bearing trees at each quarter section, noting the species, diameter, azimuth (direction), and distance from the points. A lack of bearing trees was interpreted to mean that the land was un-forested, and thus either grassland or shrubland. Each of the 175 points was assigned a vegetation classification based on the information contained in the GLO survey field notes. To ensure that the classification based on the GLO survey notes was as comparable as possible with classification of the same points from the aerial imagery (see below), only information pertaining to vegetation cover in the immediate vicinity of the point (rather than descriptions of broader landscape character) was used in assigning a vegetation classification. Points were classed as either one of the four broad terrestrial vegetation types (grassland, shrubland, hardwood forest, conifer forest), water, or developed/disturbed. It was not possible to consistently distinguish riparian forest based on the survey notes; possible riparian forest features were thus generally classified as hardwood forest, consistent with the aerial

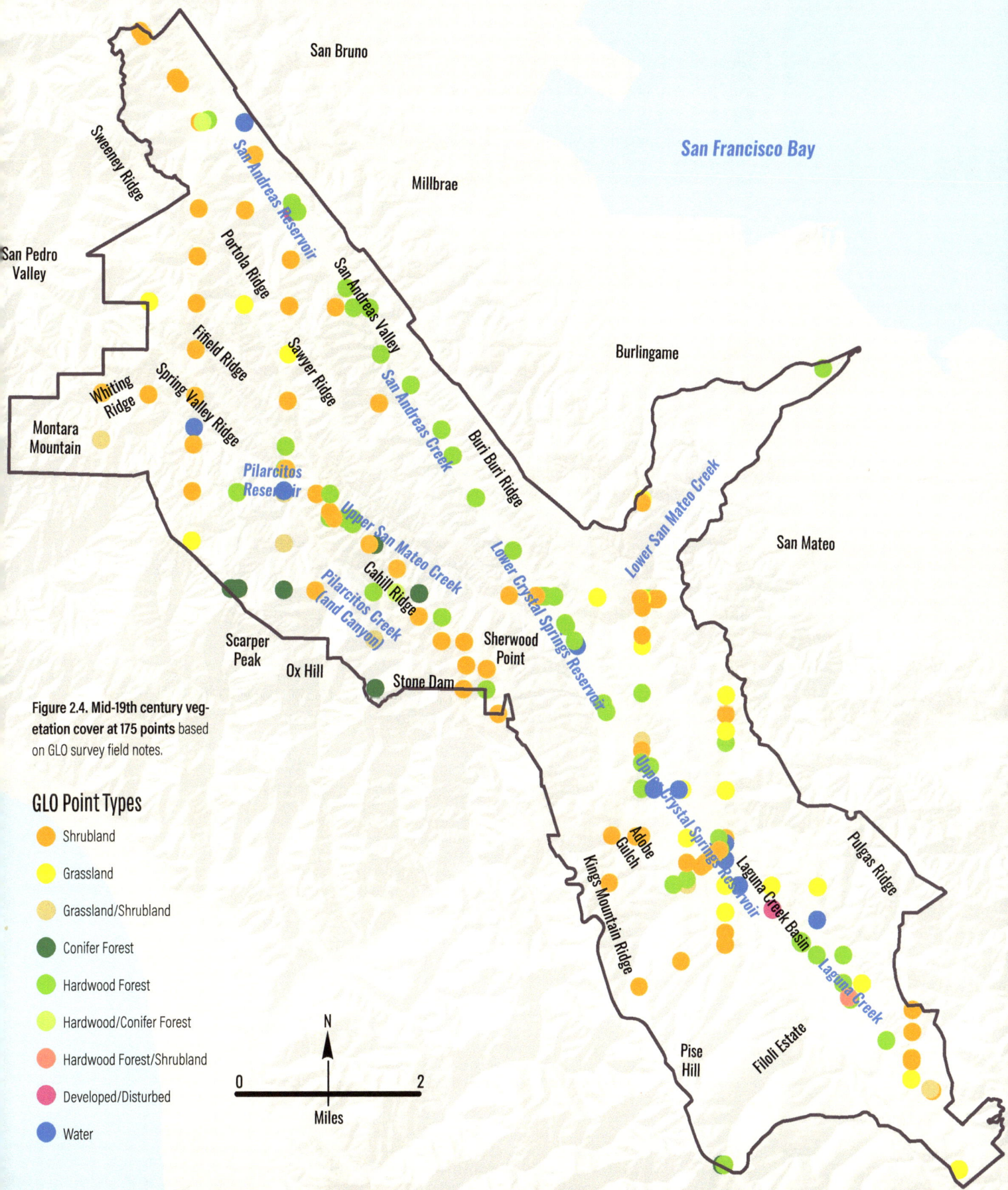

Figure 2.4. Mid-19th century vegetation cover at 175 points based on GLO survey field notes.

GLO Point Types

- 🟠 Shrubland
- 🟡 Grassland
- 🟤 Grassland/Shrubland
- 🟢 Conifer Forest
- 🟢 Hardwood Forest
- 🟢 Hardwood/Conifer Forest
- 🔴 Hardwood Forest/Shrubland
- 🔴 Developed/Disturbed
- 🔵 Water

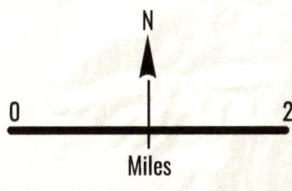

San Bruno

Millbrae

San Francisco Bay

Sweeney Ridge

San Pedro Valley

Portola Ridge

San Andreas Reservoir

San Andreas Valley

San Andreas Creek

Burlingame

Fifield Ridge

Sawyer Ridge

Buri Buri Ridge

Whiting Ridge

Spring Valley Ridge

Lower San Mateo Creek

San Mateo

Montara Mountain

Pilarcitos Reservoir

Upper San Mateo Creek

Lower Crystal Springs Reservoir

Cahill Ridge

Pilarcitos Creek (and Canyon)

Scarper Peak

Ox Hill

Sherwood Point

Stone Dam

Upper Crystal Springs Reservoir

Adobe Gulch

Laguna Creek Basin

Pulgas Ridge

Kings Mountain Ridge

Laguna Creek

N

0   Miles   2

Pise Hill

Filoli Estate

imagery point analysis. In some cases the field notes were ambiguous and it was not possible to assign a single vegetation type; several points were thus classified as one of two possible vegetation types (e.g., grassland/shrubland, shrubland/hardwood forest, hardwood/conifer forest).

In order to examine landscape changes since the mid-19th century, vegetation cover at each of the 175 points was also classified using the 1946-48 aerial images and the 2015-17 aerial images. The approach was the same as that used for the aerial imagery point analysis (see page 27): each point was assigned a vegetation type based on the dominant cover within a 7.5 m buffer.

One major limitation of the GLO survey analysis is that the 175 points used in the analysis are not evenly or randomly distributed throughout the watershed, and thus there is substantial uncertainty in extrapolating from the GLO data to draw conclusions about watershed-wide vegetation patterns. Sixty-seven percent of the study area fell within the boundaries of private land grants (see pages 58-59), and for the most part GLO survey data are limited to the boundaries, rather than the interiors, of these land grants. The remaining 33% of the study area, which was publicly owned and thus surveyed at regular intervals according to the rectangular grid established by the Public Land Survey System, was concentrated in the northwestern portion of the watershed (and to a lesser extent along Kings Mountain Ridge immediately south of present-day Highway 92). However, the aerial imagery point analysis can be used to help validate the results of the GLO analysis: because vegetation cover at each of the 175 points used in the GLO analysis was also classified using the historical and modern aerial imagery, comparison of the relative cover of each vegetation type between the GLO analysis and aerial imagery point analysis for the 1946-48 and 2015-17 time periods provides an estimate for the overall degree of bias in the GLO dataset.

Synthesis

Results from the quantitative analyses described above were integrated with information from qualitative data sources to develop a more robust understanding of historical vegetation patterns and trajectories of vegetation change over time. Several approaches were used to synthesize the diverse data sources and visualize the results; a summary of the quantitative results and the synthesis framework is presented in Chapter 4, while detailed results for each vegetation type (synthesizing quantitative results and qualitative data) are discussed in chapters 5-9.

VEGETATION CHANGE TRAJECTORIES

Simple diagrams were created for each of the four major vegetation types (grassland, shrubland, hardwood forest, conifer forest) to illustrate temporal changes in extent at the watershed scale. These trajectory diagrams present a visual summary of the key findings from the quantitative analyses described above (aerial imagery point analysis, vegetation mapping comparison, and GLO analysis). The proportion of GLO points occupied by a particular vegetation type was used to estimate its mid-19th century (ca. 1857) extent; the assumptions/limitations of this analysis are described above. Wieslander VTM mapping provided an estimate of ca. 1930 cover, and the aerial imagery analysis provided an estimate of both ca. 1947 and ca. 2016 cover. Modern vegetation mapping provided an additional estimate of ca. 1995 cover.

VEGETATION CHANGE HOTSPOTS

Maps showing vegetation change "hotspots" were created to highlight parts of the watershed where particular types of vegetation change (or lack of change) have been prevalent. The results of the aerial imagery point analysis were used as the starting place for creating the hotspot maps: clusters of vegetation loss, gain, or persistence since the 1940s were identified based on visual inspection of the 5,000 points used in this analysis, and polygons were drawn around these clusters in ArcGIS. These initial hotspots were refined based on a review of other spatially comprehensive historical (ca. 1930-40) and modern datasets, including the raw aerial imagery, Wieslander VTM mapping, and modern vegetation mapping.

The hotspot mapping then served as a framework for synthesizing spatially explicit qualitative historical data, such as maps, GLO field notes, textual descriptions, landscape photographs, and SVWC court case data. Where early (i.e., 18th, 19th, and early 20th century) sources were consistent with the type of vegetation change represented by a particular hotspot (e.g., grassland to shrubland conversion), those sources were attributed to that hotspot as supporting evidence. Thus, the earliest documented evidence varies between hotspots: in some cases, there is no direct evidence about vegetation cover prior to the 1930s, while in other cases there is evidence dating back to 1769. Where early sources were not consistent with the type of vegetation change represented by a particular hotspot, a new overlapping hotspot was created to indicate that the pre-1940s vegetation cover differed from the 1940s vegetation.

Vegetation change hotspot maps are presented in Chapters 5-8. The maps are intended to call out major zones of vegetation change based on the available evidence, but should be interpreted with caution. The hotspots are generalized representations meant to highlight dominant trends of vegetation change at the watershed scale, but they simplify more complex changes that have occurred at finer scales. For example, a "grassland to shrubland conversion" hotspot highlights a general area where this type of conversion is believed to be the dominant vegetation trajectory since the 1940s (and possibly before), but within that hotspot there may be some areas that have persisted as grassland, other areas that have converted from grassland to hardwood or conifer forest, and other areas that were a non-grassland vegetation type historically. In addition, while the vegetation change hotspots cover the majority of the watershed, there are gaps that do not fall within any of the identified hotspots. Thus, the hotspot maps should not be interpreted as precise maps of vegetation change or used to quantify change in vegetation extent; they are most useful as a tool for visualizing major patterns of vegetation loss, gain, and persistence.

Historical Stream Network

A GIS layer representing the approximate historical channel network (Fig. 2.5) was created to provide context for understanding baseline conditions of historical wetland and riparian features. Development of the historical stream network focused on areas beneath current reservoir features, as these are the areas where the configuration has been most heavily modified; streams in the upper reaches of the watershed are relatively confined topographically, and have experienced comparatively minor changes in channel alignment, and thus contemporary stream data (e.g., USGS topographic maps and modern LIDAR data) were used for these areas. Because hydromodification occurred early on in many areas (see pages 67-69), data sources available for determining historical channel locations were limited in some cases. Channels on the valley floor were synthesized from historical maps (Wackenreuder 1855, Stevens 1856, Easton 1868, Scowden 1875; see Fig. 9.1 on page 217), cross-referenced with elevation data and aerial imagery. Though Lower San Mateo Creek (below the dam) is less topographically confined than streams in the upper watershed, the channel configuration shown in historical maps (e.g., Wackenreuder 1855, Easton 1868) is generally consistent with the contemporary configuration, and thus the modern stream mapping was used for this reach.

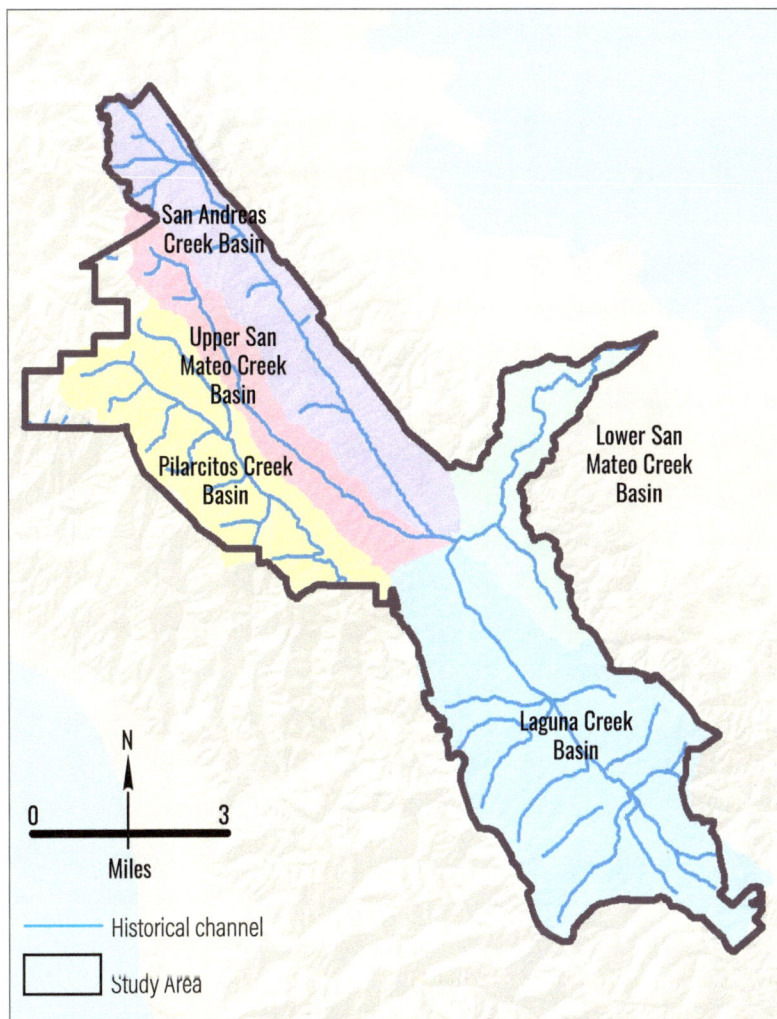

Figure 2.5. This map represents the approximate historical stream network within the Peninsula Watershed, and shows the general locations of major subwatersheds, or creek basins, within the watershed. Note that, at this scale, creeks under the Crystal Springs reservoirs appear straighter than they likely were.

San Andreas Creek Basin

Upper San Mateo Creek Basin

Pilarcitos Creek Basin

Lower San Mateo Creek Basin

Laguna Creek Basin

N

0   3
Miles

— Historical channel

☐ Study Area

Stream channel. *(Photo courtesy SFPUC)*

Above: View east across Peninsula Watershed from the Scarper Peak area. Below: Serpentine grassland flowers in the Peninsula Watershed. *(Photos courtesy SFPUC)*

CHAPTER 3

Drivers of Landscape Change

Overview

The distribution and assembly of vegetation communities is dictated by a wide range of factors. At global and regional scales, climatic variables such as precipitation and temperature, and topographic variables such as elevation, drive broad patterns of vegetation occurrence. At more local scales, variables such as slope, aspect, soil type, and groundwater depth further structure vegetation distribution. Physical processes such as fire and flooding, which are themselves governed by climate and topography, and ecological processes such as herbivory are also fundamental drivers of ecosystem dynamics and distribution.[1] Within the context of these regional and local environmental drivers and processes, human activities and human-driven changes, including anthropogenic climate change, atmospheric nitrogen deposition, land and water use modifications, land management practices, and nonnative species invasions (including introduction of plant pathogens) can have profound impacts on vegetation patterns.

Succession—the theory that, following a disturbance, the vegetation community in a given area will transition through a predictable sequence from an early "pioneer" stage to a final "climax" stage—is a foundational concept in ecology (Clements 1916, 1936). However, ecologists have come to recognize that vegetation dynamics can be much more complex than suggested by this traditional model of plant succession. While some areas may undergo a relatively linear transition in response to a disturbance, others may follow different successional pathways towards "alternative stable states" (Suding et al. 2004, Beisner et al. 2003). In some cases, state changes can occur suddenly when a critical threshold is crossed (Suding and Hobbs 2009). The trajectory of change may be dependent on multiple drivers interacting at different scales, including highly localized environmental conditions. For example, fire suppression may encourage

---

1 In this report, the terms "physical or ecological processes" and "disturbance regimes" are often used interchangeably to refer to fires, flooding, herbivory, and other intrinsic ecosystem drivers.

recruitment of woody vegetation, but simultaneous grazing may limit or prevent this recruitment, and the ultimate effect on vegetation change may depend on the relative magnitude of these and other factors such as topographic aspect. Consequently, the direction, rate, and scale of vegetation community transitions can be very difficult to predict (Soranno et al. 2014). Identifying the drivers of past landscape change is similarly challenging: information about historical disturbance regimes and local land use history is often incomplete, and linking vegetation change to particular drivers is often based on inference rather than direct evidence of causality.

Nevertheless, there is a large body of literature on succession and vegetation community response to disturbance and land use drivers in central coastal California that makes it possible to predict broad patterns of vegetation change with some degree of confidence. In the absence of major land use modifications or regular disturbances like fire or grazing, vegetation communities in the region have frequently been observed to transition from more open vegetation types, such as grasslands, to more closed vegetation types, such as shrublands or woodlands (Keeley 2005, Sandel et al. 2012). Changes within a particular vegetation type, such as increases in shrub or tree density or changes in species dominance, are also possible. Numerous studies have documented conversion of grasslands to coastal scrub (often dominated by coyote brush [*Baccharis pilularis*]) or, less frequently, to chaparral (McBride and Heady 1968, Hobbs and Mooney 1986, Williams et al. 1987, Callaway and Davis 1993, Russell and McBride 2003). Coastal scrub and chaparral can facilitate transition to either hardwood forest (McBride 1974, Callaway and Davis 1993, Mensing 1998, Zavaleta and Kettley 2006) or conifer forest (Greenlee et al. 1983, Horton et al. 1999, Dunne and Parker 1999, ABI et al. 2003, Ford and Hayes 2007). Established shrub species can facilitate survival of tree seedlings through a number of mechanisms, including enhanced soil moisture, reduced predation, and sharing of mycorrhizae, thus promoting the transition from shrubland to woodland (Callaway and D'Antonio 1991, Callaway and Davis 1998, Horton et al. 1999, Dunne and Parker 1999). Direct conversion from grassland to Douglas-fir (*Pseudotsuga menziesii*)-dominated conifer forest (likely facilitated by existing trees or saplings) has also been documented, particularly during wet climatic periods (Kennedy and Sousa 2006), though evidence suggests that grassland conversion to oak-dominated hardwood forest generally transitions through an intermediate shrubland stage (McBride 1974, Callaway and Davis 1993, Zavaleta and Kettley 2006). Among hardwood forest types, coast live oak (*Quercus agrifolia*) woodland may be outcompeted by California bay

40 • Peninsula Watershed Historical Ecology Study
Drivers of Landscape Change

(*Umbellularia californica*) woodland (Safford 1995). Hardwood forest may also transition to Douglas-fir-dominated conifer forests (Steinberg 2002, Schriver et al. 2018).

Disturbances can interrupt these successional pathways, and may be needed to maintain certain vegetation types. In many parts of the state, coastal prairies and other grasslands require periodic fire or grazing to prevent transition toward shrubland or woodland (Tyler at al. 2007, Ford and Hayes 2007), except where other environmental conditions (such as serpentine soils) inhibit the transition to woody vegetation types (Harrison and Viers 2007). In the East Bay hills, for example, McBride (1974) found that grasslands exposed to cattle grazing had minimal coyote brush cover, while adjacent ungrazed grasslands experienced rapid coyote brush invasion. Recurrent fire in coyote brush-dominated coastal scrub interrupted conversion to woodland by killing oak and California bay saplings (McBride 1974). Similarly, repeated prescribed burns were found to be an effective means of controlling coyote brush encroachment into coastal prairie at Point Pinole Regional Shoreline (Bartolome et al. 2012, Hopkinson et al. 2020). Many authors have inferred that indigenous use of fire was responsible for maintaining native grasslands in areas that experience a very low incidence of lightning ignitions and would likely otherwise have been dominated by woody vegetation (Greenlee and Langenheim 1990, Mensing 1998, Keeley 2002, Anderson 2005, Cuthrell 2013a). Prior to intensive native land management, large native herbivores such as tule elk (*Cervus elaphus nannodes*), mule deer (*Odocoileus hemionus*), pronghorn antelope (*Antilocapra americana*), and a variety of Pleistocene megafauna likely played a fundamental role in maintaining coastal prairies and other grasslands (Ford and Hayes 2007, Edwards 2007, Jackson and Bartolome 2007).

Within the context of these environmental drivers and processes operating at various spatial and temporal scales, human activities and human-induced ecosystem changes have strongly influenced vegetation trajectories within the Peninsula Watershed. Removal of indigenous communities and several centuries of fire suppression have caused a substantial departure from likely historical fire regimes, contributing to expansion of woody vegetation types. Intensive grazing and cultivation in parts of the watershed during the 19th and early 20th centuries likely reduced rates of shrub and forest encroachment, contributed to nonnative species introductions, and increased erosion rates. By the 1930s grazing was excluded from the watershed, and thus no longer served as a control on the expansion of woody vegetation. Logging activities in the mid-19th century removed old-growth redwoods and other trees, while deliberate planting of eucalyptus (*Eucalyptus* spp.), acacia (*Acacia* spp.), Monterey pine (*Pinus radiata*), Monterey cypress (*Hesperocyparis macrocarpa*), and other species resulted in persistent populations of these species. Reservoir construction and other hydromodifications beginning in the mid- to late 19th century resulted in direct habitat loss as well as altered hydrologic regimes. Urban development has resulted in substantial habitat loss as well, particularly on the eastern side of the watershed.

Grounded in these concepts of disturbance and community succession in coastal California, the remainder of this chapter summarizes each of the key environmental and land use drivers believed to have influenced vegetation trajectories in the watershed over the past two centuries, and discusses the likely impacts of each of these drivers in terms of vegetation extent, distribution, and composition. Because the objective is to understand drivers of landscape *change*, the focus of the chapter is on factors that have varied substantially over time, such as climate conditions, physical processes, and land use activities; less focus is given to factors that have remained relatively stable over time, such as topographic and edaphic controls. Hypothesized vegetation community changes resulting from these drivers are presented in Table 3.1 and Figure 3.1. These hypotheses serve as the basis for interpreting the observed vegetation shifts discussed in chapters 4–8. A concise graphical summary of the land use history of the watershed is shown in Figure 3.2.

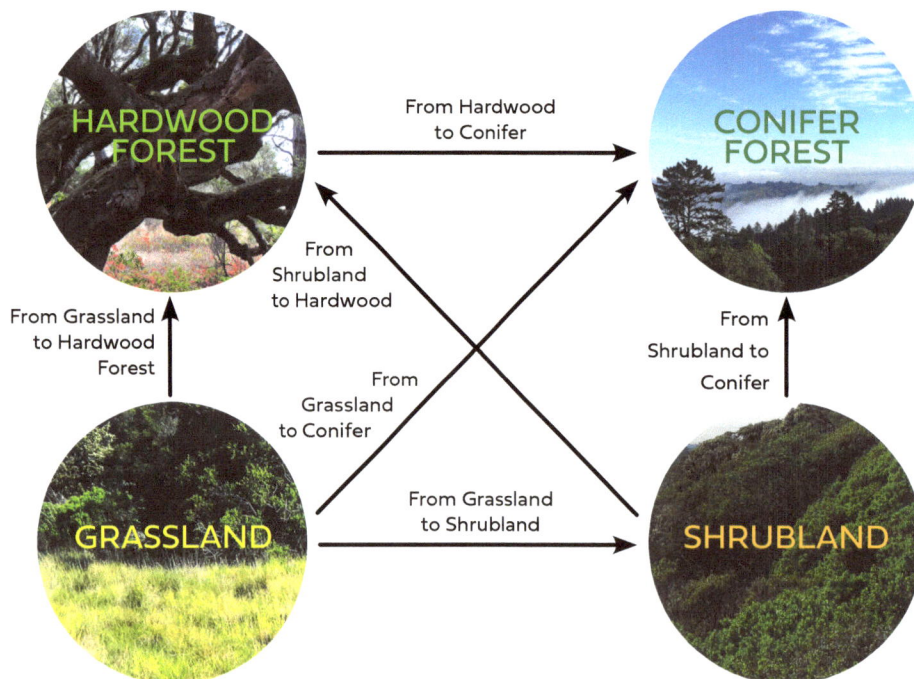

Figure 3.1. Hypothesized vegetation shifts within the Peninsula Watershed following Euro-American colonization and associated changes in disturbance regimes, land uses, and other drivers. *(Photos courtesy SFPUC)*

Table 3.1. Hypothesized transitions between or within vegetation types within the Peninsula Watershed following Euro-American settlement (including type conversion, composition shift, and maintenance), and predicted drivers of each type of landscape change.

	Hypothesized trajectory		Potential driver of landscape change
	From	**To**	
Vegetation community type conversion	Grassland	Shrubland	Succession/competition Fire [suppression] Grazing [removal] Climate [more precipitation]
	Grassland	Hardwood forest	Succession/competition Tree plantings Fire [suppression] Grazing [removal]
	Grassland	Conifer forest	Climate [more precipitation] Tree plantings/spread of invasive tree species Grazing [removal] Fire [suppression]
	Shrubland	Conifer forest	Climate [more precipitation] Fire [suppression] Biotic facilitation [e.g., mycorrhizae]
	Shrubland	Hardwood forest	Succession/competition Fire [suppression] Invasive species [Eucalyptus] Biotic facilitation [e.g., mycorrhizae]
	Hardwood forest	Conifer forest	Succession/competition Fire [suppression] Invasive species [Monterey cypress, Monterey pine]
	Any habitat type (non-wetland)	Wetland	Hydromodification [flooding around margins of reservoirs]
	Wetland	Non-wetland	Hydromodification [draining wetlands]
	Any habitat type (non-open water)	Open water	Hydromodification [reservoir construction]
	Any habitat type	Development	Urban development
Composition shift	Hardwood forest (Oak savanna/woodland)	Hardwood forest (CA bay)	Invasive species [Sudden Oak Death]
	Hardwood forest (Oak savanna)	Hardwood forest (Oak woodland)	Fire [suppression]
	Grassland (native forbs, native perennial grasses)	Grassland (non-native grasses and forbs)	Invasive species [non-native grasses and forbs] Grazing [active] Substrate [e.g., nutrient rich soils]
Maintenance	Grassland	Grassland	Fire [active] Grazing [active] Substrate [e.g. serpentine soils]
	Shrubland	Shrubland	Grazing [removal] Fire [active]
	Hardwood forest	Hardwood forest	Fire [suppression; may also lead to conversion to conifer forest]
	Conifer forest	Conifer forest	Fire [suppression]

FIRE

12,000 BCE to late 1700s: Indigenous peoples likely practiced burning of landscape.

Late 1700s to early 1800s: Spanish colonization and subjugation of native cultures; burning prohibited.

Early to late 1800s: General policy of fire prohibition/suppression.

GRAZING

1790s to early 1800s: Mission, Presidio livestock grazing on perimeter of watershed and on the eastern side of San Andreas Valley.

Early- to mid-1800s: Mexican land grants, Anglo-American dairies.

CULTIVATION

1790s – 1810s: San Mateo outpost established. Peak harvest at San Mateo in 1812.

1810s – 1860s: Cultivation in watershed interior begins; valley and ridgetop farms established by mid-century. Peak harvest in valleys by the 1860s.

LOGGING

1790s to mid-1800s: Spanish and early Mexican/American use of timber/clearing of land for settlements.

1850s - 1860s: Commercial redwood logging in southern portion of watershed.

SPECIES INVASIONS

Early 1800s: Earliest records of non-native species in central coastal California.

Mid-1800s to early 1900s: Eucalyptus, Monterey pine, Monterey cypress, Douglas-fir and other species planted for aesthetic, cultural, tax, timber and windbreak purposes. Earliest planted groves expand in coverage.

HYDROLOGIC MODIFICATIONS and URBAN DEVELOPMENT

Pilarcitos, San Andreas, Upper Crystal Springs, and Lower Crystal Springs dams constructed in 1867, 1868, 1877, and 1888 respectively.

<1769	1800s	1850s

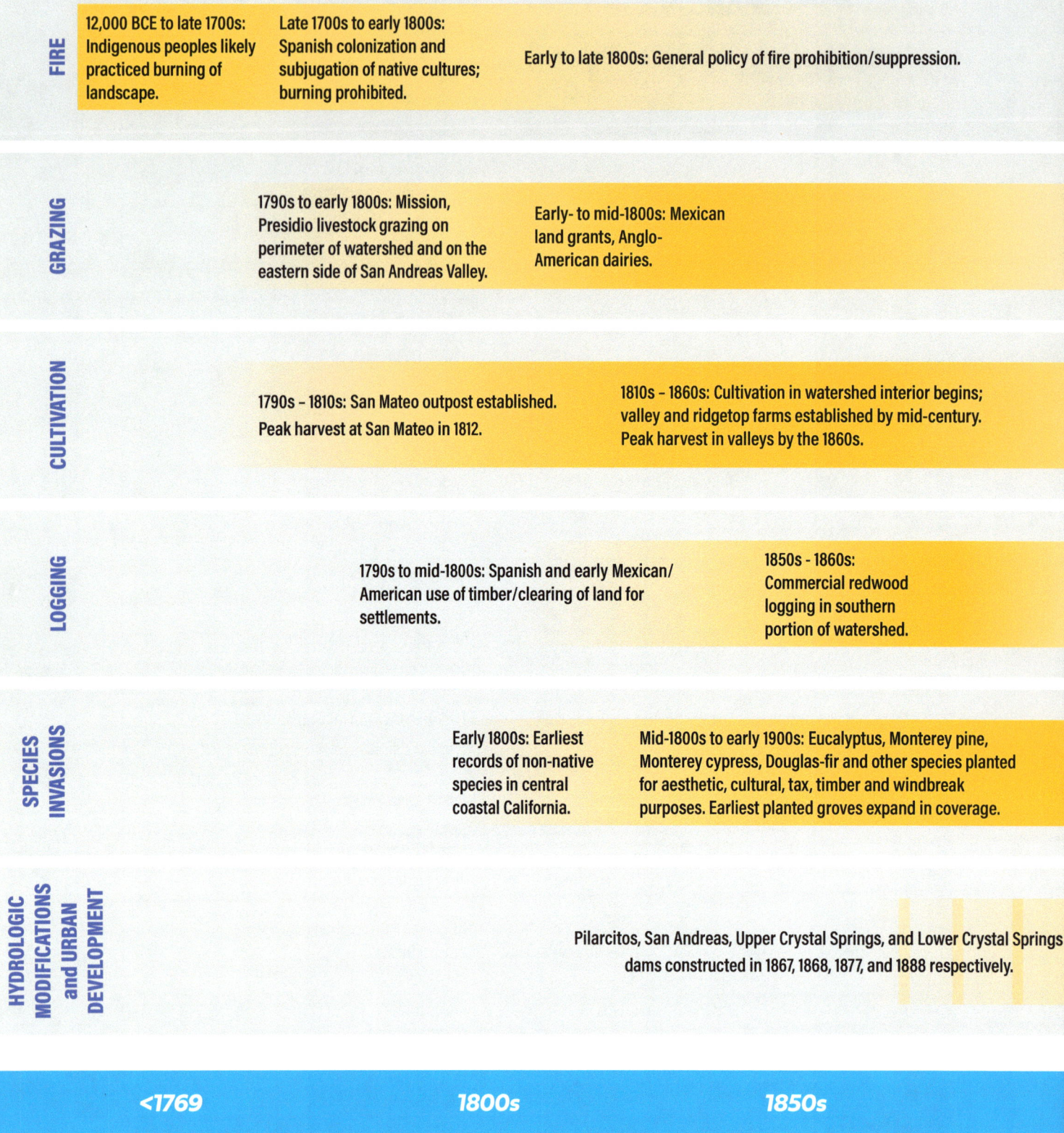

Figure 3.2. Graphical summary of major land use impacts within the watershed, showing the relative intensity of each land use at different points in time (darker orange represents higher intensity).

1889 and 1929: Large watershed fires.

1860s – 1930: SVWC land acquisition, grazing phased out within watershed.

1930 - present: SVWC watershed protection; grazing limited (except for small-scale goat grazing for fuels reduction).

1860s - 1930: SVWC acquires lands for watershed protection and phases out cultivation.

1930 - present: SVWC watershed protection.

1930 - present: Watershed protection; limited logging for fire breaks.

Early 1900s: Most contemporary non-native invasive species documented on the watershed via herbarium records.

Early 2000s: Introduction of *Phytophthora ramorum* and other plant pathogens.

Dams constructed across Upper San Mateo Creek in 1898.

Urban development intensifies within Lower San Mateo Creek Canyon and along the eastern margin of the watershed.

| 1900s | 1950s | Present |

Climate Variability

Climate, along with elevation, is a key driver of vegetation patterns at a regional scale. Vegetation distribution is influenced by a number of climatic variables, such as mean annual precipitation, seasonality of precipitation, temperature range and duration, periods of extreme rainfall or drought, fog and cloud cover, and wind patterns. The four major terrestrial vegetation types present in the Peninsula Watershed (grasslands, shrublands, hardwood forests, and conifer forests) have different tolerances to climatic conditions, but their tolerances are largely overlapping within the range of variability found within the watershed. Mean annual precipitation in California grasslands, for instance, ranges from less than 6 in to more than 78 in; sites with higher rainfall are generally characterized by greater species richness and greater dominance of perennial species than sites with lower rainfall (Bartolome et al. 2007). Among shrubland communities, annual precipitation in northern coastal scrub varies from 40-80 in (Holland and Keil 1995), and summer fog may be an important water source for coastal scrub species such as coyote brush (Emery et al. 2018). Chaparral occurs across a broad range of climatic conditions, but is most prevalent in areas receiving from 10-30 in of rainfall annually (Keeley and Davis 2007). Foothill oak woodlands typically receive 16-31 in of winter rainfall (Davis et al. 2016), while Douglas-fir forests in coastal settings occur in areas receiving 34-134 in of rain annually (Hermann and Lavender 1990).

Local precipitation trends from 1918-2007 were analyzed by Dingman (2014), who used daily rainfall records from four SFPUC gauges located throughout the Peninsula Watershed as well as downscaled monthly rainfall data from Oregon State University's PRISM Group. These analyses indicate that total annual precipitation within the watershed has increased by 1.5-2 mm/yr on average over the past 100 years (Table 3.2). A separate analysis was performed by SFEI-ASC staff for the period 1920-2008 using data from Point Blue's ClimateSmart Watershed Analyst Tool (analyzing the four sub-watersheds extending across the study area), which found a similar (though not statistically significant) trend. Findings for the Peninsula Watershed are consistent with an observed trend of increasing annual precipitation in the San Francisco Bay Area and northern California over the last century (Russo et al. 2013, Killam et al. 2014).

The historical temperature data from the Peninsula Watershed generally suggest a warming trend, with observed increases in both average minimum and maximum temperatures over time. Dingman (2014) analyzed PRISM temperature data from the watershed for the period 1918-2007, and found that average minimum temperature has increased by approximately 1.19 °C (2.14 °F) per century while average maximum temperature has increased by about 0.63 °C (1.13 °F) per century. SFEI-ASC staff analyzed average monthly temperature data for the four sub-watersheds in the study area using Point Blue's ClimateSmart Watershed Analyst Tool and found a similar (though not statistically significant) trend.

The changes in precipitation and temperature within the Peninsula Watershed over the past century may have contributed to observed changes in vegetation distribution over time. A study of grassland invasion by coyote brush at Jasper Ridge Biological Preserve, for example, found that shrub establishment was much higher in years with above average annual and spring precipitation (Williams et al. 1987). Similarly, a species distribution model for Douglas-fir in the Peninsula Watershed developed by Dingman (2014) found that the modeled distribution of Douglas-fir based on the historical increase in precipitation over the 20th century was consistent with an observed expansion in Douglas-fir distribution on Cahill Ridge since the 1920s. However, that study did not assess the relative importance of other potential factors (e.g., fire suppression) in driving the expansion of Douglas-fir. Spatial variability in climate may also influence the susceptibility of an area to this sort of vegetation type conversion: for example, Hsu et al. (2012) found that transitions from shrubland to forest ecosystems in central coastal California between 1985 and 2010 occurred more frequently in areas with higher annual precipitation and higher rates of summer evapotranspiration.

Table 3.2. "Thirty year average precipitation, minimum and maximum temperature" for the Peninsula Watershed. *(from Dingman 2014)*

Model	Average Annual Precipitation (mm)		Average Annual Minimum Temperature (°C)		Average Annual Maximum Temperature (°C)	
Time Period	1918-1947	1978-2007	1918-1947	1978-2007	1918-1947	1978-2007
PRISM	811	922	8.1	8.8	18.4	18.8
SFPUC	779	857	-	-	-	-

Native Land Management and Fire History

INDIGENOUS PRESENCE, HISTORY, AND CONTEXT

Indigenous people have lived on the San Francisco Peninsula for at least 13,500 years, and long-standing presence in or near watershed lands has been documented at multiple archaeological sites (Panich et al. 2009). At the time of European contact, at least eleven distinct local tribal groups, collectively recognized as speakers of the San Francisco Bay Ohlone/Costanoan language, occupied lands from the Golden Gate south to Point Año Nuevo and Los Altos. The indigenous population from the Peninsula to the Santa Cruz mountains at the time of European contact is estimated at around 1,400 - 2,145; estimated population densities range from 1.35 - 4.27 individuals per square mile (Babal 1990, Milliken 2009).

Within the Peninsula Watershed, the Ssalson local tribal group, numbering around 210 individuals, controlled the largest area (Fig. 3.3). The Ssalson occupied at least three primary village sites along San Mateo Creek and in the San Andreas Valley. The Lamchin group, estimated at 350 individuals, held lands on the southern end of the Watershed, likely including the present-day Phleger Estate just to the south of the watershed boundary. The Aramai, a group of approximately 50 people centered around two villages in the present-day Pacifica and Rockaway Beach areas, occupied lands extending up to Sweeney Ridge on the northwest side of the Watershed. To the south, the Chiguan group consisted of approximately 50 people centered around present-day Half Moon Bay; their territory may have encompassed a small area on the western side of the Watershed. Several other local tribal groups inhabited lands around the periphery of the Watershed, including the Urebure (around San Bruno Creek) and the Cotegen (in the Purissima Creek watershed; Milliken et al. 2009).

In the late 18th century, Spanish explorers encountered numerous villages along their routes across the Peninsula, traversing the lands of most of these groups (Stanger and Brown 1969, Milliken et al. 2009). For example, members of the Palóu-Rivera Expedition, traveling north through the Laguna Creek Basin and San Andreas Valley in November 1774, encountered "five large" villages in the southern portion of the valley (south of present-day Lower Crystal Springs Dam; Palóu and Bolton 2011). Further north, along San Andreas Creek, they encountered "women with their infants" who "came out from among the thickets, where they have their little huts on the bank" (Rivera et al. 1969). Friar Pedro Font, traveling northwards with the Anza Expedition along the shore of San Francisco Bay in March of 1776, described a "good-sized village situated on the banks of the arroyo of San Matheo [sic]" (Font and Brown 2011); numerous shell mounds have also been documented along Lower San Mateo Creek (Vanderlip 1980). Noted historian Alan Brown wrote that "probably the most heavily inhabited spot of any in the county in aboriginal times is now under water behind Crystal Springs Dam" (Brown in Vanderlip 1980).

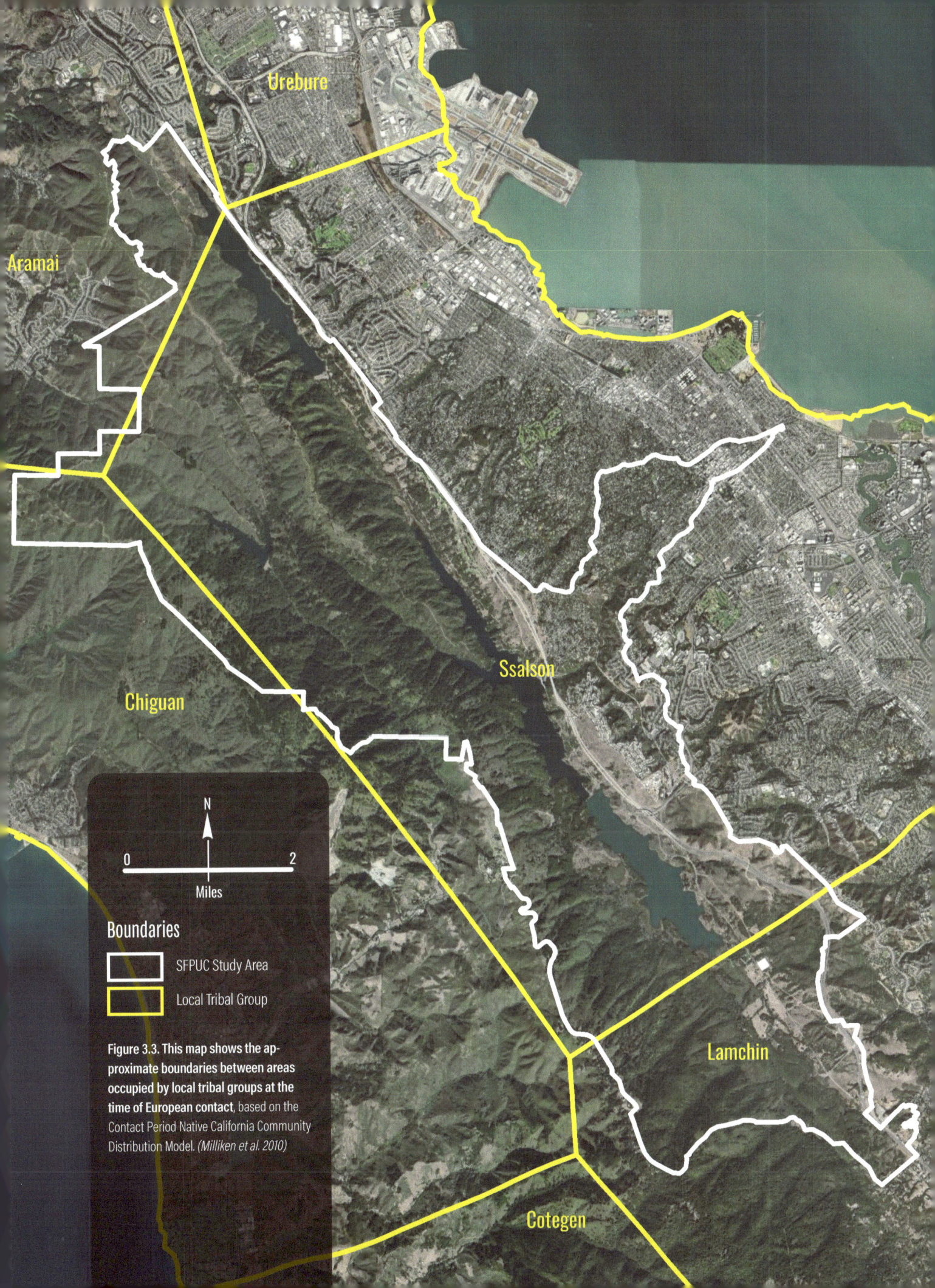

Urebure

Aramai

Chiguan

Ssalson

Lamchin

Cotegen

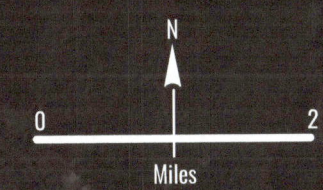

N

0 — 2
Miles

Boundaries

SFPUC Study Area

Local Tribal Group

Figure 3.3. This map shows the approximate boundaries between areas occupied by local tribal groups at the time of European contact, based on the Contact Period Native California Community Distribution Model. *(Milliken et al. 2010)*

With the arrival of the Spanish and establishment of the missions, indigenous populations experienced rapid declines precipitated by disease, violence, and forced loss of territory. By the early 1800s, nearly all indigenous peoples of the area had migrated to or were forced into enslavement at the Missions, though armed rebellions occurred at Mission Dolores in the 1820s (Leventhal et al. 1994, Milliken et al. 2009). Descendents of the Ohlone/Costanoan peoples in the San Francisco Bay region have undertaken efforts to seek Federal recognition and to revive many aspects of indigenous language and culture.

HISTORY OF FIRE MANAGEMENT

Indigenous burning and land management

Indigenous peoples on the central California coast were generally hunters and gatherers, foraging acorns and seeds as diet staples, supplemented with mammals and fish (Babal 1990). Like many native tribes throughout California, the Ohlone/Costanoans used fire as a land management tool to create mosaics of open habitat and promote favorable plants and game animals (Anderson 2005, Lightfoot and Parrish 2009, Cuthrell 2013a). A combination of evidence from Spanish explorers accounts, macrobotanical records, dendrochronology, and phytolith studies suggests that indigenous communities regularly burned grasslands and other habitat types across much of the Peninsula. Lightning-ignited fires in the region are rare (though potentially signfiicant; Keeley 2002, Lightfoot et al. 2013b), and thus anthropogenic burning was likely a significant factor in structuring vegetation distribution and maintaining grassland habitats. Though direct evidence for anthropogenic burning in the Peninsula Watershed is limited, reasonable extrapolations can be made based on evidence from surrounding areas.

Early evidence for indigenous use of fire comes from Friar Juan Crespí, a member of the Portolá Expedition of 1769. While ascending towards Sweeney Ridge from San Pedro Valley, for instance, Crespí observed that "the grass… [was] all burnt off by the heathens" (Crespí and Brown 2001). The following day, while traveling south between San Mateo Creek and San Francisquito Creek, Crespí observed that the valley "was all level land, seemingly of many leagues' extent, the entire plain being very good, dark, very grass-grown soil—although most of the grasses had been burnt off—and the entire plain much grown over with a great many large white oaks and live oaks" (Crespí and Brown 2001). Accounts from other early Spanish explorers corroborate Crespí's observations (Lightfoot 2006).

In addition to grasslands, intentional burning was also conducted in other habitat types, including oak woodlands, pine forests, hazelnut (*Corylus cornuta*) thickets, and coast redwood (*Sequoia sempervirens*) forests (Crespí and Brown 2001, Palóu and Bolton 1966, Cuthrell 2013b, Stephens and Fry 2005, Striplen 2014). Intervals of burning likely varied by habitat type and throughout time and space. For instance, fire scar dendrochronology of coast redwoods in the Santa Cruz Mountains (focusing on the Whitehouse, Waddell, and Scotts creek watersheds) was used to estimate a mean fire return interval of ~7 years during the period from 1600-1850 (Striplen 2014; see also Johnson et al. 2010). Additional dendrochronology research in redwood forests in the northern Santa Cruz Mountains (including at Huddart Park, not far from the West Union Creek watershed at the southern tip of the study area) suggests a mean fire return interval of 12-16 years between 1615-1884 (Stephens and Fry 2005). Phytolith analysis, archaeobotanical data, and pollen and charcoal records provide evidence of indigenous use of fire to maintain grasslands and prevent encroachment of woody vegetation at Quiroste Valley, in the Whitehouse Creek watershed in the Santa Cruz Mountains (Evett and Cuthrell 2013; Cuthrell et al. 2013; Cowart and Byrne 2013; Lightfoot et al. 2013a,b). Similarly, phytolith analysis, archaeological evidence, and archival records indicate that indigenous burning was used to maintain grasslands at McCabe Canyon in Pinnacles National Park (Johnson et al. 2010). In the greater Monterey Bay area, Greenlee and Langenheim (1990) estimate that prior to European settlement, mean fire return intervals were 1-15 years for prairies and coastal sage, 18-21 years for chaparral, 1-2 years for oak woodland, 50-75 years for mixed evergreen forest, and 17-82 years for redwood forest. In general, anthropogenic burning of grasslands likely occurred in late summer/early fall (Cuthrell 2013b); fires in redwood forests likely occurred between late summer and late winter, based on fire scar data (Striplen 2014).

The frequency of anthropogenic burning likely decreased rapidly following Spanish colonization. In 1793, Governor Arrillaga proclaimed an edict to prohibit traditional burning in Alta California:

> With attention to the widespread damage which results to the public from the burning of the fields, customary up to now among... Indians in this country... I see myself required to have the foresight to prohibit for the future... all kinds of burning, not only in the vicinity of the towns but even at the most remote distances... I order... presidios... to... take whatever measures... necessary to uproot this very harmful practice of setting fire to pasture lands... [to] stop fire, or failing that, to direct it into another direction which may result in less damage, apprehending the violators... [and seeing them] punished. (Arrillaga in Timbrook et al. 1982)

Fire and fire-management history post-1800

The prohibition of traditional fire management was succeeded by a period of land settlement, though fires still broke out sporadically. While traveling from Santa Cruz to Santa Clara in 1841, for instance, diplomat Eugène Duflot de Mofras wrote, "Occasionally the traveler is amazed to observe the sky covered with black and copper colored clouds... to see a fine cloud of ashes fall. Such extraordinary spectacles are caused by prairie or forest fires started by careless Indians or white men... frequently such fires smolder for several months, and spread from one end of the province to the other. These in fact often seriously handicap travelers who are overtaken by fires out on the plains where the grass is nine or ten feet" (de Mofras in Clar 1959).

Despite a general policy of fire suppression, a number of major fires were recorded in and around the Peninsula Watershed in the late 19th and early 20th centuries. A fire in September 1877 burned portions of the Sawyer Tract west of San Andreas Reservoir, including areas forested with oak, California bay, and Pacific madrone (*Arbutus menziesii*; Lawrence 1925). Following the fire, the *Times-Gazette* (in Lawrence 1925) reported that "the soil is covered with ashes nearly two feet deep. A bird's eye view of the burned district presents a picture of desolation."

The most severe wildfires in the recorded history of the area occurred in September 1889, when a series of blazes swept through large portions of the watershed and the hills to the south and west, extending nearly to Halfmoon Bay and Woodside (Fig. 3.4). Fires burned large areas to the west and south of San Andreas Lake, in the southern portion of Pilarcitos Canyon, along Cahill Ridge, and to the south of Upper Crystal Springs Lake, affecting many different vegetation types (*San Francisco Chronicle* 1889a,b; *Petaluma Courier* 1889, *Daily Alta California* 1889). The *Times-Gazette* (in Lawrence 1925) reported, "The heavens seemed to be on fire in that portion of this country lying west of San Andres [sic] Reservoir," while the San Francisco Chronicle wrote, "Fires... raged in the valley and on the mountain sides in all directions." In San Andreas Valley, the fires burned through "grass and brush" (*Petaluma Courier* 1889). West of Upper Crystal Springs Lake, the fire burned through agricultural fields, chaparral, and hardwood forest, leaving "here and there the ragged branches of an oak or madrone denuded of its foliage" (*San Francisco Chronicle* 1889b). Along Kings Mountain Road, observers stated that "flames roared up fifty feet higher than the top of the tallest [redwoods]," and that "the trunks of many of the large redwoods were partially burned through" (*San Francisco Chronicle* 1889b, McFarland in *San Francisco Chronicle* 1889b).

After the enormity of the 1889 fires, SVWC established a formal fire prevention management program in the watershed. This program consisted

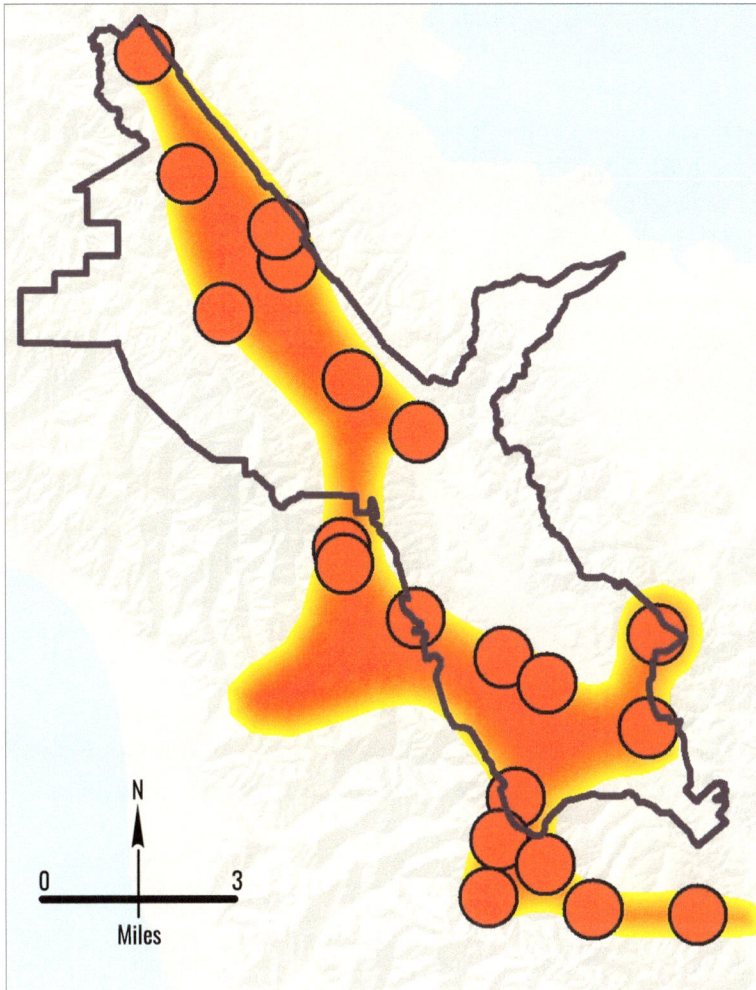

Figure 3.4. Conceptual reconstruction of the footprint of the September 1889 fires (red/yellow polygon). Areas where explicit evidence exists of fire impacts during the 1889 event (e.g., references to specific landowners affected) are shown in red dots. Note that there is considerable spatial uncertainty associated with this reconstruction.

at first of burning grasses on the east of San Andreas and Crystal Springs Lakes, plowing fireguards in timber stands, and burning all grasses between fireguards and trees (Fig. 3.5). After some damage to younger trees, this program was modified to use a disk plow and harrows for a strip 10 ft wide along all roads, planting areas and managed timber stands. Along all wooden flumes, timber from 50 ft on either side and understory vegetation within 6 to 10 ft were removed, every two years. In 1909, fire breaks of 10 to 15 ft along all high ridges in the watershed were installed, totaling 32 miles. This fire prevention program was evidently effective, as SVWC Water Division Superintendent W.B. Lawrence wrote in 1925 that there had been "no serious fires" since the program began and the "San Mateo County fires are not allowed to devastate as of yore" (Lawrence 1925).

From that point forward, documented fires in the watershed were rare. Several notable fires did occur in 1929 on the periphery of the watershed, with fires flaring up in San Pedro Valley, around Skyline Blvd. near the summit of Half Moon Bay Road, near Pilarcitos Canyon and Stone Dam, and between Pedro and Scarper Peaks. Ultimately, fires that year burned approximately 1,100 ac in the Pilarcitos Creek watershed, as well as 400 ac of company property north of the Pilarcitos Creek watershed in the San Pedro Mountains. The burned areas were predominantly "dense brush," with "very few trees" destroyed (Davis 1929). The last major recorded fires were in 1946, when "large intense fires" were documented in the "northern San Francisco Watershed" (Poinsot and Wong 2008). A number of small lightning-caused fires have occurred in the watershed and adjacent areas in recent decades, including areas to the west of Pilarcitos Dam and Stone Dam (J. Fournet pers. comm.).

Contemporary fire management has continued this fire prevention program and focused on fire suppression and fuels reduction. Fireguards are generally maintained via mowing on the watershed perimeter as well as along roadsides, fencelines, gates and other areas. Disking is conducted along the eastern boundary of the watershed (S. Simono pers. comm.). Vegetated fuelbreaks are created and maintained by using a hydraulic masticator (Brontosaurus), mounted on a tracked excavator or skid steer chassis, to mulch vegetation in place. The unit is capable of producing a small chip size which facilitates decomposition (J. Fournet pers. comm.). Goats are also occasionally used to reduce fuel loads (Ciardi 2019). Prescribed burns have been planned to reduce fuels and potential wildfire severity and to augment and enhance grasslands on the watershed (E. Natesan pers. comm.). Pruning, removal of hazard trees, and management of invasive species also contribute to fuel reduction efforts (Ciardi 2019, J. Avant personal comm.).

As a result of the cessation of indigenous burning and implementation of fire suppression over the past two centuries, the frequency of fire has decreased throughout the watershed. These changes in fire frequency have major consequences for the watershed's vegetation communities, and have likely been a primary driver of many of the observed vegetation shifts discussed in subsequent chapters.

Firebreaks in the hills. *(Photo by KQED Quest, courtesy CC BY 2.0)*

Figure 3.5. Early images of fire prevention and prescribed burning efforts on the watershed. (top) The caption for this image reads, "Fire on the watershed. The forest rangers started it as part of their field work, in order to test the most expeditious method of putting it out." (bottom) The caption for this image reads, "To put out flames in burning logs and dead trees, the forest rangers use a hand force-pump attached to a metal water-container" (Sweeley 1925). *(Top: ark:/13960/t00z8g243; Bottom: ark:/13960/t00z8g243 (bottom); both courtesy San Francisco History Center, San Francisco Public Library)*

Grazing and Agriculture

MISSION ERA (1769-1833)

Following the establishment of the Presidio military outpost and Mission San Francisco de Asís (Mission Dolores) in San Francisco in 1776, growth of these institutions was accompanied by increased demand for food, water, and timber (Alley 1883). The climate, soils, and native vegetation of San Francisco proved insufficient to supply these resources, and thus the Spanish looked southward for suitable grazing and farming lands. Initially, the Mission's focus was on San Pedro Valley, to the northwest of the Peninsula Watershed (around present-day Pacifica), where a Mission outpost was established in the 1780s (Stanger 1963). Cattle grazed in the hills around the valley, and up to 40 ac of wheat and 10 ac of corn were cultivated (Hynding 1982). An epidemic struck the area in 1791, however, and the outpost was deserted entirely by 1794 (Stanger 1963, Babal 1990).

In 1793, a new Mission outpost was established on Lower San Mateo Creek, which quickly became the center of agricultural activity (Babal 1990). Mission crop yields on the Peninsula increased rapidly, from approximately 1,200 bushels (consisting primarily of wheat, barley, corn, beans, and peas) in 1783 to a peak of over 11,100 bushels in 1814 (Stanger 1938, Postel 1994). Cultivated areas were likely concentrated around the Mission outpost on Lower San Mateo Creek. The size of the Mission herds likewise grew considerably, from less than 1,000 head in 1783 to a peak of over 23,000 head (including 11,240 cattle and 11,000 sheep) in 1811 (Engelhardt 1924). The Presidio also maintained a sizable herd, known as the "King's cattle," which numbered 1,215 head in 1790 and approximately 2,000 head of cows and several hundred head of horses in 1815 (Martinez 1852, Stanger 1963). The Mission and Presidio

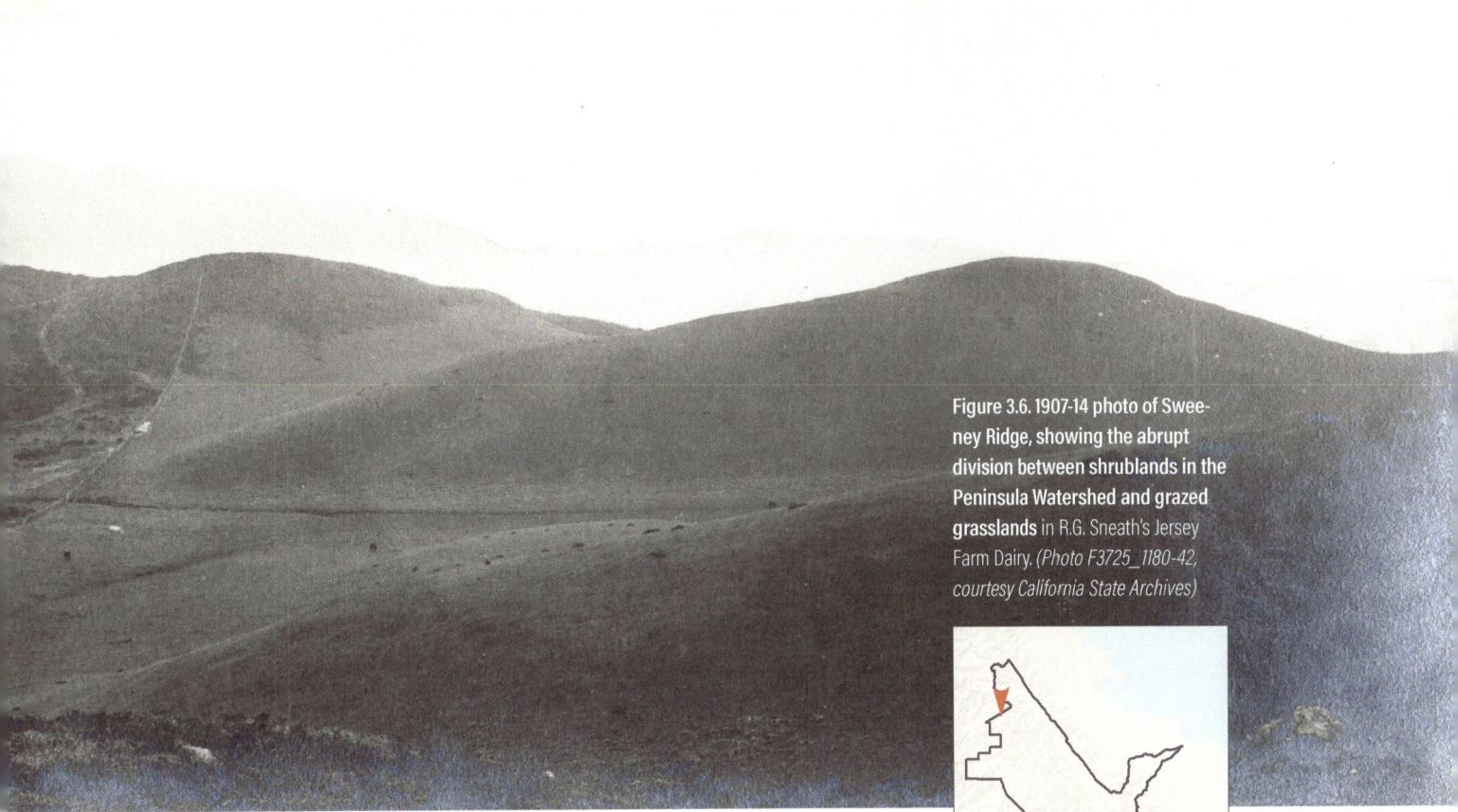

Figure 3.6. 1907-14 photo of Sweeney Ridge, showing the abrupt division between shrublands in the Peninsula Watershed and grazed **grasslands** in R.G. Sneath's Jersey Farm Dairy. *(Photo F3725_1180-42, courtesy California State Archives)*

PHOTO LOCATOR MAP

competed for prime grazing lands, and over the course of the late 18th and early 19th centuries their herds alternately grazed areas extending from present-day South San Francisco south to San Francisquito Creek and west to the San Andreas Valley (Stanger 1938, 1969).[2]

The impact of Mission-era grazing on the Peninsula Watershed is uncertain, but was likely mostly limited to portions of San Andreas Valley, the Laguna Creek Basin, and areas on the eastern margin of the watershed. Dense woody vegetation in many parts of San Andreas Valley precluded widespread grazing, though did not prevent the movement of stray cattle: early accounts describe how cows would hide in "thickets and ravines" (Jones 1853), reverting to a semi-wild state (Viader 1853, Stanger 1938). Mission and Presidio herds grazed areas that later became the Buri Buri and Pulgas land grants, which include much of the eastern portion of the watershed (see pages 58-59; Jones 1853, *San Francisco Daily Herald* 1853, Stanger 1938). Presidio herds were ultimately taken over by the Mission or sold off, and Mission herds dwindled after secularization of the missions in the 1830s, finally disappearing by 1844 (Englehardt 1924).

2 Herds maintained by Mission Santa Clara, to the south, only ranged as far north as San Francisquito Creek (Martinez 1852, Bowman 1947).

RANCHO AND AMERICAN ERAS (1834 TO EARLY 20TH CENTURY)

As the mission system collapsed, the Mexican government granted large tracts of land to private citizens. Portions of six land grants, or ranchos—San Pedro, San Mateo, Buri Buri, Feliz, Pulgas, and Cañada de Raymundo—extended into the study area and supported large numbers of livestock as well as some cultivated areas (Fig. 3.7). The Pulgas Rancho, granted to Maria Soledad Ortega in 1835 (though provisionally granted to Don Jose Dario Arguello by the Spanish governor in 1795), supported 4,000 cattle and 2,000 horses in 1838 (Hynding 1982, Babal 1990, Fredricks 2008). The Feliz Rancho, granted to Domingo Feliz in 1844, supported 500-600 head of cattle, as well as 15-20 ac cultivated in barley and wheat (Brown 1853, Babal 1990). Rancho Buri Buri, granted to Jose Antonio Sanchez in 1835, supported 8,000 cattle and 1,000 horses in 1838 (Wyatt 1947). The San Mateo Rancho, granted to Cayetano Arenas in 1846 and resold to William Howard and Henry Mellus shortly thereafter, supported horses, cattle, and other livestock (Teschemaker 1857, Hoover and Kyle 2002). Rancho Cañada de Raymundo, granted to John Coppinger in 1840, was used primarily for logging (see below) rather than raising livestock (Hynding 1982, Babal 1990).

Following the U.S. annexation of California in 1848 and the start of the Gold Rush in 1849, settlers began to buy up and subdivide the Mexican land grants, and a number of smaller farms and ranches sprang up throughout the Peninsula. By the late 1850s and 1860s, agriculture had expanded throughout the San Andreas Valley and the Laguna Creek Basin, and also occupied substantial areas along San Mateo Creek and Kings Mountain Ridge (Easton 1868, *Santa Cruz Weekly Sentinel* 1871, Moore and DePue 1878, Lawrence 1922a, Stanger 1938, Johnson 1955, Babal 1990). In 1860, for example, the *San Mateo County Gazette* (in Babal 1990) reported, "Throughout the extent of the cañada [Raymundo], the sturdy farmer's thrift is seen in the well-cultivated fields of grain and meadow land"; another observer in 1869 stated that "thriving farmers cultivate every available acre in this Cañada [Raymundo]" (Davidson 1869). Resident William Burke, recalling the area around San Feliz Station (near Upper Crystal Springs Dam) in the late 19th century, stated, "To the north were fine farms and country estates... To the south was a long stretch of hay-fields and pastures in which dairy herds grazed" (Burke 1926).

Figure 3.7. Two-thirds of the study area falls within the boundaries of private land grants established during the early 1800s. These ranchos, granted to wealthy citizens by the Mexican government, included portions of the San Pedro, San Mateo, Buri Buri, Feliz, Pulgas, and Cañada de Raymundo land grants.

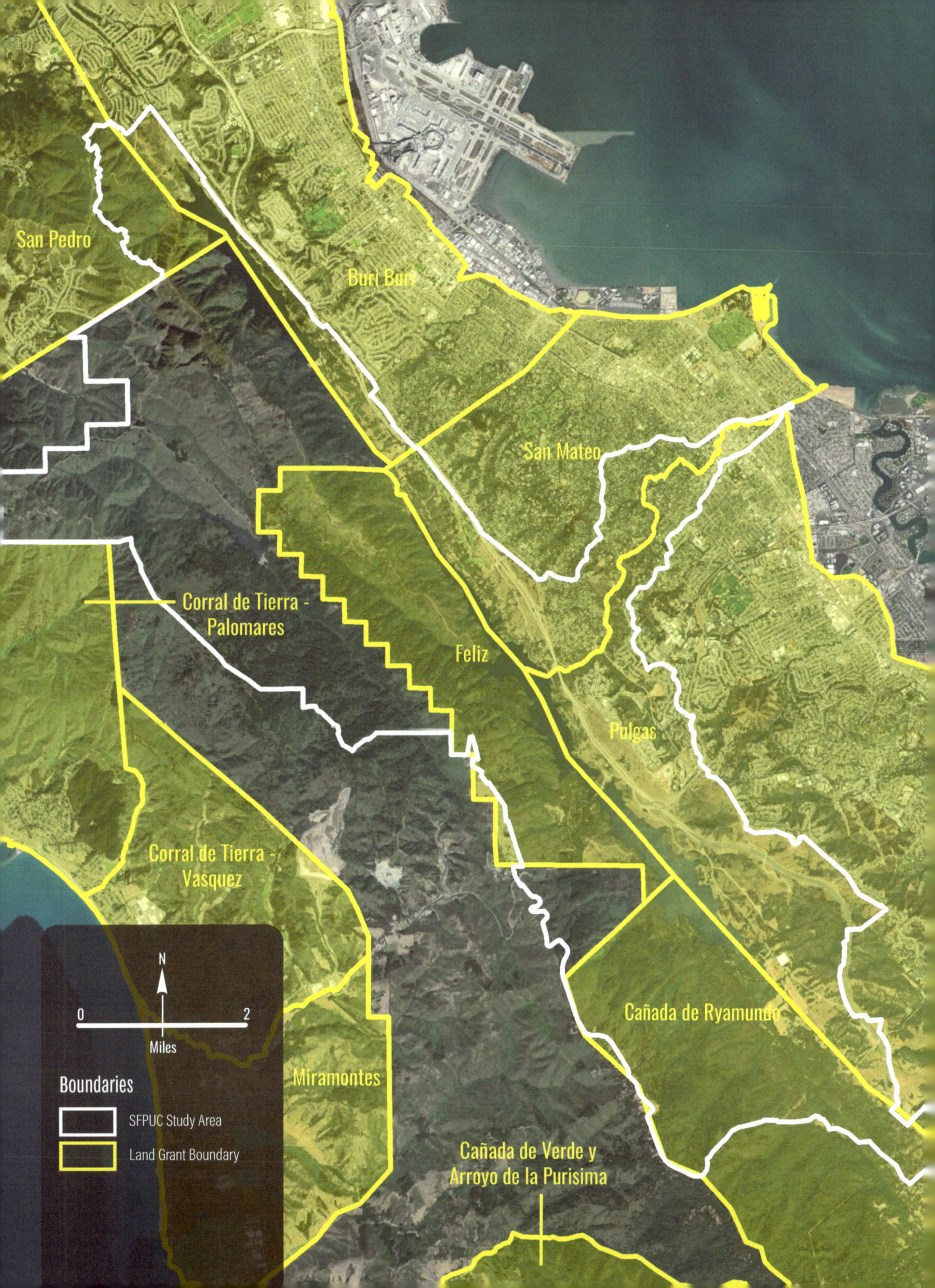

San Pedro

Buri Buri

San Mateo

Corral de Tierra -
Palomares

Feliz

Pulgas

Corral de Tierra -
Vasquez

Cañada de Ryamundo

Miramontes

Cañada de Verde y
Arroyo de la Purisima

N

0   2

Miles

Boundaries

SFPUC Study Area

Land Grant Boundary

Common crops included barley, flax, and other grains; legumes; orchards; and to a lesser extent watermelon, corn and tobacco; dairy farms were also prevalent (*California Farmer* 1864, *Santa Cruz Weekly Sentinel* 1871, Moore and DePue 1878, Lawrence 1922b, Babal 1990). One of the largest dairies in the area, the Jersey Farm Dairy (established in 1875), occupied almost 3,000 ac and extended into the northernmost part of the watershed around Sweeney Ridge (Bromfield 1894, Babal 1990; Fig. 3.6 on page 57). This dairy remained in operation until the 1930s, much longer than many of the other farms in the valley (Babal 1990). To the west, the Fifield Dairy (which operated from 1864 to 1905) occupied 1,066 ac between Fifield and Sawyer Ridges, and also included large areas planted to barley (California Farmer 1864, Babal 1990; Fig. 3.8; see also Fig. 5.13 on page 132). Another large farm and horse ranch, operated by Leander Sawyer, occupied 2,200 ac along San Andreas Creek; it was purchased by SVWC in 1876 (Babal 1990). The area just west of Lower Crystal Springs Dam supported a vineyard and orchard owned by Agoston Haraszthy (Johnson 1955). To the south and west of present-day Upper Crystal Springs Dam, Michael Casey ran a 327 ac farm and ranch with cattle and horses until about 1900 (Fig. 3.9). On the eastern side of the watershed, along Lower San Mateo Creek, William Howard grew barley and raised a small herd of cattle (Maynard ca. 1850). Further south, the Bollinger dairy (which operated 1854-1874) occupied 628 ac, extending from San Andreas Valley up the western slope of Pulgas Ridge (Easton 1868, Babal 1990). Numerous other smaller farms and ranches occupied lands within the valley prior to reservoir construction. Grazing and agriculture were also common on the upper portion of Kings Mountain Ridge, near present-day Skyline Blvd (Fig. 3.11; Bromfield 1894, Unknown 1912).

Figure 3.8. (right top) "Residence of W.J. Fifield... San Mateo County, Cal." Fifield's cattle can be seen grazing well into the hills. *(Moore and DePue 1878, courtesy Gilbert Richards publications)*

Figure 3.9. (right bottom) "Ranch and Residence of Michael Casey, Cañada Ramonda [sic], San Mateo Co. Cal." The newly constructed Upper Crystal Springs Reservoir can be seen in the background. *(Moore and DePue 1878, courtesy Gilbert Richards publications)*

LOCATOR MAP

LOCATOR MAP

Drivers of Landscape Change

Figure 3.10. "San Felix Station M. Carey, Prop. San Mateo Co. Cal." Cattle can be observed grazing to the extreme lower right of the illustration. (*Moore and DePue 1878, courtesy Gilbert Richards publications*)

LOCATOR MAP

Overall, the historical record reveals a trend of increasing impacts from grazing and agriculture in the watershed between the late 18th and early 20th centuries. Sizeable livestock herds roamed large areas of San Andreas Valley, the Laguna Creek Basin, and the eastern side of the watershed during the Mission and Rancho eras, while crops and dairy cattle dominated many of the same areas during the late 19th and early 20th centuries. More remote parts of the watershed, such as Pilarcitos Canyon, were not cultivated (Lawrence 1922b). While the cumulative impacts of grazing and cultivation in the watershed are a matter of some speculation, they likely included soil erosion and compaction, reduced rainfall infiltration, reduced water quality, altered plant biomass, reduction in the extent of woody vegetation communities, and introduction of non-native plant species (*San Francisco Chronicle* 1897, Burcham 1961, Trimble and Mendel 1995, Bilotta et al. 2007). In the absence of frequent fires, livestock grazing (as well as active clearing) was likely a dominant driver in limiting the establishment of woody vegetation and maintaining grasslands and other early successional habitats. Early observers noted that cattle grazing was sufficient in areas to limit woody vegetation growth (Davy 1895, Hoag 1973). Ultimately, grazing was essentially discontinued entirely in watershed lands due to water company policy, as SVWC acquired watershed lands and excluded grazing to promote water quality.

> As we near the crest of the divide we are astonished by the number of gardens, reaching as they do, to the very tops of the hills...at the top of the mountains... we have never before saw such fields of flax as here border the road.
>
> —*SANTA CRUZ WEEKLY SENTINEL* 1871, describing agriculture along Kings Mountain Ridge on the road from San Mateo to Half Moon Bay

Figure 3.11. "Grazing" and "cultivation" near summit of Kings Mountain Ridge. *(Bromfield 1893, courtesy San Mateo County Public Works and Recorder)*

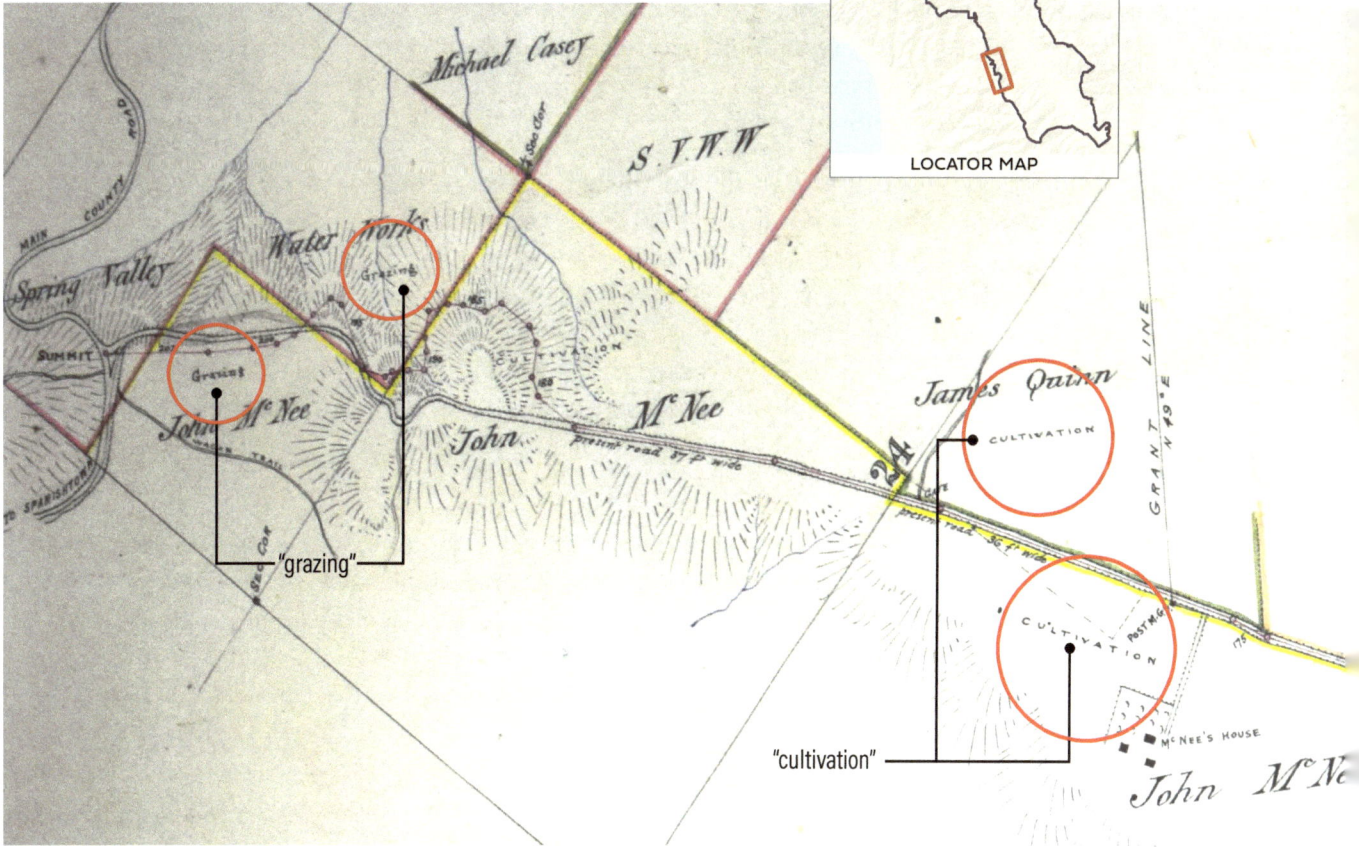

Logging and Tree Plantings

The timber resources of the Peninsula's forests and woodlands have been utilized by generations of settlers. During the Mission era (late 18th and early 19th centuries), modest amounts of redwood timber from the Peninsula were used in the construction of the Presidio, the Franciscan missions, and other structures (Stanger 1967, Carranco and Labbe 2003, Postel 2010). Langellier and Rosen (1992), for instance, report that during construction of an esplanade at the Presidio in 1793, "woodchoppers went into the hills west of San Mateo... to secure redwood." Timbers were dragged by oxen to San Francisco or loaded on schooners at nearby embarcaderos (Postel 2010). Because it was so labor intensive, however, the extent of redwood logging was relatively limited during the Mission era.

Beginning in the 1830s and 40s, an increasing number of settlers began to exploit the timber resources of the region's vast redwood forests for commercial gain. Bill "The Sawyer" Smith, reported to be the first sawyer in the region, arrived in the early 1830s and began logging the area around Woodside (Alley 1883, Postel 2010). John Coppinger, the grantee of the Cañada de Raymundo Rancho, was another early settler, and conducted extensive logging on his land in the 1840s. The National Park Service Historic Resource Study for San Mateo County, for instance, reports that "according to Mexican government records of 1841, 100,000 board feet of wood were ready at the embarcadero near Santa Clara for export... it is presumed that most or all of this was logged on Copinger's [sic] *rancho*" (Postel 2010). Operating before the development of industrial sawmills, these early sawyers relied on whipsaws, adzes, and other handtools, and thus the output of timbers was relatively small (Stanger 1967).

With the onset of the Gold Rush in 1849, demand for redwood lumber swelled. The first sawmill in the region was constructed in 1849 along Alambique Creek, in the San Francisquito Creek watershed south of the study area (Stanger 1967). By 1853, the *Daily Alta California* (1853) reported that there were ten sawmills operating on the Peninsula, producing 67,000 feet of lumber each day. The number of operating mills increased to approximately fifty by the 1850s-1860s (Stanger 1967). The center of this lumbering activity was in the heart of the redwood forest to the south of the Peninsula Watershed, in areas around Kings Mountain and Woodside. Redwood City was the principal port used to export lumber from the Peninsula.

Because redwood forests only extended into the southernmost portion of the Peninsula Watershed (see pages 196-199), the impacts of commercial logging in the watershed were relatively small compared with other parts of the Peninsula. Nevertheless, at least two sawmills existed in the southern part of the Peninsula Watershed: Albert Smith's mill (which operated from 1853 to 1854, when it burned down), and Pinckney's Mill (which began operation in 1855; Fig. 3.12). Stanger (1967) suggests that "there is evidence that the area [around Smith's and Pinckney's mills] was thoroughly logged." Oberlander (1953) offers a similar assessment, stating that at the extreme southern end of Kings Mountain Ridge, "with the exception of a few virgin trees 6-9 feet in diameter, the entire grove is a secondary one, the trees having come from sprouts around the original cut stumps. The trees were cut approximately one hundred years ago and trees 1-3 feet in diameter have come up in secondary growth." Indeed, according to Alley (1883), by 1883 the old growth redwoods on the bayward side of the Peninsula had "almost wholly disappeared."

In addition to commercial extraction of redwoods, smaller-scale tree cutting by homesteaders and other settlers occurred elsewhere in the watershed. In the area around Sawyer Camp (between present-day San Andreas and Lower Crystal Springs reservoirs), for instance, Leander Sawyer harvested trees from the nearby hills and "dragged [them] by oxen-sleds down the ravine," where they were turned into firewood (Hoag 1973). A reporter in 1860 described how in Cañada de Raymundo, "clearings, where needed, are... being made, the open or less wooded lands being mostly occupied, having been the first to be taken up" (*San Mateo County Gazette* in Babal 1990). Trees were also removed in preparation for dam construction: for instance, one reporter in 1874 noted that "three hundred men are to be put to work immediately, scrubbing out stumps and cutting all trees of sufficient size" to make way for Crystal Springs Dam (Vanderlip 1980). As SVWC gradually acquired lands throughout the watershed, the company prohibited logging to protect water quality (Schussler 1875).

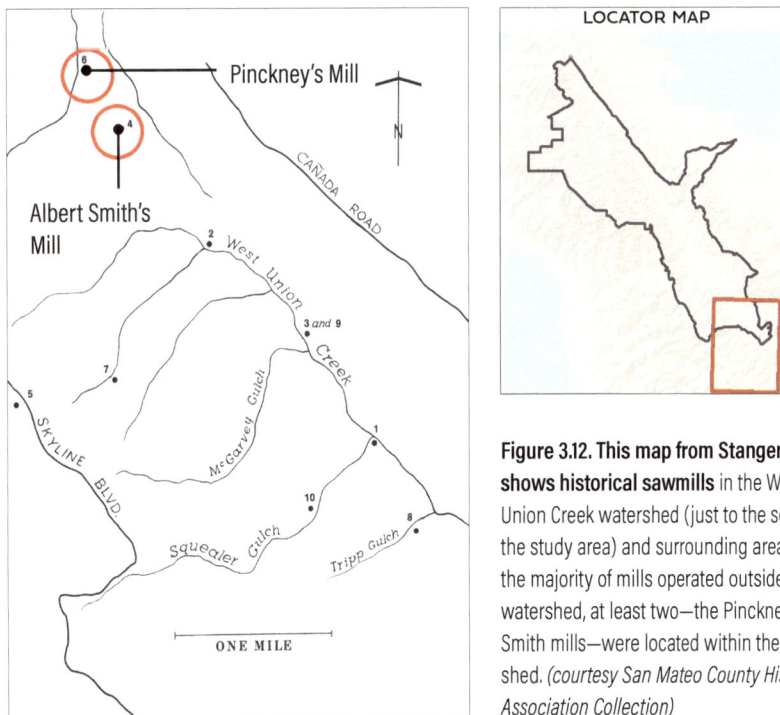

Figure 3.12. This map from Stanger (1967) shows historical sawmills in the West Union Creek watershed (just to the south of the study area) and surrounding areas. While the majority of mills operated outside the watershed, at least two—the Pinckney and Smith mills—were located within the watershed. (courtesy San Mateo County Historical Association Collection)

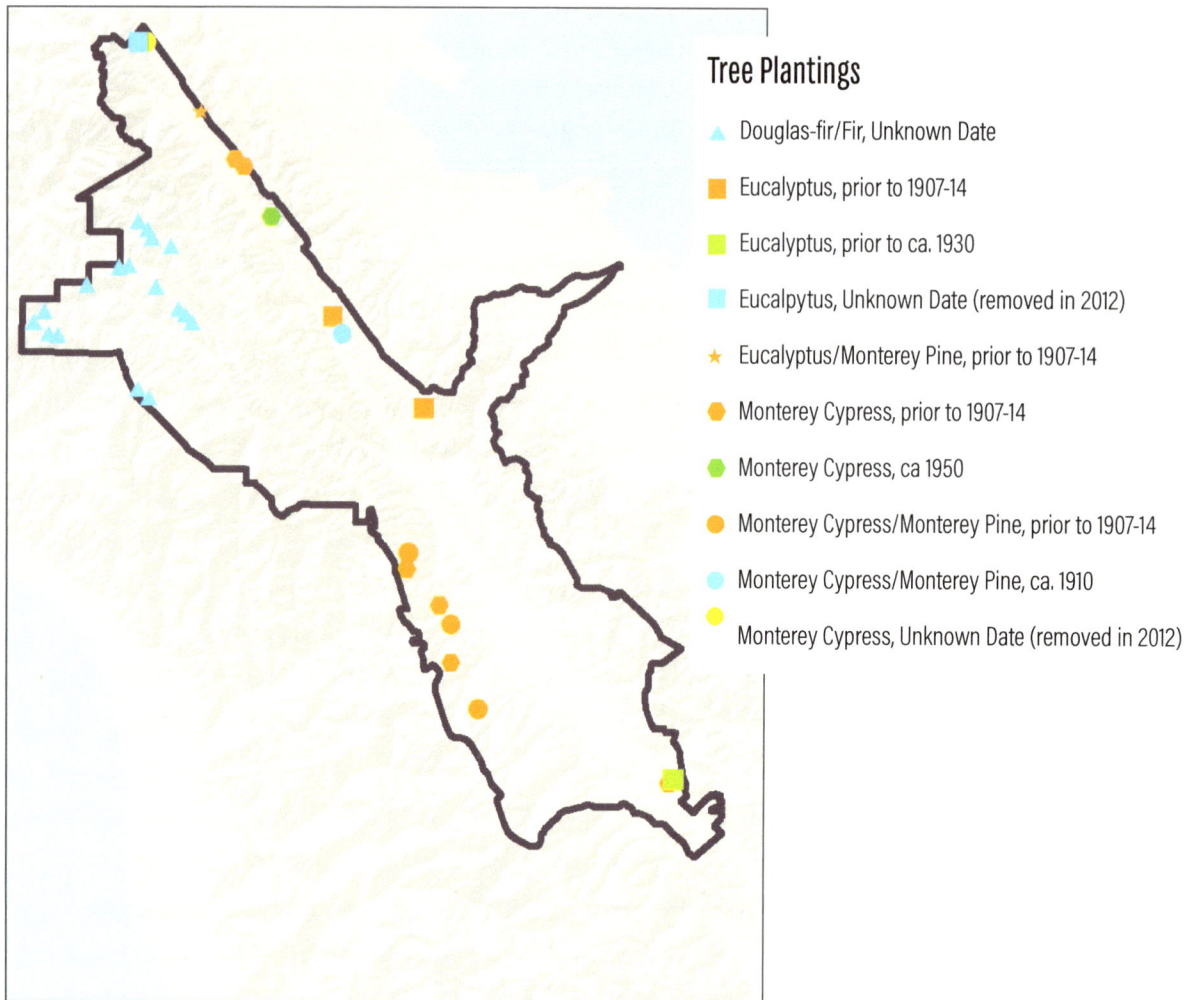

Fig. 3.13. Locations of known or suspected tree plantings in watershed. Planted species include non-indigenous trees such as Monterey cypress, Monterey pine, and eucalyptus, as well as native species such as Douglas-fir.

Tree Plantings

▲ Douglas-fir/Fir, Unknown Date

■ Eucalyptus, prior to 1907-14

■ Eucalyptus, prior to ca. 1930

■ Eucalpytus, Unknown Date (removed in 2012)

★ Eucalyptus/Monterey Pine, prior to 1907-14

⬤ Monterey Cypress, prior to 1907-14

⬤ Monterey Cypress, ca 1950

⬤ Monterey Cypress/Monterey Pine, prior to 1907-14

⬤ Monterey Cypress/Monterey Pine, ca. 1910

⬤ Monterey Cypress, Unknown Date (removed in 2012)

Intentional planting of trees has also occurred in a number of locations throughout the watershed. Trees were planted to serve as windbreaks, to reduce land taxes (wooded areas were generally taxed at a lower rate), to mark the site of homesteads, or for other reasons (Oberlander 1953). Favored trees included non-indigenous species such as Monterey pine (*Pinus radiata*), Monterey cypress (*Hesperocyparis macrocarpa*), fir (*Abies* spp.), eucalyptus (*Eucalyptus* spp.), and acacia (*Acacia* spp.). Douglas-fir and other native trees were also planted in some locations. Figure 3.13 documents locations of tree plantings throughout the watershed; the 1995 vegetation mapping (Schirokauer et al. 2003) shows approximately 650 ac dominated by planted or naturalized populations of Monterey cypress and eucalyptus. Further discussion of tree plantings is provided on pages 210-213.

Hydrologic Modifications

Rapid growth and development of San Francisco and the greater Bay Area during the second half the 19th century drove increasing demand for water, resulting in substantial hydromodification of the Peninsula Watershed. SVWC, established in the 1850s, played the most significant role in driving changes to the watershed. The company's first efforts involved construction of an earthen dam across Pilarcitos Canyon, which was completed in 1863 and captured water from the upper portion of Pilarcitos Creek watershed. The larger Pilarcitos Dam, 75 ft high, was constructed further downstream in 1867, and heightened to 95 ft in 1874. A pair of tunnels, constructed in the 1860s, conveyed water from Pilarcitos Reservoir under Cahill and Sawyer ridges, where it then flowed through pipes and flumes to San Francisco. Stone Dam, located along Pilacitos Creek two miles downstream of the much larger Pilacitos Dam, was constructed in 1871 (Lawrence 1922b, Babal 1990).

Shortly after Pilarcitos Dam was built, construction began on San Andreas Dam. The dam was completed in 1868, capturing water from the San Andreas Creek watershed and inundating the sag ponds, marshes, and forests within the northern portion of San Andreas Valley (see pages 218-220). A tunnel was constructed through Buri Buri Ridge soon after, and by 1870 San Andreas Reservoir was providing water to San Francisco. The height of San Andreas Dam was subsequently increased several times, to 95 ft in 1874 and 105 ft in 1928 (SVWC 1923, Babal 1990, Hanson 2005).

During this same period, a series of flumes, aqueducts, and diversion dams was constructed to capture water from the mountains west of Pilarcitos Creek, including the drainages of Lock's Creek, Aponolio Creek, and Frenchman's Creek (outside of the Peninsula Watershed). These conveyances, part of what was known as the Lock's Creek Line, transported water to Pilarcitos Canyon, where it joined another flume from Stone Dam. The Lock's Creek Tunnel (Stone Dam Tunnel Number One), completed in 1870, carried the combined waters through Cahill Ridge to the San Mateo Creek Canyon, and from there the water was carried by flume across San Mateo and San Andreas creeks to San Andreas Reservoir (San Francisco Board of Supervisors 1875, SVWC 1923, Babal 1990). In 1898, a pair of small dams were constructed across Upper San Mateo Creek, and a tunnel (Stone Dam Tunnel Number Two) was constructed through Sawyer Ridge; water in the Lock's Creek Line was rerouted to flow through this tunnel and then into San Andreas Reservoir (Babal 1990).

Upper Crystal Springs Dam was constructed in 1877, impounding water from Laguna Creek and inundating Laguna Grande and other wetlands in the Laguna Creek Basin. The height of the dam was raised in 1891 (after completion of Lower Crystal Springs Dam downstream), to allow the road to Half Moon Bay (now Highway 92) to cross over the reservoir. Construction of Lower Crystal Springs Dam, the largest of the Peninsula Watershed's dams (and at the time one of the largest concrete dams ever built), began in 1886, and the first phase of construction was completed in 1888 (Cummings 1972, Babal 1990; Fig. 3.14). Located at the head of Lower San Mateo Creek Canyon, just downstream of the historical confluence of Laguna, San Andreas, and San Mateo creeks, the dam inundated a broad swath of San Andreas Valley and the Laguna Creek Basin. The height of the dam was increased in 1890 and again in 1911 and 2012 (Babal 1990; M. Ingolia pers. comm.). Culverts linking Upper and Lower Crystal Springs reservoirs were constructed in 1924. Water from other parts of SFPUC's water supply system, including the Tuolumne River (Hetch Hetchy Reservoir) and the Alameda Creek Watershed, are also stored in San Andreas and Crystal Springs reservoirs (San Francisco Planning Department 2008).

In addition to dam construction, channel stabilization projects were conducted in a number of areas of the watershed to prevent erosion or channel migration. The most heavily modified channel in the study area is Lower San Mateo Creek within the City of San Mateo, which has been straightened and armored, and in some areas flows through underground culverts (San Francisco Planning Department 2008). At the mouth of San Mateo Creek, an earthen dam and dikes were constructed in the 1880s to regulate water levels and facilitate reclamation of former marshlands (*San Francisco Chronicle* 1882, Mendell 1885). Less drastic modifications have occurred in other parts of the watershed: for instance, along Laguna Creek south of Upper Crystal Springs Reservoir, SVWC Superintendent W.B. Lawrence (1917) wrote, "The main stream... has... been cared for by the clearing out of the creek channel, and the construction of willow or rock protection walls and the placing of bundles of willow branches in the bottom of the stream, properly anchored with boulders where necessary, to prevent further deepening of the channel."

Collectively, hydrologic modifications have profoundly altered both physical processes and habitat availability in the Peninsula Watershed. Reservoir construction inundated large areas of wetland, riparian, and aquatic habitats, and cut off upstream spawning habitat for steelhead, coho salmon, and other fish. Dam construction and water diversion also drastically altered hydrogeomorphic patterns within the study area, impounding much of the upstream freshwater and sediment supply and reducing or eliminating both dry season and flood flows within the creeks downstream. The altered flow and sediment dynamics have impaired channel functioning and habitat quality downstream. In Lower San Mateo Creek, for instance, the lack of periodic high flows has led to a build up of fine sediment and riparian vegetation in the channel (NMFS 2019). Aside from periodic spills and releases and a small amount of seepage, there is typically no streamflow in San Andreas Creek immediately downstream of San Andreas Dam (though tributary and groundwater inputs provide some streamflow further downstream; San Francisco Planning Department 2008). For many years the situation was similar in San Mateo Creek downstream of Lower Crystal Springs Dam, though in 2015 SFPUC began releasing of 3-17 cubic feet per second (cfs) from Lower Crystal Springs Dam to support steelhead habitat downstream (San Francisco Planning Department 2008, NMFS 2019). In the Pilarcitos Creek Watershed, water is periodically released from Pilarcitos Reservoir during the summer months, and then diverted again from Pilarcitos Creek at Stone Dam for use by the Coastside County Water District. Since 2006, SFPUC has experimented with summer and fall releases from Stone Dam in order to augment instream flows in Pilarcitos Creek to support steelhead and other aquatic resources (PWA 2008, WWR 2010).

Figure 3.14. Circa 1887-8 photo of Lower Crystal Springs Dam construction and the directors of the Spring Valley Water Company. *(Photo D-1245, courtesy SFPUC)*

Drivers of Landscape Change

Urban Development

For the most part, early land uses within the Peninsula Watershed were pastoral, though several small communities had emerged by the mid-1800s. The Crystal Springs Hotel was established in the 1850s just west of present-day Lower Crystal Springs Dam, and became a popular resort destination in the region (see Fig. 7.3 on page 175). The Byrnes Store, a stage stop located on the west side of present-day Upper Crystal Springs Dam, was established around the same time. San Feliz Station, with a hotel, stage stop and stable, was established in 1869 on the west side of San Andreas Valley (Fig. 3.10 on page 62; Lawrence 1922a, Burke 1926, Babal 1990, Edmonds 2014).

As more settlers arrived, a network of roads was constructed through the watershed. A horse trail connecting San Francisco Bay with Half Moon Bay, bisecting the watershed, likely existed at least as early as 1846. A formal roadway was constructed in the mid-1850s to carry stagecoaches from Burlingame to Half Moon Bay. A second roadway, following San Mateo Creek to the Crystal Springs Hotel and finally to Half Moon Bay, was opened in 1860. An early alignment of Cañada Road, connecting Crystal Springs to Woodside, was completed in 1863. Another road from Woodside was built in 1894, running west around Upper Crystal Springs to connect with the Half Moon Bay road (Richards 1973).

As SVWC constructed reservoirs (see pages 67-69) and bought up land for watershed protection (see pages 72-73), the historic settlements were submerged along with the other farms and homesteads in the valley. Several of the roads were also decommissioned as a result of reservoir construction or other challenges in maintaining the roads. In 1877, for example, the original Cañada Road was submerged with the development of Upper Crystal Springs Reservoir, and the 1894 route was abandoned in 1939 (Richards 1973). In other cases, modern roads and highways were developed along the original rights-of-way. Present-day Highway 92, for instance, follows the approximate route of the historic San Mateo-Half Moon Bay road. In the interior of the watershed, most of the structures built by early settlers have disappeared, though several watershed keeper houses still exist on the watershed to house its watershed stewards.

Several quarries—including Skyline Quarry along present-day Highway 92, Casey Quarry along Lower San Mateo Creek, Donovan/Cañada Quarry along Cañada Road at the southern end of the watershed, San Andreas Quarry southeast of San Andreas Dam, and Spring Valley Quarry on Spring Valley Ridge—were operated within the study area to supply raw materials for the construction of roads, dams, and other structures. Ongoing operations and mitigation activities are conducted at all of the quarry sites (J. Appel and J. Fournet pers. comm.).

Urban development on the margins of the watershed accelerated during the latter half of the 20th century. Low-density residential housing was constructed throughout much of Lower San Mateo Creek Canyon, and along the eastern side of the watershed in Burlingame, Millbrae, Hillsborough, and San Mateo (e.g., the Lakeview neighborhood just north of Crystal Springs Rd and the Highlands-Baywood Park neighborhood just south of Crystal Springs Rd). Highway 280 was constructed in the mid-1970s, and runs through the eastern side of the watershed. In addition to directly displacing grasslands and other habitat types, the highway also dramatically reduced connectivity between plant and wildlife populations and significantly altered the hydrology of many spring-fed small tributaries that drained westward into San Andreas Valley. In addition, sediment basins were built to divert runoff from the freeway into Lower San Mateo Creek.

Highway 280, San Andreas Reservoir, and suburban streets. (courtesy of Google Earth)

Drivers of Landscape Change

Watershed Protection

Between the 1860s and 1930, SVWC gradually acquired land within the Peninsula Watershed for water supply infrastructure and watershed protection. Private ranches, farms, and homesteads were purchased, and agriculture, livestock grazing, and other existing land uses were discontinued. In many cases these parcels were subsequently submerged following reservoir construction. Logging was also prohibited on watershed lands to protect water quality in the reservoirs. As W. B. Lawrence, Superintendent of SVWC's Water Division put it, "In the end, the entire thirty-five square miles of catchment area were swept clean of all human habitation" (Lawrence 1922a).

The first areas purchased by SVWC, in the 1860s, were around present-day Pilarcitos Reservoir (constructed 1863-67). Over the following several decades, SVWC purchased additional large tracts of land in the Pilarcitos Watershed, as well as large tracts in San Andreas Valley and the Laguna Creek Basin. By 1868, SVWC owned approximately 2,600 ac in the northern part of the watershed (Easton 1868). This number had increased to approximately 11,500 ac by 1877 (Cloud 1877), 16,500 ac by 1894 (Bromfield 1894), and 23,520 ac by 1927 (Kneese 1927; Fig. 3.15). SVWC was purchased by the City of San Francisco in 1930.

One of the primary ways that SVWC ownership impacted vegetation patterns in the watershed was through the removal of prior land uses, most notably grazing, agriculture, and logging. Removal of grazing represented a substantial shift in this disturbance regime; in the absence of regular fires, the lack of grazing was likely a key factor driving widespread encroachment of woody vegetation into grasslands. In the 1880s, for instance, A.L.C. Potter observed that, near the current Jepson Laurel Tree (in San Andreas Valley), "these small flats were more open years ago, as cattle kept small brush out" (Potter 1923). In 1895, botanist J. Burtt Davy noted that the area around San Andreas Reservoir was "fine botanical ground... owing to the exclusion of cattle and preservation of timber" (Davy 1895).

Though SVWC owned a substantial portion of the watershed by the turn of the century, there were still a number of active dairies and other private enterprises in portions of the watershed that continued well into the 20th century. In at least one case, hundreds of cattle were allowed to graze down to the water's edge of San Andreas Reservoir (*San Francisco Call* 1895). The Fifield Dairy continued operating until 1906, and the Jersey Farm Dairy kept up operations until 1930 (Babal 1990). Following the City of San Francisco's purchase of SVWC in 1930, farms and grazing lands in the whole Peninsula were observed to be quite hard to come by. One observer commented that "the only land I knew that could be used for cattle ranching is west of the Skyline Boulevard, south of the Half Moon Bay summit" (Davis 1951).

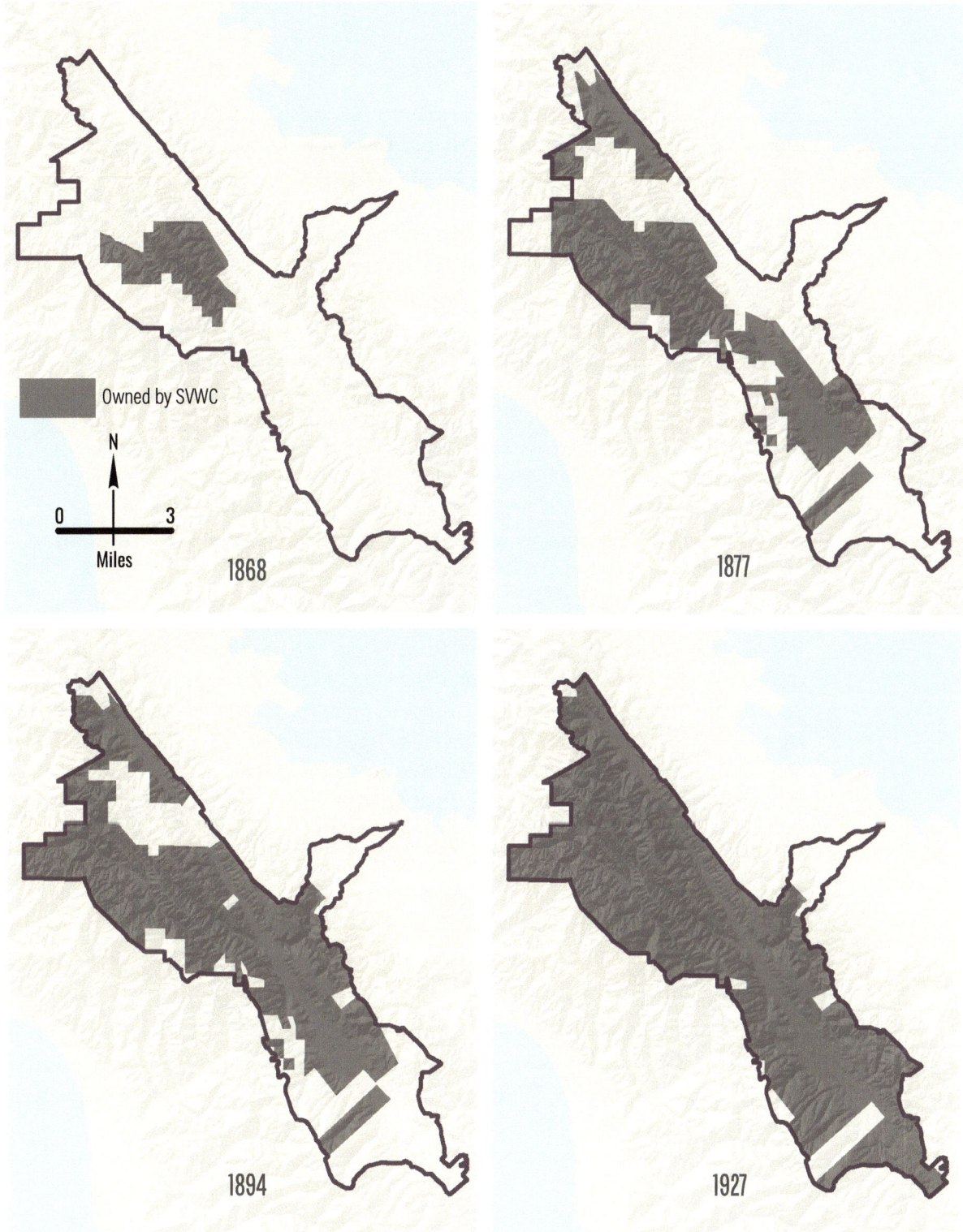

Figure 3.15. The Spring Valley Water Company (SVWC) progressively bought up lands within the Peninsula Watershed over the span of approximately 70 years, from ca. 1860 to 1930. Lands owned by SVWC were protected from grazing, logging, development, and other land uses, and thus serve as an indicator of when different portions of the watershed were removed from intensive land uses. These images show lands owned by the SVWC at intermediate time periods, based on early maps of San Mateo County (Easton 1868, Cloud 1877, Bromfield 1894, Kneese 1927).

Drivers of Landscape Change

Invasive Species

The introduction and spread of non-native species, including invasive plants, animals, and microorganisms, has been widely documented to affect populations of native species and overall ecosystem functioning. Invasive plant species can alter the flora within a vegetation community type, or in some cases result in vegetation type change (e.g., grassland to shrubland). Grasslands, for example, represent one of the most highly invaded ecosystems in California, with a drastically altered species composition dominated by Mediterranean annual grasses (Jackson and Bartolome 2007, Caziarc 2012). The successful establishment of invasive species can be dependent on many biotic or abiotic factors, including local disturbance or management regimes, and can be uneven in space and time (Williamson and Harrison 2002, Thomsen et al. 2005). In some cases, invasive species-dominated vegetation communities represent an "alternative stable state," such that reintroduction of historical disturbance regimes and ecological processes will not necessarily result in the restoration of the historical plant community (Suding et al. 2004, D'Antonio et al. 2007).

Many of the non-native plants that colonized coastal California were intentionally or unintentionally introduced from the Mediterranean region during the Mission era. Agricultural operations at the missions introduced a variety of exotic agricultural plants: Father Longinos-Martinez noted in 1792 that "all the trees and seeds that have been introduced [to California] from the country multiply with the same abundance and quality as in that climate [of Spain]" (Simpson in Minnich 2008). Parish (in Minnich 2008) similarly observed that "there is a persisting tradition among the Spanish-speaking population that the mission fathers were accustomed to carry the seed with them... to sow by the way side." Livestock grazing has also been identified as a major vector for invasive species introductions, though in many cases the establishment of non-native species appears to have preceded widespread livestock grazing (Minnich 2008).

A variety of Mediterranean non-native annual plants proliferated in grasslands and valleys in central coastal California by the mid-1800s. Wild oat (*Avena fatua*) and black mustard (*Brassica nigra*) were noted by numerous observers in the region. In 1868, Cronise noted that San Jose had a "thick crops of wild oats," uncultivated parts of Santa Cruz County coastal marine terraces "produce enormous crops of wild mustard," and northern Monterey Bay had "bottom-land, which produces fine crops of wild oats, bunch grass, and a variety of clover and native grasses" (Cronise in Minnich 2008). A later wave of introductions resulted in the spread of species such as bromes (*Bromus* spp.) and summer mustard (*Hirschfeldia incana*; Minnich 2008), and more recently species such as jubata grass (*Cortaderia jubata*). Woody vegetation (including eucalyptus, Monterey pine, and Monterey cypress) also spread throughout central coastal California, mostly through active planting and subsequent expansion.

In the Peninsula Watershed, many non-native plant species have invaded native ecosystems, driving changes in vegetation community structure and composition and potentially leading to vegetation type changes. Annual and perennial Mediterranean grasses have invaded native grasslands in some areas (particularly non-serpentine grasslands), though large areas dominated by native perennial grasses still remain (see Chapter 5). Particularly concerning invasives currently include stinkwort (*Dittrichia graveolens*), harding grass (*Phalaris aquatica*), and yellow star thistle (*Centaurea solstitialis*; M. Ingolia pers. comm.). Woody plant species such as eucalyptus, acacia, Monterey cypress, Monterey pine, and French broom (*Genista monspessulana*) have overtaken areas of former grassland and shrubland as well (Nomad Ecology 2009). Table 3.3 documents the earliest observations of many of the most pervasive non-native plant species in central coastal California, and where more resolution is possible, on the watershed lands.

Native ecosystems in the Peninsula Watershed have also been impacted by invasive animal species such as Argentine ant (*Linepithema humile*), New Zealand mudsnail (*Potamopyrgus antipodarum*), American bullfrog (*Lithobates catesbeiana*), and carp (Family Cyprinidae; E. Natesan and J. Avant pers. comm.). In some cases these species were intentionally introduced to the watershed for recreational or other purposes (see page 80), while in other cases their spread has been unintentional. Invasive animals may affect native wildlife species directly, through competition or predation, or indirectly, by altering food web dynamics or abiotic conditions.

In addition to non-native plants and animals, novel plant pathogens have spread into the Peninsula Watershed and surrounding areas, in some cases resulting in major vegetation community changes at the landscape-scale. Most notable among these pathogens is *Phytophthora ramorum*, the organism that causes Sudden Oak Death. *P. ramorum* is an oomycete (or "water mold") that was likely introduced through international trade of ornamental plants around 1995 (California Oak Mortality Task Force 2019). It flourishes in cool, wet climates, and its spores can spread through water, wind-driven rain, movement of plant material, or human activity (such as equipment transport or planting of infected plants). Species most vulnerable to Sudden Oak Death include tanoak (*Notholithocarpus densiflorus*) and oak species such as coast live oak, California black oak (*Quercus kelloggii*), and Shreve oak (*Quercus parvula*). California bay laurel (*Umbellularia californica*) is one of its most common host plants. *P. ramorum* has spread to a number of locations throughout the Peninsula Watershed, and the SFPUC has taken steps to monitor the spread of the pathogen, slow its spread, and evaluate eradication alternatives. However, the continued spread of *P. ramorum* and other plant pathogens may result in widespread mortality of certain oaks and other woody plant species, potentially resulting in significant changes in species composition or even vegetation type change in some areas in the future.

Table 3.3. Historical evidence of local non-native species, based on compiled historical records from Minnich 2008 and participants from the Consortium of California Herbaria.

Species	Common name	Selection of earliest general evidence in central coastal California*	Earliest confirmed observation in study area**
Brassica nigra	Black mustard	1848: "Fertile valleys [of coastal California] are overgrown with wild mustard" (Fremont 1848)	1946: Near junction of Belmont-las pulgas roads, close to Crystal Springs (Bacigalupi)
Erodium cicutarium	Filaree	Early 1800s: "according to much testimony it was as common throughout California early in the present century as now" (Brewer and Watson 1876-80) 1811: evidence in San Jose adobe brick (Hendry 1931, Hendry and Bellue 1925, Frenkel 1970).	-
Malva parvifolia	Cheese mallow	1811: Evidence in San Jose adobe brick (Hendry 1931, Hendry and Bellue 1925, Frenkel 1970).	-
Avena fatua	Wild oat	1835: "The wild oat in the year 1835 was found south of the bay of San Francisco... this grain being sown in a natural way by horses and cattle... grows both on the plains and the hills" (Cronise 1868 and/or Hittel 1874)	1950: San Andreas Valley, San Francisco Watershed Reserve (Oberlander)
Poa annua	Annual blue grass	1860s: Widespread in the state (Brewer and Watson 1876-80)	1894: Santa Cruz Mountain Peninsula. Crystal Springs Lake (Dudley)
Hordeum murinum	Wall barley	1811: Evidence in San Jose adobe brick (Hendry 1931, Hendry and Bellue 1925, Frenkel 1970).	1950: San Francisco Watershed Reserve (Oberlander)
Medicago polymorpha	Bur clover	1841: Recorded on botany report on area between San Francisco and Monterey (Hooker and Arnott 1841 / Burcham 1957).	-
Melilotus indica	Sour clover	1860s: Common in the state (Brewer and Watson 1876-80)	1903: Crystal Springs Lake (Elmer)

Filaree (*Erodium cicutarium*) at Sweeney Ridge.
(Photo by Tom Hilton, courtesy CC by 2.0)

Species	Common name	Selection of earliest general evidence in central coastal California*	Earliest confirmed observation in study area**
Bromus diandrus	Ripgut brome	1862: Near Mission Dolores at San Francisco (Bolander)	1949: San Francisco Watershed Reserve, San Andreas Valley (Oberlander)
Hirschfeldia incana	Summer mustard	-	1949: San Francisco Watershed. Grassland about San Andreas Lake (Oberlander)
Acacia longifolia	Golden wattle	-	1948: Residence, San Mateo (Hoyt)
Cytisus scoparius	Scotch broom	-	1922: Lake San Andreas (Smith)
Dipsacus fullonoum	Teasel	-	1923: Near Pulgas tunnel (Spalding)
Hirschfeldia incana	Shortpod mustard	-	1949: Grassland about San Andreas Lake (Oberlander)
Hypochaeris radicata	Hairy cat's-ear	-	1903: Lake San Andreas (Elmer)
Cotula australis	Brass buttons	-	1902: Lake San Andreas (Abrams)
Conium maculatum	Poison hemlock	-	1903: San Mateo (Elmer)
Centaurea calcitrapa	Purple star thistle	-	1896: San Mateo (Dudley)
Centaurea solstitialis	Yellow star thistle	-	1949: San Francisco watershed reserve, south of San Andreas Lake (Oberlander)
Plantago lanceolata	Ribwort	-	1949: San Andreas Lake (Oberlander)
Pseudognaphalium luteoalbum	Jersey cudweed	-	1935: Lake San Andreas (Howell)
Phalaris minor	Mediterranean canarygrass	-	1896: San Mateo County: Crystal Springs Lake (Elmer)

*based on Minnich 2008. **based on historical Consortium of California Herbaria records.

Ripgut brome (*Bromus diandrus*). (Photo by Tom Benson, courtesy CC by 2.0)

Drivers of Landscape Change

Hunting and Wildlife Management

Fish and wildlife populations in the Peninsula Watershed have changed considerably since the time of Spanish contact. Large-scale landscape changes, such as agricultural development and reservoir construction, have indirectly affected the populations of many species by changing the availability and quality of habitat. Hunting, predator control programs, introduction of non-native fish and wildlife species, and other activities have further impacted wildlife populations both directly (e.g., hunting) and indirectly (e.g., trophic cascades). In some cases, changes in wildlife populations themselves may have driven further landscape change.

As Euro-American settlers increasingly migrated to the region, recreational game hunting surged, as did human-wildlife conflicts. Hunting was popular in the watershed lands until its prohibition by SVWC in the late 1800s. For instance, quail (presumably either California Quail [*Callipepla californica*] or Mountain Quail [*Oreortyx pictus*]) were a local delicacy: Lawrence (1922a) reported that "the fashionable guests from San Francisco... thronged the [Crystal Springs] hotel grounds every Sunday [and] always demanded quail on toast with their vintage champagne." Such hunting evidently had an impact, as quail had nearly become extinct in the watershed by the 1920s (Lawrence 1923). Even after instituting hunting prohibitions and a predator control program in 1912, a biological survey of the area in 1929 reported finding only "in the vicinity of 2000 birds," and pointed out "that is certainly not a heavy population of breeding birds... far short of a normal population for such an area that is well fitted to a quail's habits" (McClean 1930). Furthermore, Mountain Quail, once prevalent in higher elevations parts of the county, appear to have been locally extirpated by the early 1900s (Alexander and Hamm 1916). Greater Roadrunner (*Geococcyx californianus*), too, was once observed in the watershed, though it has also been extirpated (Evans 1873, eBird 2017).

Hunting led to local extirpation of other species as well. Grizzly bears (*Ursus arctos*) were once common in the region, and were noted for their unusually large body size (Alexander and Hamm 1916). Fray Pedro Font, for instance, traveling through San Andreas Valley with the Anza Expedition in March of 1776, observed that "there are a great many bears throughout these woods" and reported an encounter with a "very large bear, which the men succeeded in killing" (Font and Brown 2011). A formidable predator that drove ranchers to relinquish their cattle enterprises in the watershed (Evans 1873, Lawrence 1922b, San Mateo County Historical Association 1967, Svanevik and Burgett 2002), grizzly bears were hunted to extinction in the region by the 1870s or 80s, and black bears (*Ursus americanus*) were killed off soon after (*Sacramento Daily Union* 1885, Alexander and Hamm 1916). American beaver (*Castor canadensis*) is believed to have been present throughout much of coastal California historically, but was extirpated from much of its historical range during the fur-trade of the early to mid-19th century (Lanman et al. 2013).

In addition to recreational hunting, SVWC "vigorously" instituted a predator control program from 1912 to at least 1923 in order to eliminate undesirable "varmints" that preyed on ground

Figure 3.16. This photograph shows the results of the SVWC predator control program. The caption on the image reads, "In voluntary co-operation with federal and state authorities, Spring Valley wages war on predatory animals. These 'varmints' were trapped on the peninsular watershed during the past season." In the image, labels of furs can be read, including "raccoon," "skunk," "wild cat" (likely bobcats), "fox," "coyote," and other species. *(Photo ark:/13960/t7gq86n1t, courtesy San Francisco History Center, San Francisco Public Library)*

Until its discovery as a promising water source, [the Peninsula watershed] has remained a secluded and safe retreat for coyotes, pumas, big wild cats and the formidable grizzly bear... Its whole native fauna in variety and abundance befitting to the wild solitude of the locality although so near to the largest city of the State.

—DESCRIPTION OF PILARCITOS DAM SITE BY ENGINEER CALVIN BROWN, 1863

Photo by Don DeBold, courtesy CC by 2.0

nesting birds or otherwise posed a nuisance, including bobcats (*Lynx rufus*), opossums (*Didelphis virginiana*), raccoons (*Procyon lotor*), gray fox (*Urocyon cinereoargenteus*), coyotes (*Canis latrans*), skunks (*Mephitis mephitis* or *Spilogale gracilis*) and ringtail (*Bassariscus astutus*; Lawrence 1923; Fig. 3.16). The program was developed in response to observed declines in quail and other ground nesting bird populations, and was focused on the Pilarcitos Reservoir and Upper San Mateo Creek areas, encompassing an area of fifteen square miles in some seasons. Lawrence (1923) reported that, apparently as a result of the hunting prohibitions and predator control efforts—but also perhaps due to increased protection of habitat by that time—the watershed properties "have become a great animal refuge." Populations of "desirable" species such as deer and tree squirrels boomed, though predator control did not appear to help quail, the program's main intended beneficiary (Lawrence 1923).

A number of non-indigenous (and in some cases invasive; see page 75) animal species were introduced to the watershed for recreational purposes or to help control the populations of other species. Largemouth bass (*Micropterus salmoides*),[3] for instance, were introduced to streams and reservoirs in the watershed at least as early as 1879 (Journals of the Legislature of the State of California 1881, *San Francisco Chronicle* 1882, *Morning Call* 1893). "Eastern trout" (likely *Salvelinus fontinalis*), carp, and a number of other non-native fish species were also introduced (*Morning Call* 1891, *San Francisco Call* 1912b). Introduced carp populations evidently became overwhelming competitors in the lakes, and sea lions (presumably *Zalophus californianus*) were introduced in the early 1900s for the purpose of controlling the carp. The sea lions persisted for some time, but disappeared from the watershed by the early 1900s (*San Francisco Call* 1912a). "Golden beavers" were introduced into the Crystal Springs reservoirs in 1939, though the outcome of this introduction is unknown (Tappe 1942). Wild turkeys (*Meleagris gallopavo*) and American bullfrogs (*Lithobates catesbeiana*) have also entered the watershed more recently, though the source of these introductions is unknown (J. Avant pers. comm.).

Today, hunting and fishing are banned in the watershed, and wildlife is carefully managed under pertinent federal and state statutes and regulations. These protections have benefitted many species: for instance, large cats that were once culled under the watershed predator control program, and large birds of prey that were hunted and exposed to a legacy of DDT, are in varying stages of recovery in the watershed (J. Avant pers. comm.). The watershed supports a rich diversity of over 190 wildlife species (iNaturalist 2019)— including listed or sensitive species such as California red-legged frog (*Rana draytonii*), San Francisco garter snake (*Thamnophis sirtalis tetrataenia*), Marbled Murrelet (*Brachyramphus marmoratus*), and several species of butterfly—and serves as an important stopover for migratory birds along the Pacific Flyway (eBird 2017). The legacy of past stressors, however, including habitat fragmentation and degradation, sustained hunting, predator control, species introductions, and hydromodification, still poses threats to wildlife in the watershed and underscores the need for ongoing management and restoration efforts to maintain and enhance the resilience of wildlife populations in the region.

---

3   Sometimes referred to in historical sources as black bass (*Micropterus nigricans*).

San Francisco garter snake (*Thamnophis sirtalis tetrataenia*). (Photo by J. Maughn, courtesy CC by 2.0)

California red-legged frog (*Rana draytonii*). (Photo by KQED Quest, courtesy CC by 2.0)

Pilarcitos Reservoir. (Photo by SFEI-ASC, August 2018)

CHAPTER 4

Quantitative Results and Synthesis Framework

Overview

The Peninsula Watershed has experienced a dynamic array of landscape changes over the last several centuries. The relative proportions of major terrestrial vegetation types within the watershed—grasslands, shrublands, hardwood forests, and conifer forests—have changed substantially (Fig. 4.1), as have their distribution, composition, and structure. Wetland, riparian, and aquatic habitats have likewise experienced substantial changes (see Chapter 9). A range of drivers, operating at various spatial and temporal scales, has contributed to these changes in habitat extent and distribution (see Chapter 3).

This chapter presents the results of the three main quantitative spatial analyses that were performed to assess terrestrial vegetation change over time: analysis of historical and contemporary aerial imagery, comparison of historical and contemporary vegetation mapping, and analysis of GLO survey data. While these datasets alone do not provide a complete picture of the historical landscape or subsequent vegetation change, they are particularly valuable because they constitute relatively spatially accurate information with watershed-wide coverage, and are conducive to quantitative analysis. As such, they are a useful starting place for discerning some of the dominant trends in terrestrial vegetation change that have occurred at the watershed scale. The findings from these quantitative analyses also serve as a conceptual framework for synthesizing all of the other more qualitative historical data (such as traveler narratives and landscape photographs) that do not lend themselves as easily to quantitative spatial analysis. The findings from this broader synthesis process, and a more thorough discussion of patterns and drivers of change for each of the major terrestrial vegetation types, are provided in chapters 5-8.

Overall, the quantitative datasets show a clear trend towards decreased grassland and shrubland cover and increased forest cover within the watershed over time. This is consistent with observed vegetation changes in many other parts of coastal California (see pages 40-42), and with the hypothesized transitions among vegetation types outlined on pages 42-43. However, closer examination shows that within this broad trend, there has been a complex trajectory of vegetation change over time and a heterogeneous pattern of change throughout the watershed. For instance, serpentine grasslands have been more persistent than non-serpentine grasslands. Coyote brush-dominated shrublands have increased in some areas and declined in others, while more diverse coastal scrub and chaparral communities have experienced considerable loss. Hardwood forests dominated by eucalyptus and acacia have expanded, while oak woodlands and other native hardwood forests have been lost in some areas and expanded in others. Conifer forests, including planted stands of Monterey pine and Monterey cypress as well as stands of Douglas-fir, have expanded in many areas. These and other trends are explored in more detail in chapters 5-8.

While not quantified as part of this study, wetland, riparian, and aquatic habitats throughout the watershed have also experienced substantial changes in extent, distribution, and composition. Particularly notable is the loss of large areas of sag ponds, freshwater marshes, and willow thickets in San Andreas Valley and the Laguna Creek Basin as a result of reservoir construction (and earlier land use changes) in the 19th century (see Chapter 9).

These vegetation changes have been driven by complex, interconnected processes and land use changes. Through the regular use of fire, indigenous communities on the Peninsula likely maintained open mosaics of grassland, shrubland, and oak savanna to increase habitat for favorable game animals, increase the abundance of favored plants, and other purposes (see pages 50-51). Spanish colonization led to a subsequent period of both fire suppression and extensive livestock grazing on the eastern side of the watershed, which had mixed effects on vegetation communities. Extensive logging in the early to mid-19th century reduced the extent of redwoods in the southwestern portion of the watershed, while clearing for agriculture and grazing altered vegetation cover within the San Andreas Valley, the Laguna Creek Basin, and other parts of the watershed. Plantings of non-indigenous conifer and hardwood species displaced grasslands, shrublands, and hardwood forests. Highways, urban development, and reservoirs have also led to additional displacement or loss of a number of vegetation types. A gradual average trend (though not statistically significant) toward a warmer and wetter climate in the watershed has been documented, potentially contributing to expansion of conifer forest.

The following sections present the results of the three quantitative analyses of terrestrial vegetation change throughout the watershed, including the aerial imagery point analysis, comparison of historical and contemporary vegetation mapping, and analysis of GLO survey data.

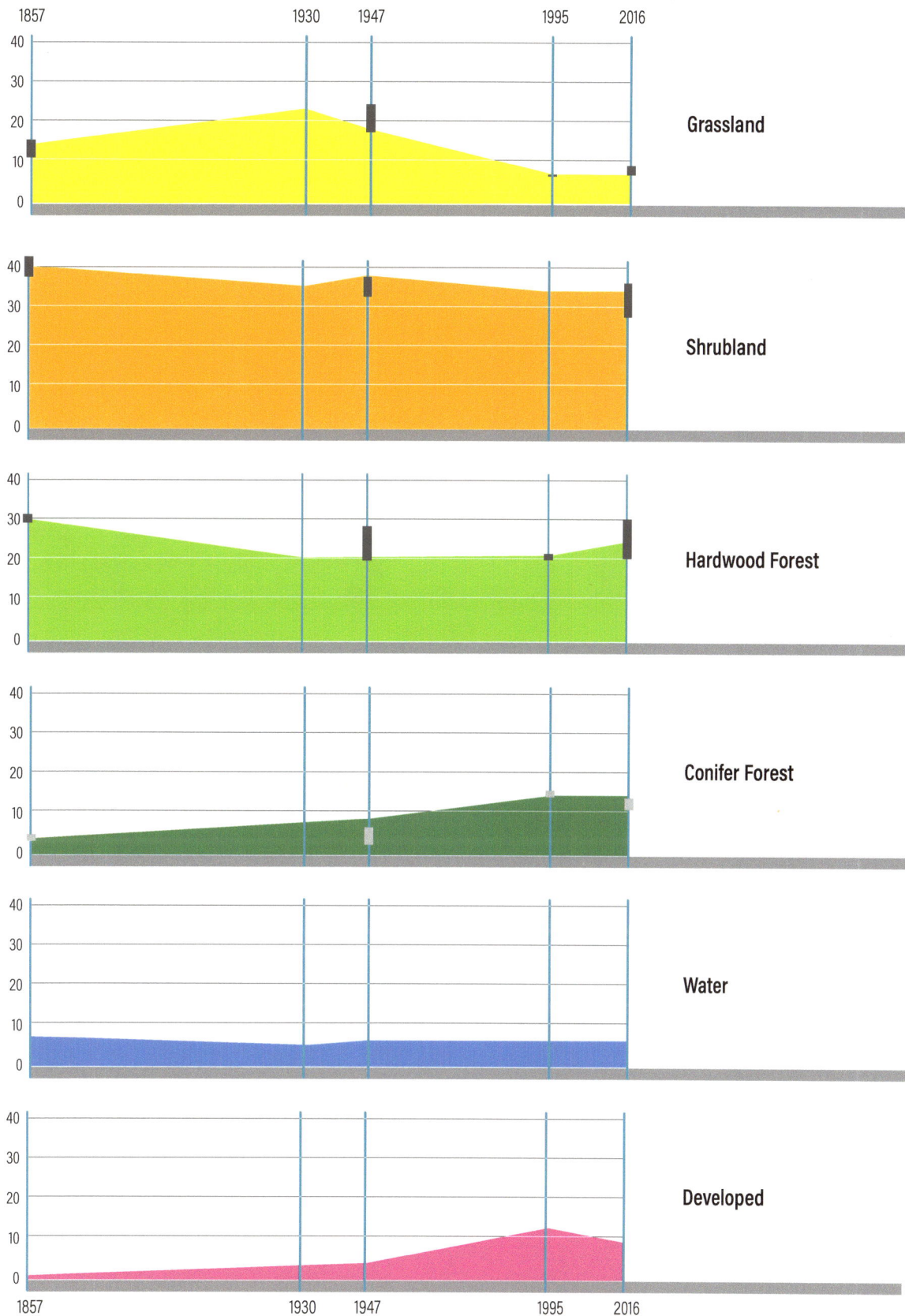

Figure 4.1. Trends in vegetation type cover within the Peninsula Watershed over time. The diagram shows the estimated coverage of each vegetation/land cover type over time. The estimates are derived from the aerial imagery point analysis (ca. 1947 and ca. 2016), vegetation mapping comparison (ca. 1930 and ca. 1995) and GLO survey data (ca. 1857), with estimated cover at intervening periods interpolated between these points. Gray boxes around certain points represent estimates of uncertainty (see Chapter 2 for further discussion). The diagram is a simplified version of the trajectory diagrams shown for each vegetation type in Figures 5.2, 6.1, 7.1, and 8.1 (chapters 5-8). Note that the GLO dataset appears to somewhat overestimate the extent of water and hardwood forest in the mid-19th century; see pages 102-104 for further discussion.

Aerial Imagery Point Analysis

TIME PERIOD: 1946-48 TO 2015-17

The aerial imagery point analysis classified land cover at 5,000 randomly distributed points in both historical (1946-48) and contemporary (2015-17) aerial imagery (Fig. 4.2; Table 4.1). The relative proportion of points in each land cover class was taken as a proxy for the estimate of percent cover of that class within the watershed for each time period (see discussion of methodology on pages 27-29).

Based on the point classification of the historical (1946-48) aerial imagery, land cover within the watershed in 1946-48 consisted of 19% grassland, 39% shrubland, 21% hardwood forest, 9% conifer forest, 7% water, and 4% developed/disturbed. Land cover in contemporary (2015-17) aerial imagery consisted of 8% grassland, 35% shrubland, 25% hardwood forest, 15% conifer forest, 7% water, and 10% developed/disturbed. Thus, change in land cover between 1946-48 and 2015-17 based on this analysis includes an estimated 56% loss of grassland, 11% loss of shrubland, 17% gain of hardwood forest, 65% gain of conifer forest, 5% loss of water, and 128% gain of developed/disturbed.

Estimates of the uncertainty in the percent cover values for each land cover class for each time period were developed based on the range of percent cover values obtained by four independent analysts classifying a subset of 200 duplicate, calibrated points (see pages 27-29). Because the uncertainty ranges are based on the subset of 200 points, rather than the full 5,000 points, in some cases the uncertainty ranges do not encompass the overall land cover estimates. For the historical imagery, the uncertainty ranges for each vegetation type are 19.5-25.5% for grasslands, 34-38.5% for shrublands, 21-29% for hardwood forest, and 3-7% for conifer forest. For the contemporary imagery, the uncertainty ranges are 8-10% for grasslands, 29-37% for shrublands, 21.5-31.5% for hardwood forest, and 12-14.5% for conifer forest. In addition to the per-class uncertainty estimates, an overall measure of reliability, calculated for both the historical and the contemporary point classification using the Krippendorff's Alpha statistic (again using the subset of 200 duplicate points), found that classification of the historical aerial imagery was suitable for drawing "tentative conclusions" (Alpha = 0.736) and classification of the contemporary aerial imagery was "reliable" (Alpha = 0.826; Krippendorff 2004).

Trajectories of land cover type conversion based on the aerial imagery analysis are shown in Figure 4.3 and the lower portion of Table 4.1. In general, these trajectories support the hypothesized transitions discussed in Chapter 3 (Fig. 4.4; see pages 42-43). Among points classified as grassland in the historical imagery, 34% were still classified as grassland in the contemporary imagery, while 19% converted to shrubland, 18% converted to hardwood forest, 5% converted to conifer forest, and 24% converted to developed/disturbed. Among points that were historically shrubland, 72% remained shrubland, while 15% converted to hardwood forest, and 9% converted to conifer forest. Among points that were historically hardwood forest, 65% remained hardwood forest, while 9% converted to shrubland, 18% converted to conifer forest, and 7% converted to developed/disturbed. Among points that were historically conifer forest, 77% remained conifer forest, while 5% converted to shrubland and 17% converted to hardwood forest. Figure 4.5 shows the spatial patterns of vegetation type conversion and persistence for each of the major vegetation types based on the aerial imagery point analysis.

Table 4.1. Results of the aerial imagery point analysis. The upper portion of the table shows the number of points (out of a total of 5,000) for both the historical (1946-48) and modern (2015-17) time periods for each vegetation/land cover class, along with the percent totals within each class. The lower portion of the table shows the percent changes between each historical class and each modern class.

POINT TOTALS	Modern (2015-17)							
Historical (1946-48)	Grassland	Shrubland	Hardwood Forest	Conifer Forest	Developed/Disturbed	Water	Historical Total (points)	Historical Total (%)
Grassland	317	179	165	45	226	5	**937**	**19%**
Shrubland	47	1416	286	167	42	0	**1958**	**39%**
Hardwood Forest	15	98	696	188	76	1	**1074**	**21%**
Conifer Forest	0	21	81	359	5	0	**466**	**9%**
Developed/Disturbed	19	24	27	8	134	0	**212**	**4%**
Water	10	8	5	1	0	329	**353**	**7%**
Modern Total (points)	**408**	**1746**	**1260**	**768**	**483**	**335**	**5000**	**-**
Modern Total (%)	**8%**	**35%**	**25%**	**15%**	**10%**	**7%**	**-**	**100%**

PERCENT CHANGE	Modern (2015-17)					
Historical (1946-48)	Grassland	Shrubland	Hardwood Forest	Conifer Forest	Developed/Disturbed	Water
Grassland	34%	19%	18%	5%	24%	1%
Shrubland	2%	72%	15%	9%	2%	0%
Hardwood Forest	1%	9%	65%	18%	7%	0%
Conifer Forest	0%	5%	17%	77%	1%	0%
Developed/Disturbed	9%	11%	13%	4%	63%	0%
Water	3%	2%	1%	0%	0%	93%

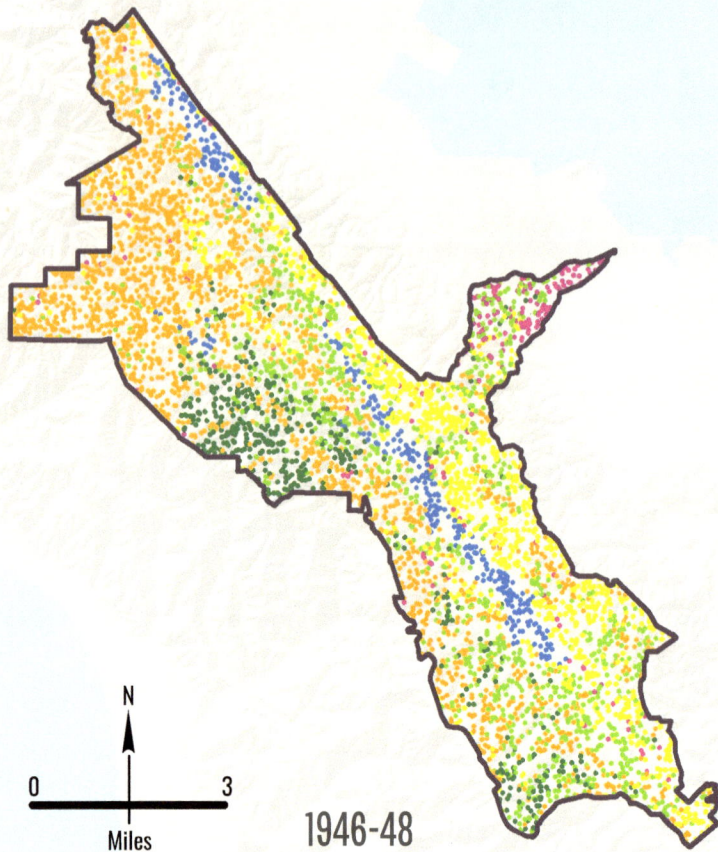

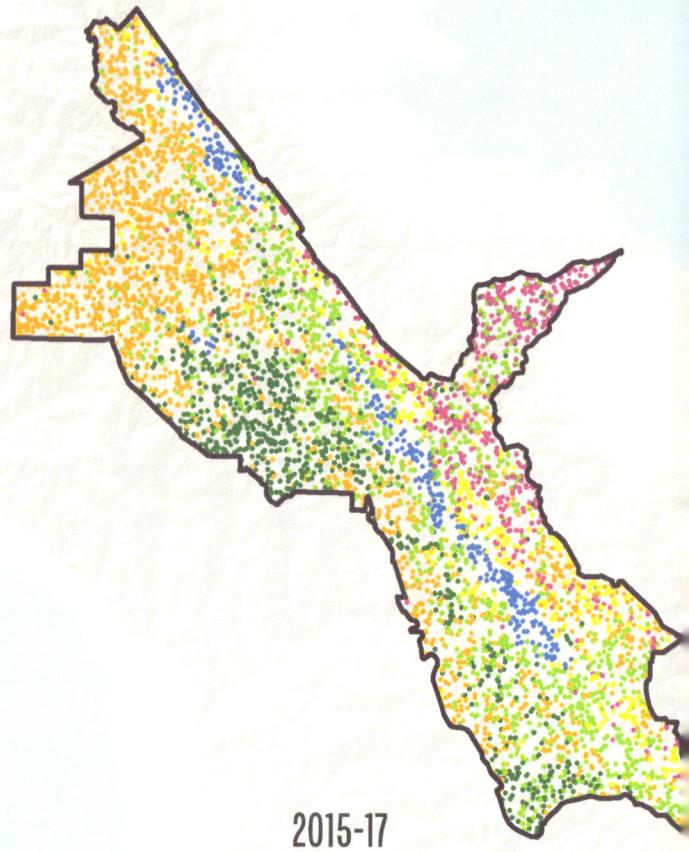

1946-48

2015-17

Change

Habitat Type

- Developed/Disturbed
- Water
- Conifer Forest
- Hardwood Forest
- Shrubland
- Grassland

Figure 4.2. Historical and contemporary vegetation cover, based on the aerial imagery point analysis.
Classified vegetation/land cover for points in historical (1946-48, above left) and contemporary (2015-17, above right) aerial imagery. The map to the left shows areas of change (blue) and persistence (gray) among the broad land cover types.

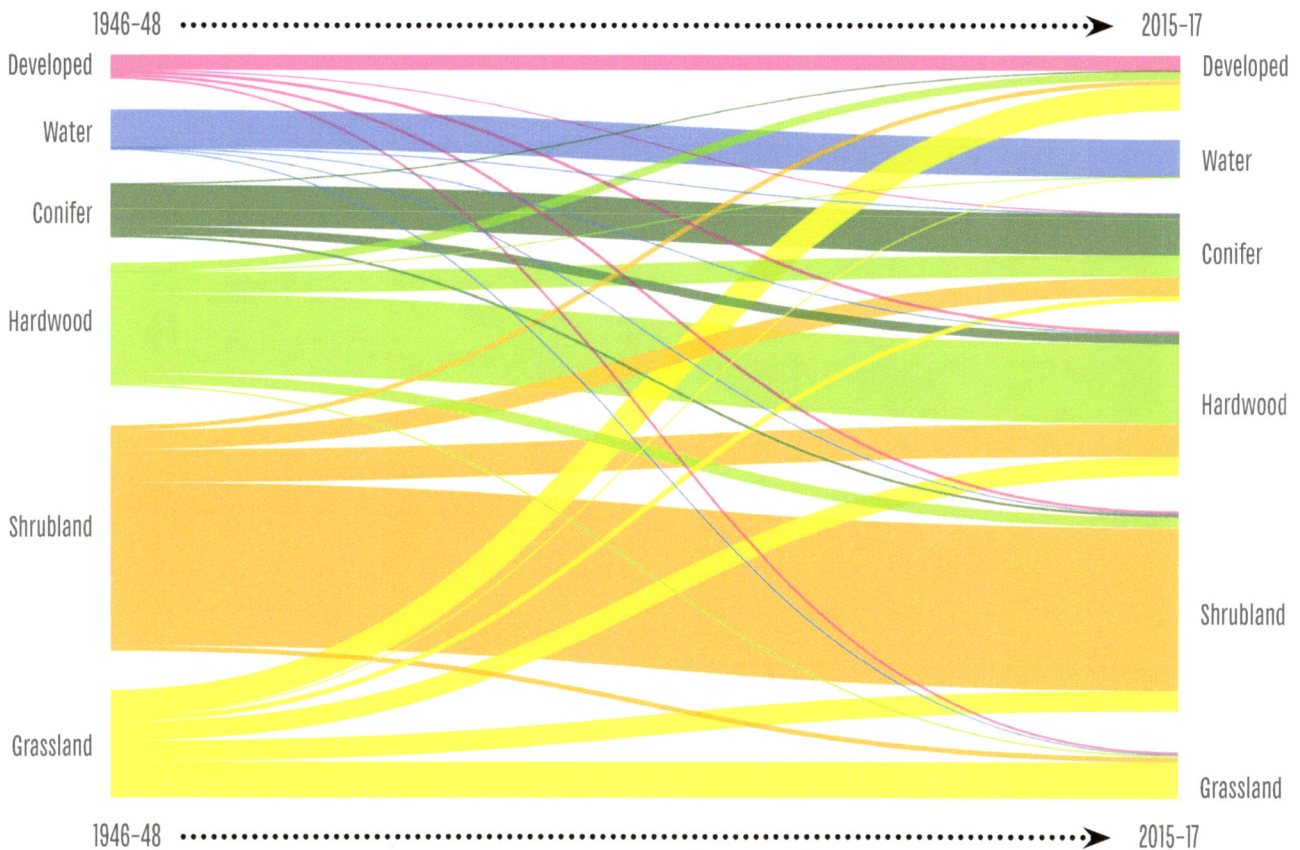

Figure 4.3. Land cover and vegetation type conversion within the study area based on the aerial imagery point analysis. The bars on the left side represent the frequency of each land cover type present within the study area in 1946-48, while the bars on the right represent the frequency of each land cover type in 2015-17. The lines connecting the two sides of the chart illustrate the conversion "pathways" that have occurred over this period; the thickness of each line corresponds to the total number of points that have undergone a given type of conversion.

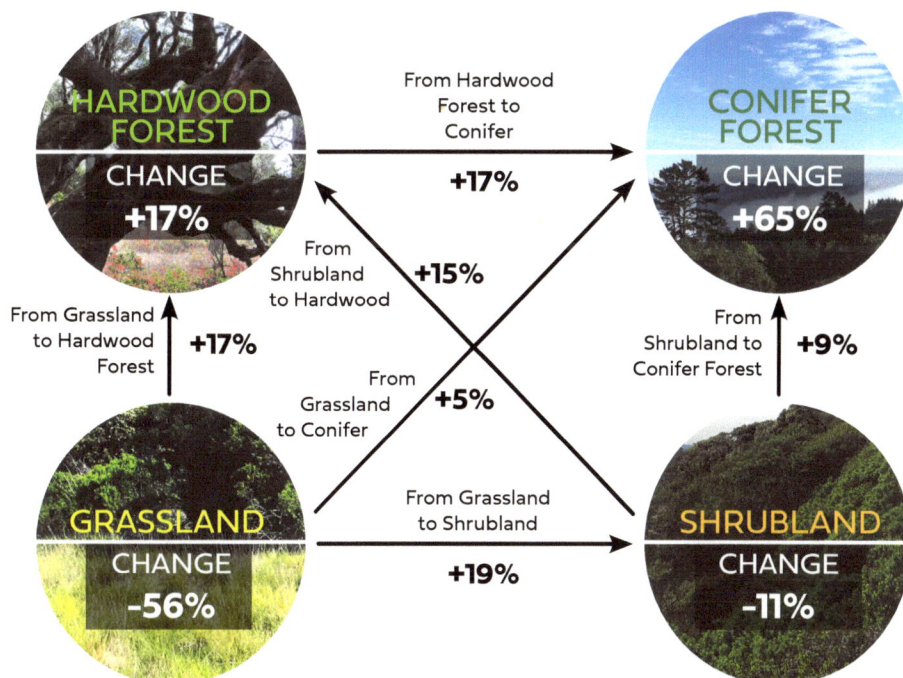

Figure 4.4.
Overall change in the distribution of major terrestrial vegetation types between 1946-48 and 2015-17 based on the aerial imagery point analysis. This diagram highlights major vegetation conversion pathways in relationship to hypothesized trends, but does not depict losses due to development; refer to Fig. 4.3 above and Table 4.1 on page 87 for complete land cover change results. *(Photos courtesy SFPUC)*

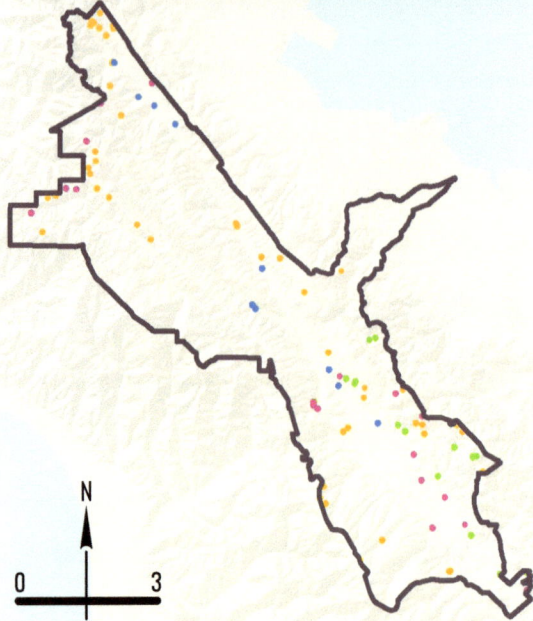

Gain

N

0 ——— 3
Miles

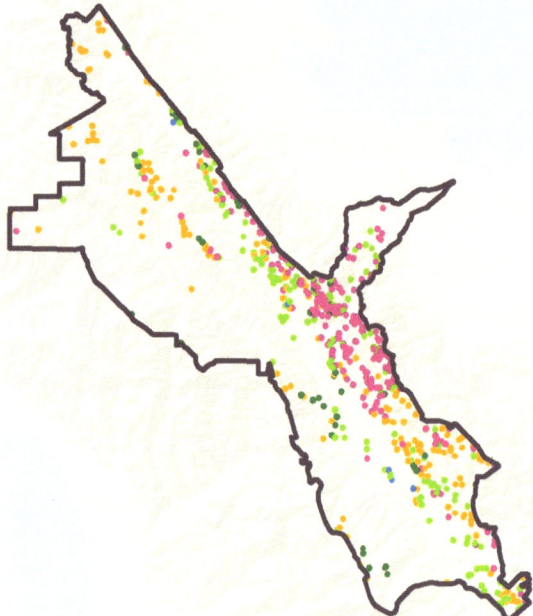

Loss

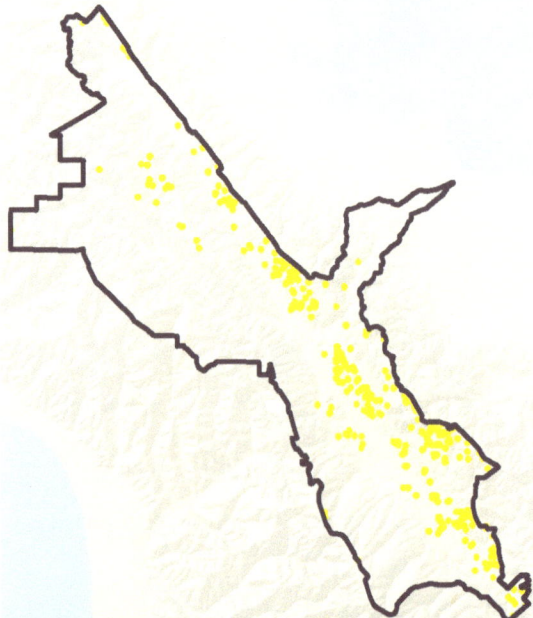

Persistence

Grassland gain, loss, and persistence: 1946-48 to 2015-17

Figure 4.5a. The maps on the left show areas of gain (top), loss (center), and persistence (bottom) of grassland between 1946-48 and 2015-17, based on the aerial imagery point analysis. Colors in the gain and loss maps indicate the vegetation/land cover types that grassland has been gained from or lost to, respectively. The bar chart indicates the number of points classified as grassland that have been gained from or lost to each vegetation/land cover type, as well as the number of grassland points that have persisted over this time period (the y-axis has been scaled proportionally).

Habitat Type

- Developed/Disturbed
- Water
- Grassland
- Shrubland
- Hardwood Forest
- Conifer Forest

Locator Map

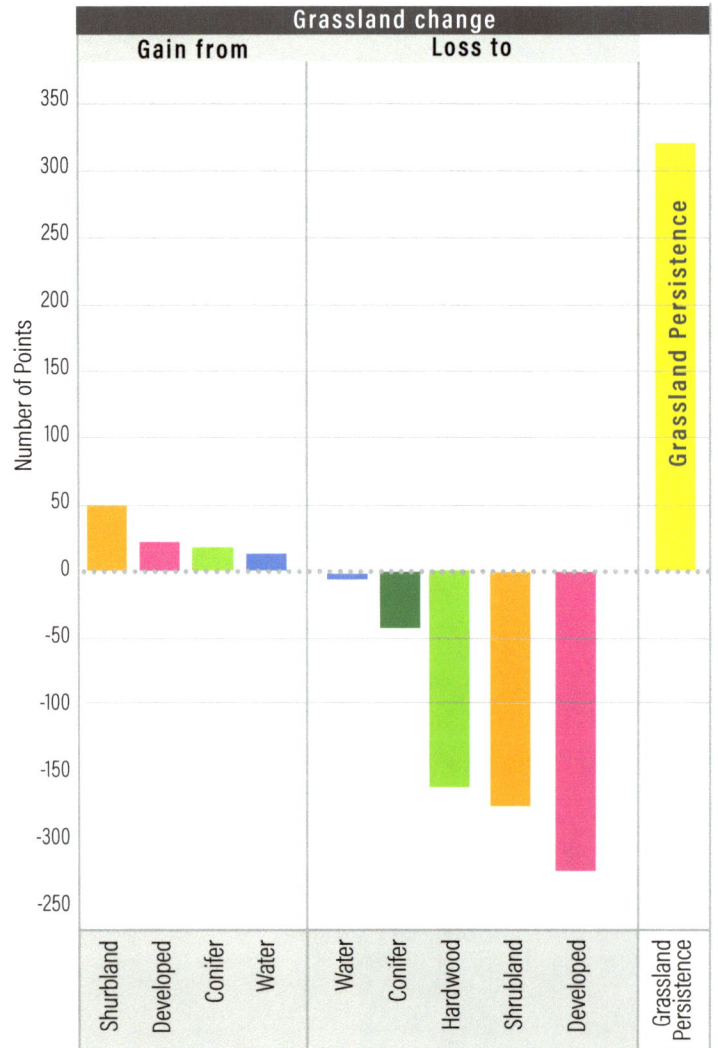

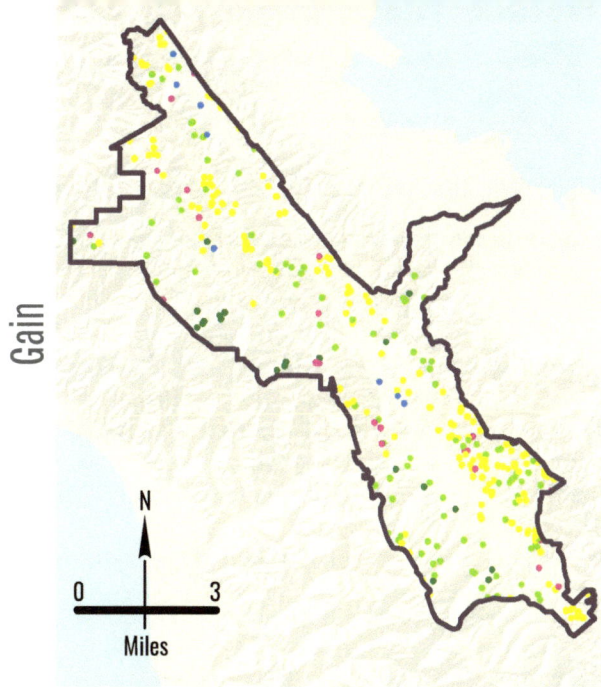

Gain

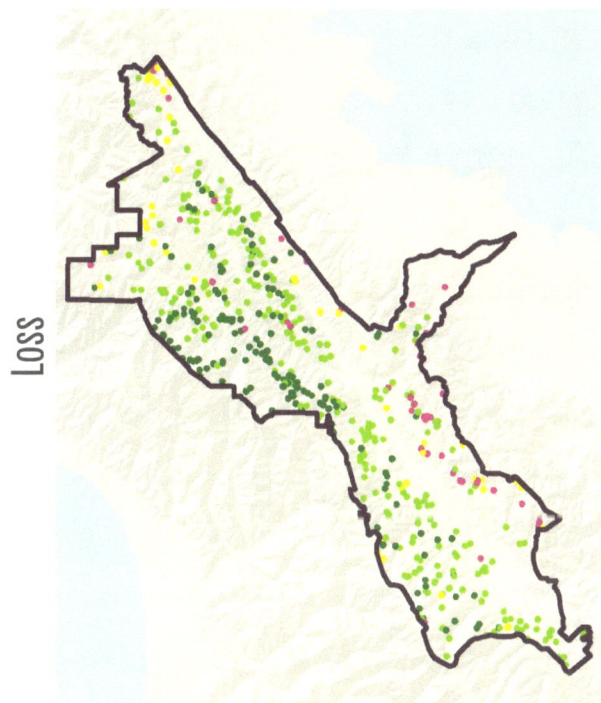

Loss

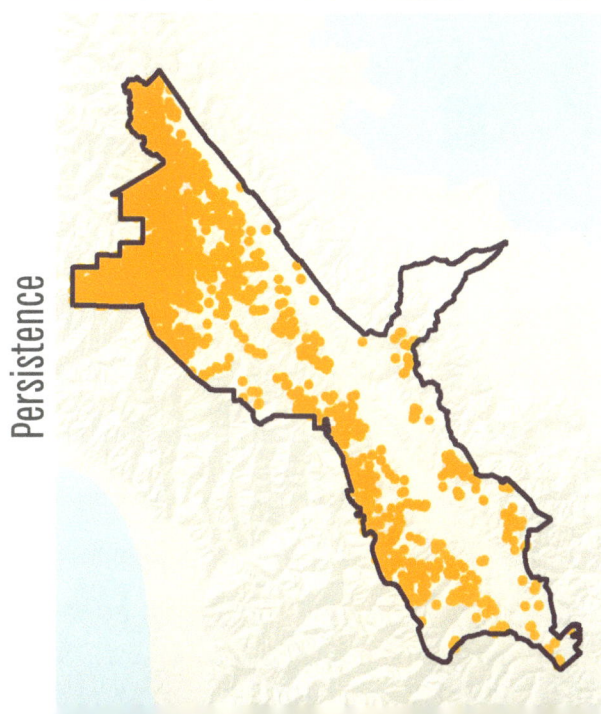

Persistence

Shrubland gain, loss, and persistence: 1946-48 to 2015-17

Figure 4.5b. The maps on the left show areas of gain (top), loss (center), and persistence (bottom) of shrubland between 1946-48 and 2015-17, based on the aerial imagery point analysis. Colors in the gain and loss maps indicate the vegetation/land cover types that shrubland has been gained from or lost to, respectively. The bar chart indicates the number of points classified as shrubland that have been gained from or lost to each vegetation/land cover type, as well as the number of shrubland points that have persisted over this time period (the y-axis has been scaled proportionally).

Habitat Type

- Developed/Disturbed
- Water
- Grassland
- Shrubland
- Hardwood Forest
- Conifer Forest

Locator Map

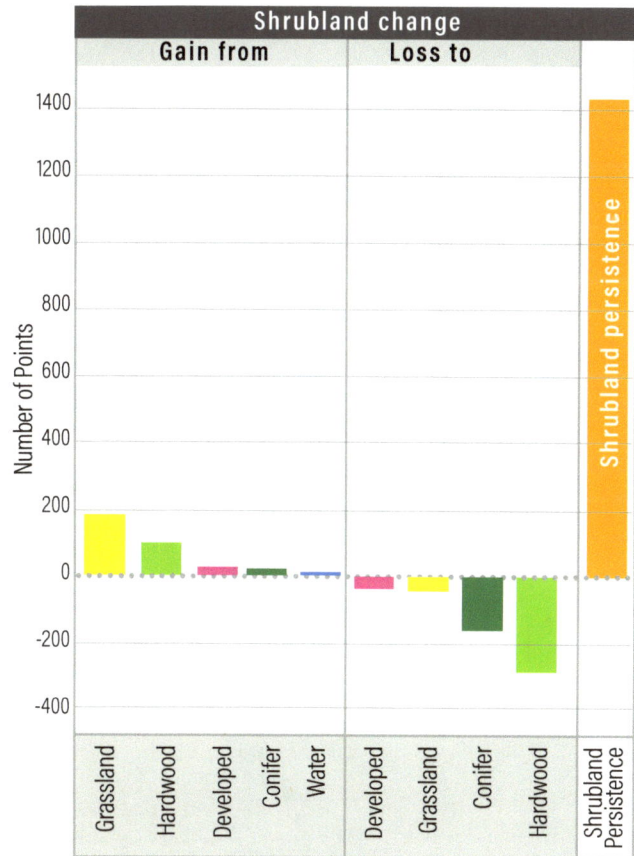

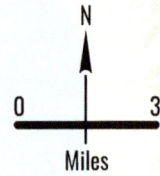

Gain

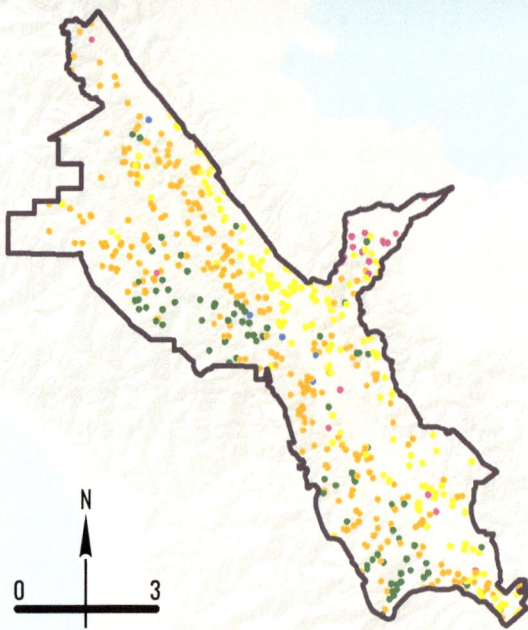

Loss

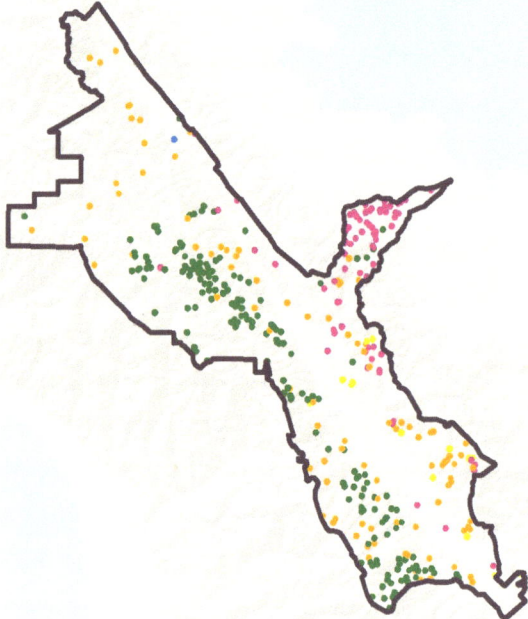

Persistence

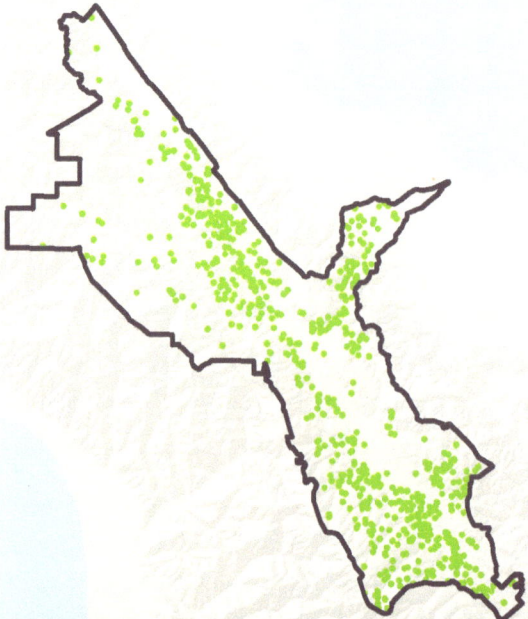

Hardwood gain, loss, and persistence: 1946-48 to 2015-17

Figure 4.5c. The maps on the left show areas of gain (top), loss (center), and persistence (bottom) of hardwood forest between 1946-48 and 2015-17, based on the aerial imagery point analysis. Colors in the gain and loss maps indicate the vegetation/land cover types that hardwood forest has been gained from or lost to, respectively. The bar chart indicates the number of points classified as hardwood forest that have been gained from or lost to each vegetation/land cover type, as well as the number of hardwood forest points that have persisted over this time period (the y-axis has been scaled proportionally).

Habitat Type

- Developed/Disturbed
- Water
- Grassland
- Shrubland
- Hardwood Forest
- Conifer Forest

Locator Map

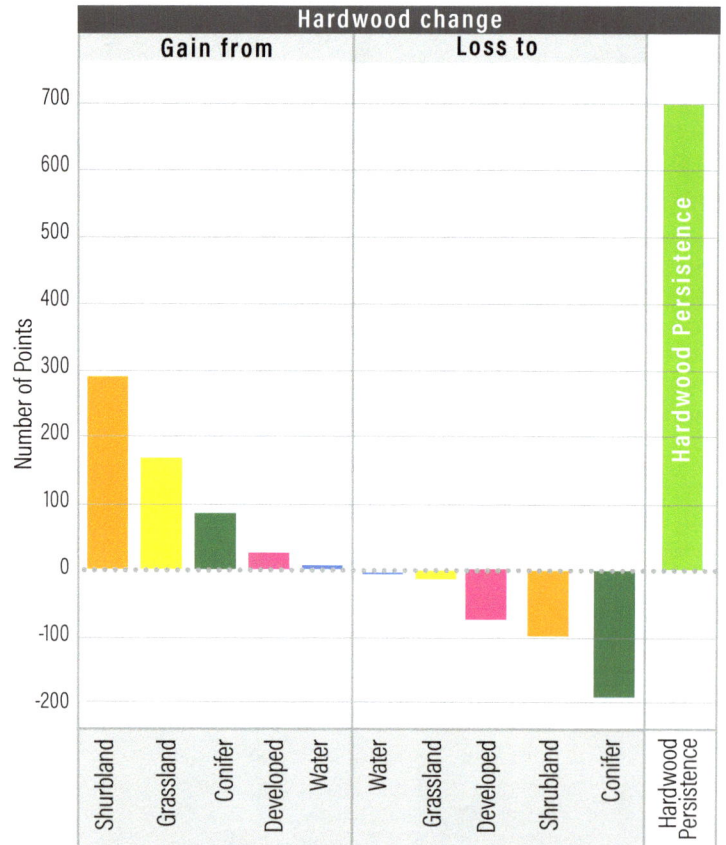

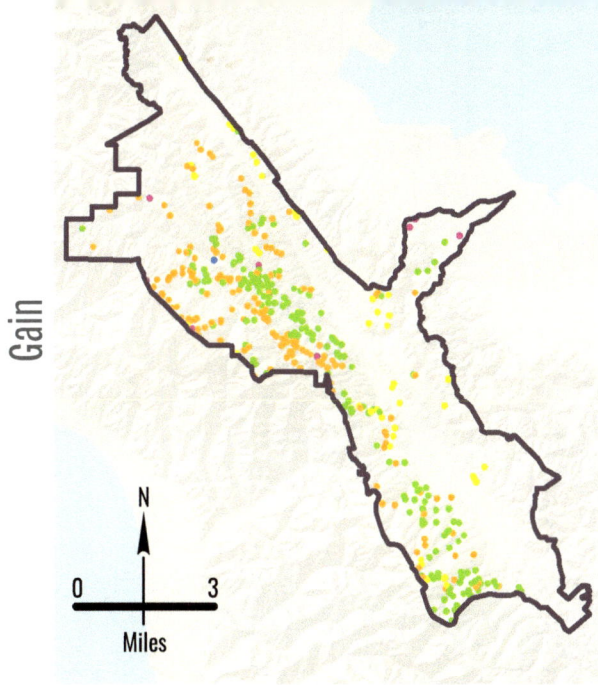

Gain

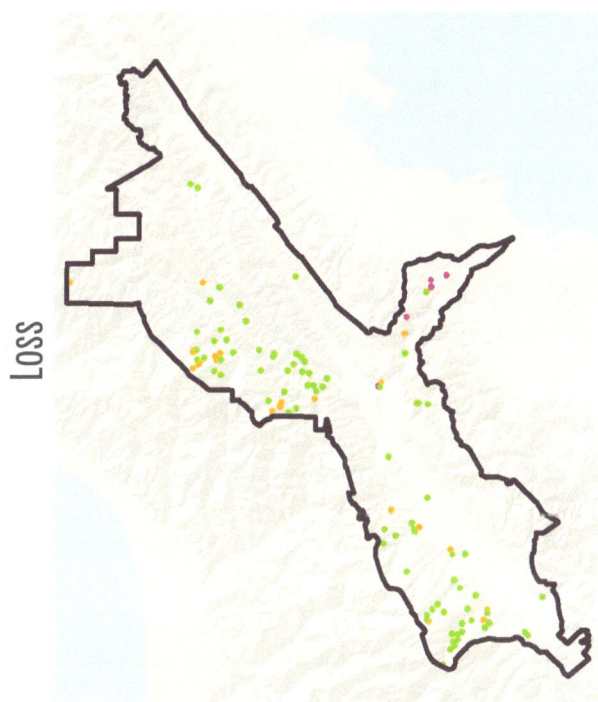

Loss

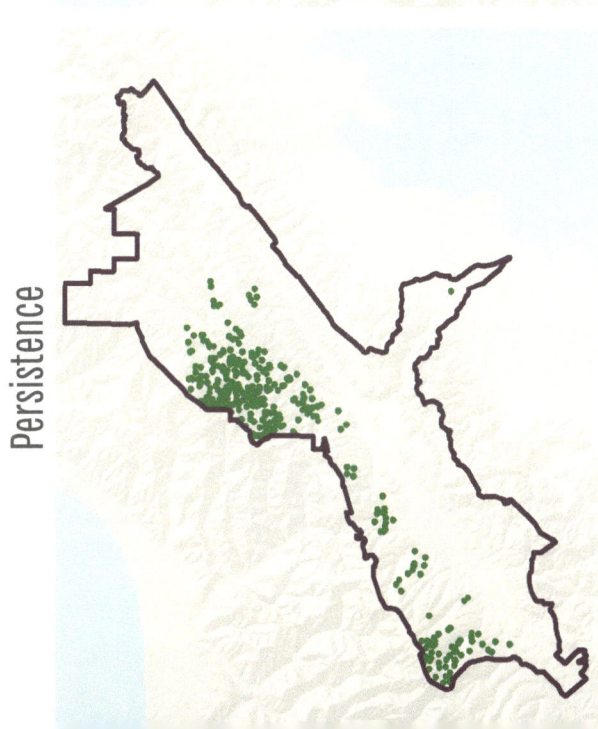

Persistence

Conifer gain, loss, and persistence: 1946-48 to 2015-17

Figure 4.5d. The maps on the left show areas of gain (top), loss (center), and persistence (bottom) of conifer forest between 1946-48 and 2015-17, based on the aerial imagery point analysis. Colors in the gain and loss maps indicate the vegetation/land cover types that conifer forest has been gained from or lost to, respectively. The bar chart indicates the number of points classified as conifer forest that have been gained from or lost to each vegetation/land cover type, as well as the number of conifer forest points that have persisted over this time period (the y-axis has been scaled proportionally).

Habitat Type

- Developed/Disturbed
- Water
- Grassland
- Shrubland
- Hardwood Forest
- Conifer Forest

Locator Map

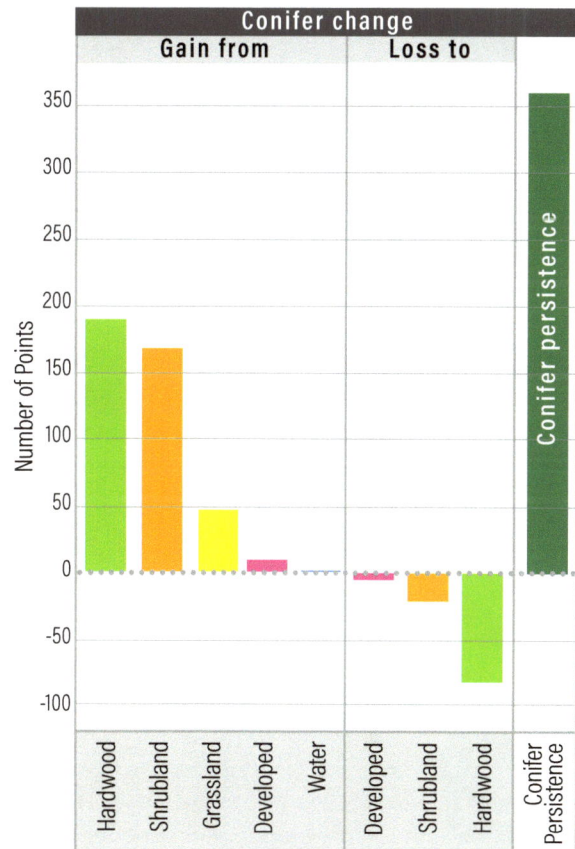

Comparison of Wieslander VTM and Modern Vegetation Mapping

TIME PERIOD: 1928-32 TO 1995-2001

Comparison of historical vegetation mapping (based on Wieslander VTM surveys) and modern vegetation mapping (MRLCC 2001, Schirokauer et al. 2003), crosswalked to the broad vegetation types used in this study, revealed many of the same trends as the aerial imagery point analysis (Fig. 4.6; Table 4.2). The time period represented by the vegetation mapping (1928-32 to 1995-2001) is slightly earlier, though largely overlaps, the period represented by the aerial imagery (1946-48 to 2015-17). Thus, taken together, the aerial imagery analysis and the vegetation mapping comparison represent two independent sources of information about vegetation change over the past 70-90 years.

Based on the Wieslander VTM mapping, land cover within the watershed in 1928-32 consisted of 24% grassland, 37% shrubland, 21% hardwood forest, 9% conifer forest, 5% water, and 4% developed/disturbed. Based on the modern vegetation mapping, land cover in 1995-2001 consisted of 7% grassland, 35% shrubland, 22-23% hardwood forest, 15-16% conifer forest, 7% water, and 13% developed/disturbed. Thus, change in land cover between 1928-32 and 1995-2001 includes an estimated 70% loss of grassland, 3% loss of shrubland, 2-7% gain of hardwood forest, and 78-91% gain of conifer forest.

Trajectories of land cover type conversion based on the vegetation mapping comparison are shown in Figure 4.7 and lower portion of Table 4.2. As with the aerial imagery analysis, in general these trajectories support the hypothesized transitions discussed in Chapter 3 (Fig. 4.8; see pages 42-43). Of the areas mapped as grassland historically, by 1995-2001 25% remained grassland, while 21% converted to shrubland, 18-19% converted to hardwood forest, 5-6% converted to conifer forest, and 24% converted to developed/disturbed. Of areas that were historically shrubland, 69% remained shrubland, while 14-15% converted to hardwood forest, and 12-13% converted to conifer forest. Of areas that were historically hardwood forest, 47-50% remained hardwood forest, while 15% converted to shrubland, 18-20% converted to conifer forest, and 12% converted to developed/disturbed. Of areas that were historically conifer forest, 69-70% remained conifer forest, while 11% converted to shrubland and 18-19% converted to hardwood forest. Figure 4.9 shows the spatial patterns of vegetation type conversion and persistence for each of the major vegetation types based on the vegetation mapping comparison.

Table 4.2. Results of the vegetation mapping comparison. The upper portion of the table shows the acreage for both the Wieslander VTM (1928-32) and modern vegetation mapping (1995-2001) for each vegetation/land cover class, along with the percent totals within each class. The lower portion of the table shows the percent changes between each historical class and each modern class.

ACREAGE TOTALS	Modern (1995-2001)								
Historical (1928-32)	Grassland	Shrubland	Hardwood Forest	Conifer Forest	Forest (Hardwood or Conifer)	Developed/Disturbed	Water	Historical Total (acres)	Historical Total (%)
Grassland	1550	1304	1118	294	51	1526	412	**6255**	24%
Shrubland	82	6587	1346	1110	86	254	23	**9488**	37%
Hardwood Forest	209	847	2607	967	130	662	85	**5508**	21%
Conifer Forest	0	254	407	1529	10	2	7	**2210**	9%
Developed/Disturbed	17	39	80	7	11	944	0	**1099**	4%
Water	4	62	69	31	0	2	1224	**1392**	5%
Modern Total (acres)	**1863**	**9092**	**5627**	**3938**	**288**	**3391**	**1751**	**25951**	-
Modern Total (%)	**7%**	**35%**	**22%**	**15%**	**1%**	**13%**	**7%**	-	**100%**

PERCENT CHANGE	Modern (1995-2001)						
Historical (1928-32)	Grassland	Shrubland	Hardwood Forest	Conifer Forest	Forest (Hardwood or Conifer)	Developed/Disturbed	Water
Grassland	25%	21%	18%	5%	1%	24%	7%
Shrubland	1%	69%	14%	12%	1%	3%	0%
Hardwood Forest	4%	15%	47%	18%	2%	12%	2%
Conifer Forest	0%	11%	18%	69%	0%	0%	0%
Developed/Disturbed	2%	4%	7%	1%	1%	86%	0%
Water	0%	4%	5%	2%	0%	0%	88%

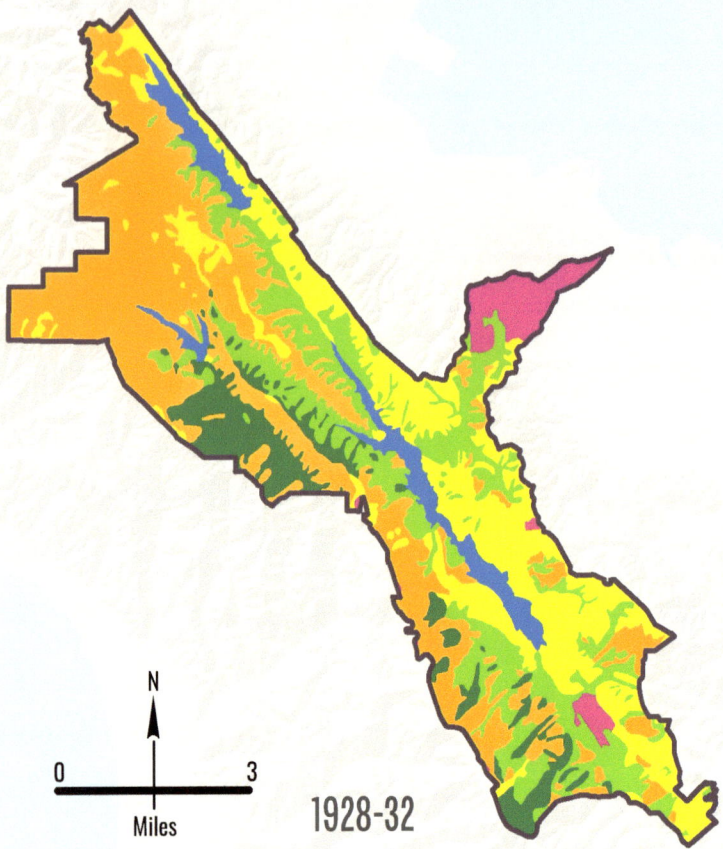

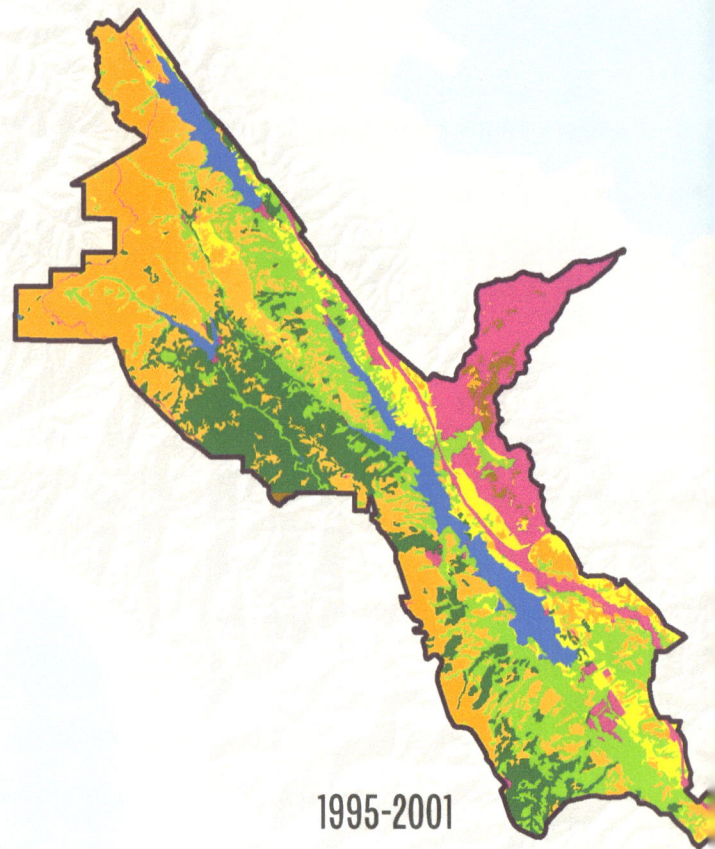

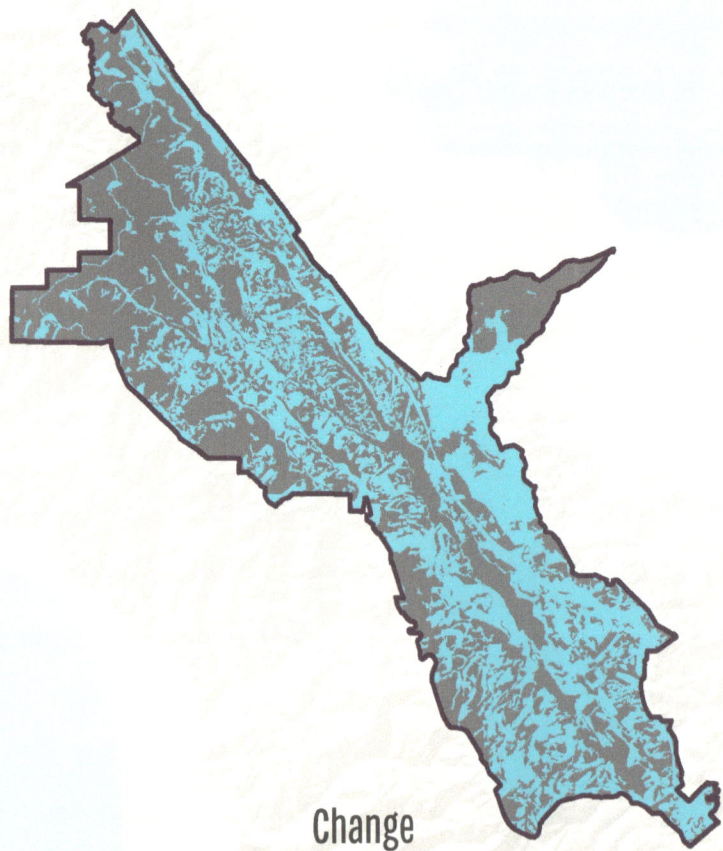

1928-32

1995-2001

Change

Habitat Type

- Developed/Disturbed
- Water
- Conifer Forest
- Hardwood Forest
- Shrubland
- Grassland
- Forest (Hardwood or Conifer)

Figure 4.6. Wieslander VTM mapping from 1928-32 (above left) and modern vegetation mapping from 1995-2001 (above right) were compared to assess change in vegetation type extent over time. The map to the left shows areas of change (blue) and persistence (gray) among the broad vegetation types.

N

0   3
Miles

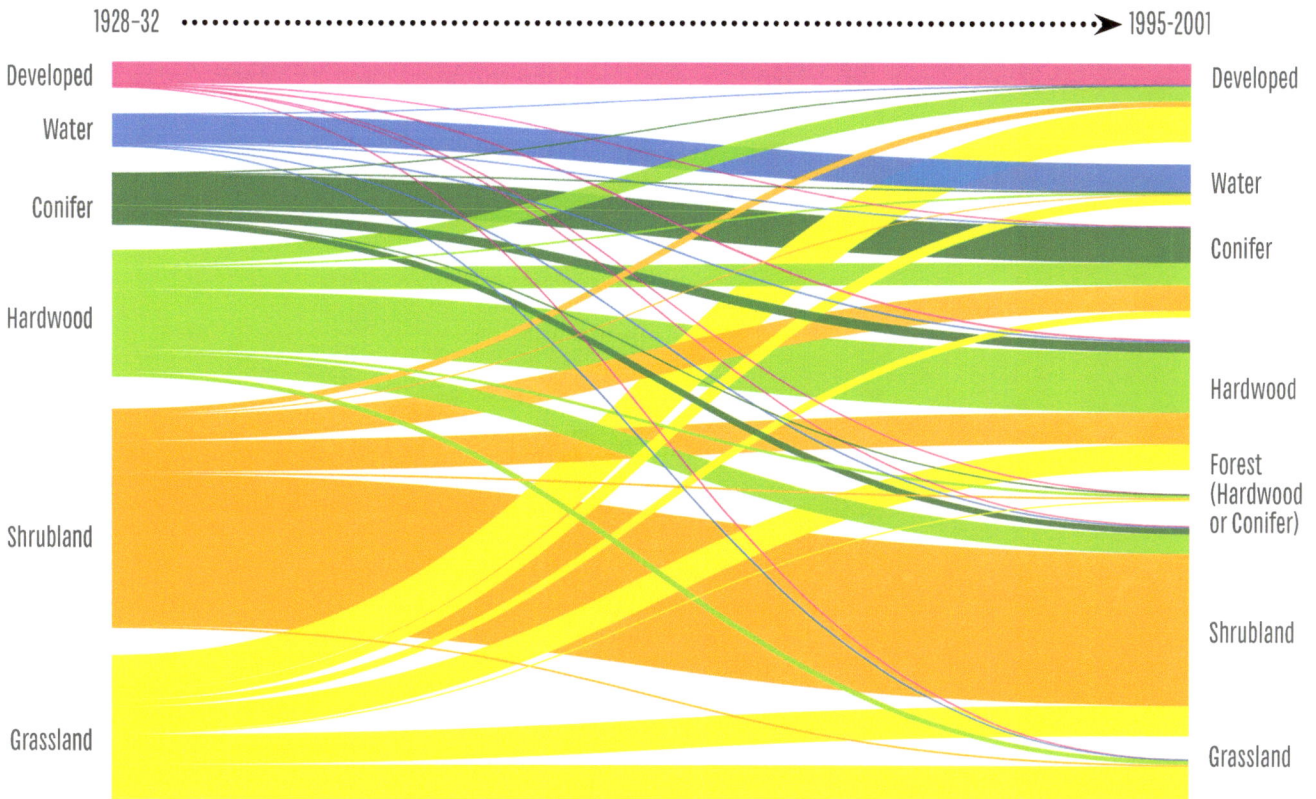

Figure 4.7. Land cover and vegetation type conversion within the study area based on the vegetation mapping comparison. The bars on the left side represent the proportion of each land cover type present within the study area in 1928-32, while the bars on the right represent the proportion of each land cover type in 1995-2001. The lines connecting the two sides of the chart illustrate the conversion "pathways" that have occurred over this period; the thickness of each line corresponds to the total area that has undergone a given type of conversion.

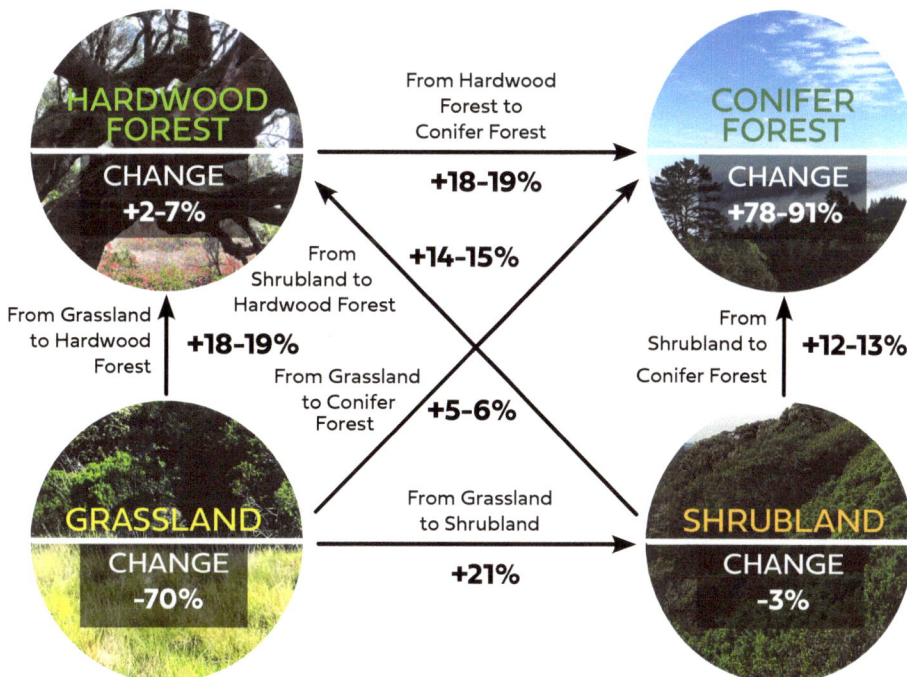

Figure 4.8. Overall change in the distribution of major terrestrial vegetation types between 1928-32 and 1995-2001 based on the vegetation mapping comparison. This diagram highlights major vegetation conversion pathways in relationship to hypothesized trends, but does not depict losses due to development; refer to Fig. 4.7 above and Table 4.2 on page 95 for complete land cover change results. *(Photos courtesy SFPUC)*

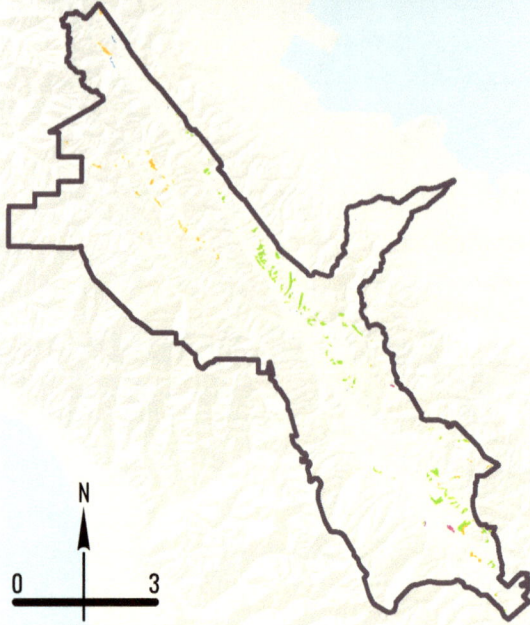

Gain

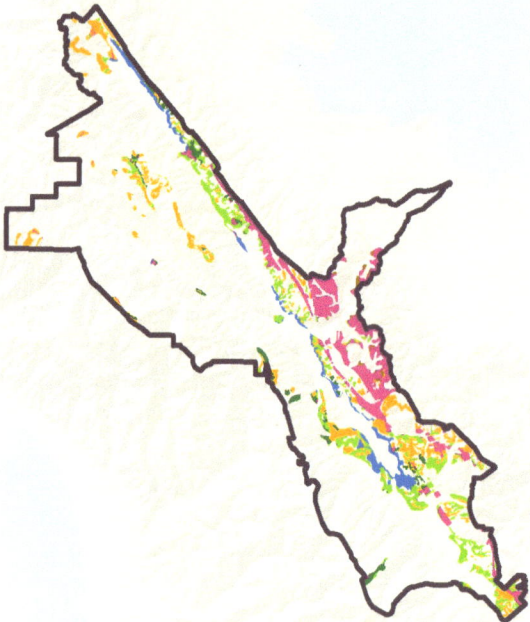

Loss

Persistence

Grassland gain, loss, and persistence: 1928-32 to 1995-2001

Figure 4.9a. The maps on the left show areas of gain (top), loss (center), and persistence (bottom) of grassland between 1928-32 and 1995-2001, based on the vegetation mapping comparison. Colors in the gain and loss maps indicate the vegetation/land cover types that grassland has been gained from or lost to, respectively. The bar chart indicates the acres of grassland that have been gained from or lost to each vegetation/land cover type, as well as the acreage of grassland that has persisted over this time period (the y-axis has been scaled proportionally).

Habitat Type

- Developed/Disturbed
- Water
- Grassland
- Shrubland
- Hardwood Forest
- Conifer Forest
- Forest (Hardwood or Conifer)

Locator Map

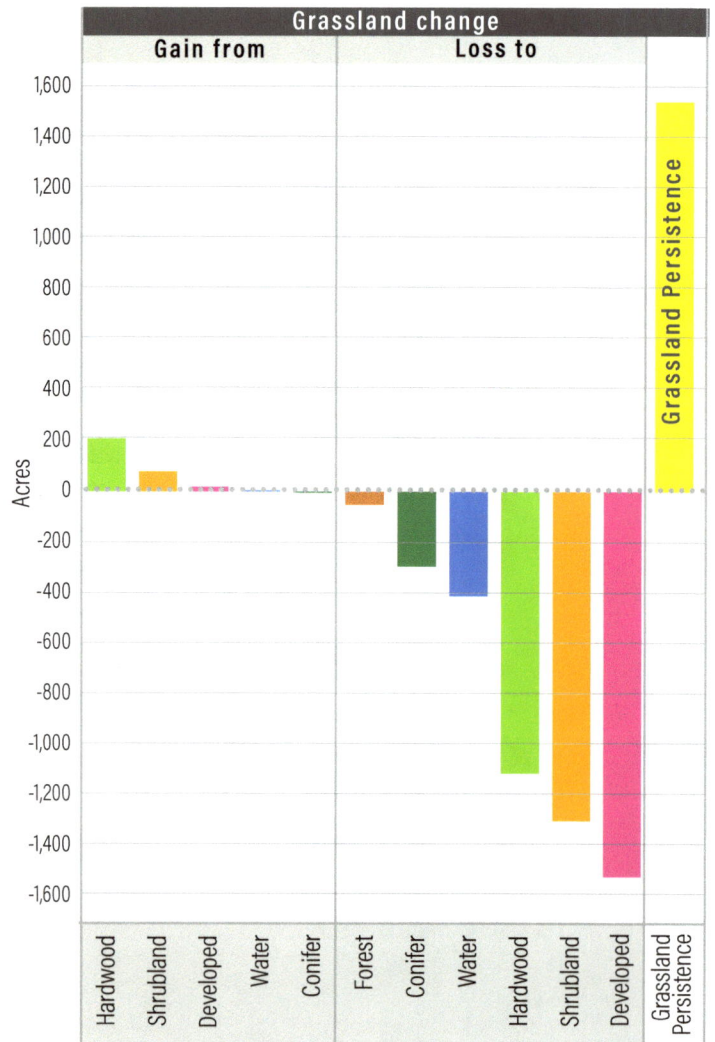

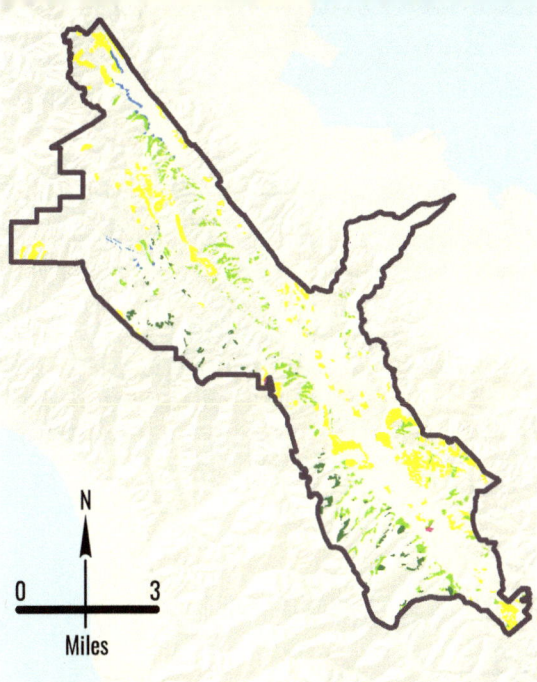

Gain

Loss

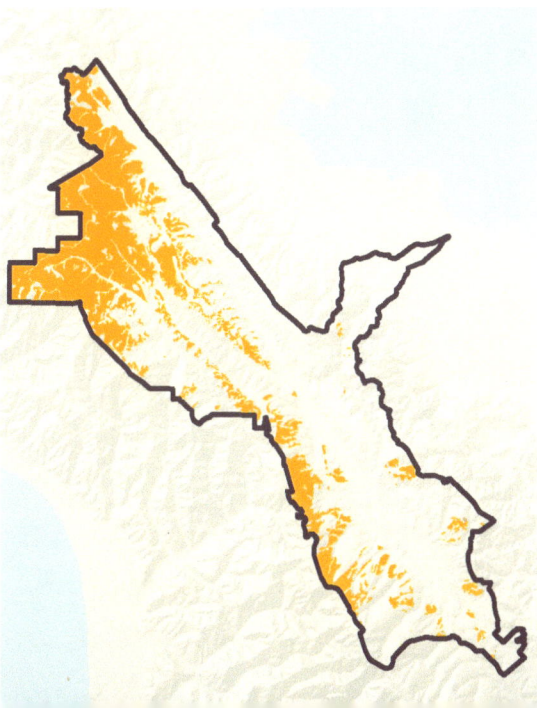

Persistence

Shrubland gain, loss, and persistence: 1928-32 to 1995-2001

Figure 4.9b. The maps on the left show areas of gain (top), loss (center), and persistence (bottom) of shrubland between 1928-32 and 1995-2001, based on the vegetation mapping comparison. Colors in the gain and loss maps indicate the vegetation/land cover types that shrubland has been gained from or lost to, respectively. The bar chart indicates the acres of shrubland that have been gained from or lost to each vegetation/land cover type, as well as the acreage of shrubland that has persisted over this time period (the y-axis has been scaled proportionally).

Habitat Type

- Developed/Disturbed
- Water
- Grassland
- Shrubland
- Hardwood Forest
- Conifer Forest
- Forest (Hardwood or Conifer)

Locator Map

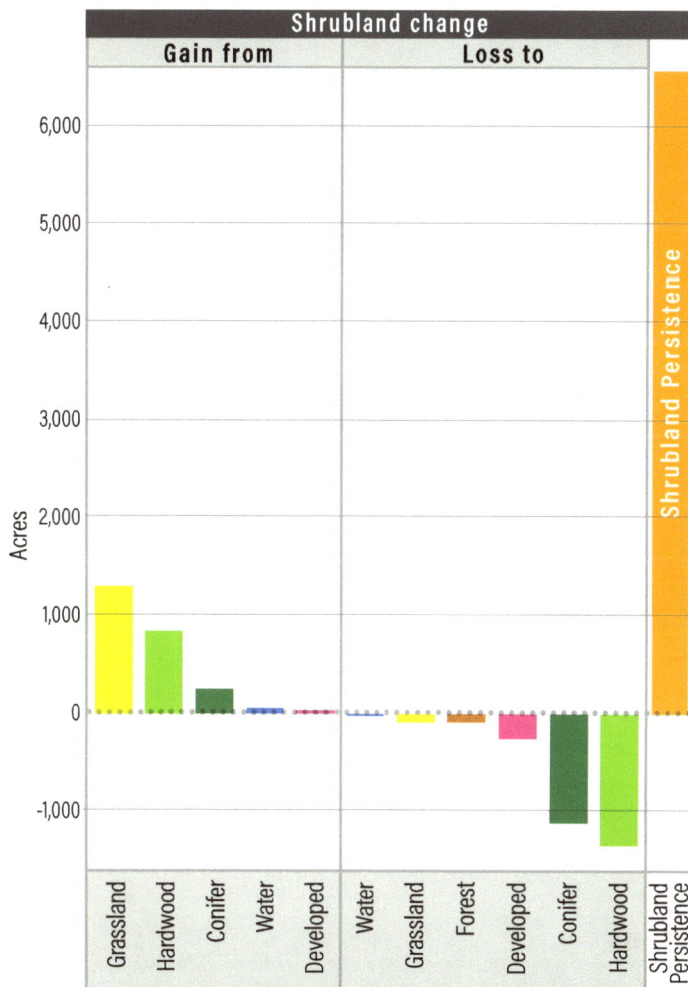

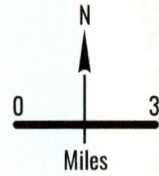

Gain

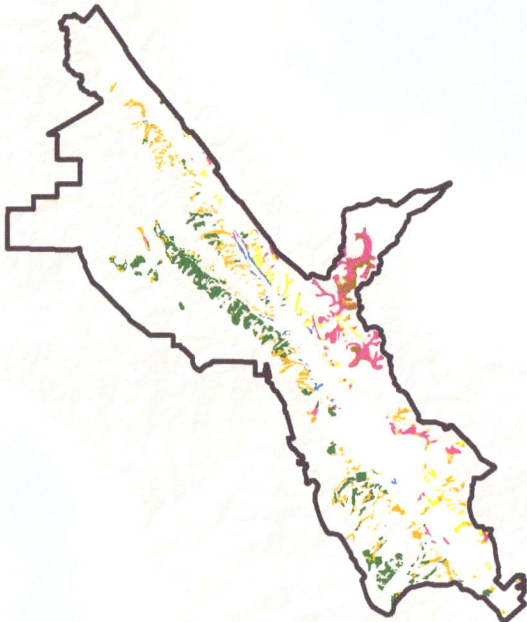

Loss

Persistence

Hardwood gain, loss, and persistence: 1928-32 to 1995-2001

Figure 4.9c. The maps on the left show areas of gain (top), loss (center), and persistence (bottom) of hardwood forest between 1928-32 and 1995-2001, based on the vegetation mapping comparison. Colors in the gain and loss maps indicate the vegetation/land cover types that hardwood forest has been gained from or lost to, respectively. The bar chart indicates the acres of hardwood forest that have been gained from or lost to each vegetation/land cover type, as well as the acreage of hardwood forest that has persisted over this time period (the y-axis has been scaled proportionally).

Habitat Type

- ▮ Developed/Disturbed
- ▮ Water
- ▮ Grassland
- ▮ Shrubland
- ▮ Hardwood Forest
- ▮ Conifer Forest
- ▮ Forest (Hardwood or Conifer)

Locator Map

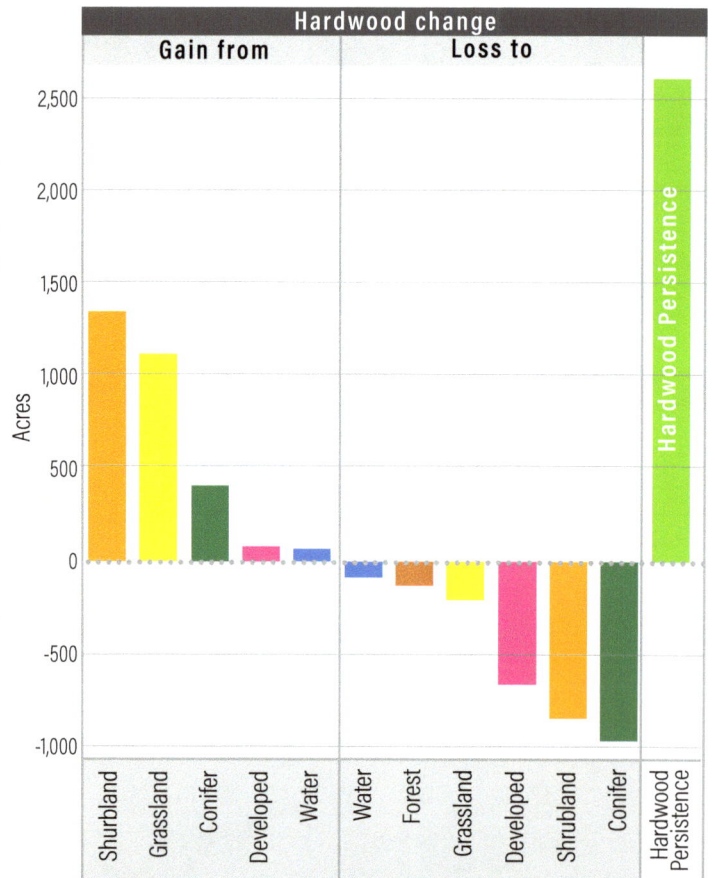

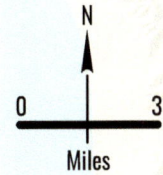

Gain

Loss

Persistence

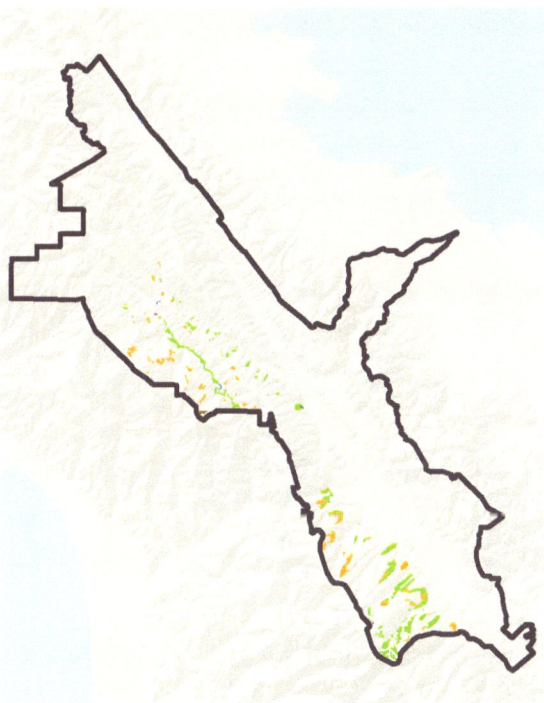

N

0 ——— 3

Miles

Conifer gain, loss, and persistence: 1928-32 to 1995-2001

Figure 4.9d. The maps on the left show areas of gain (top), loss (center), and persistence (bottom) of conifer forest between 1928-32 and 1995-2001, based on the vegetation mapping comparison. Colors in the gain and loss maps indicate the vegetation/land cover types that conifer forest has been gained from or lost to, respectively. The bar chart indicates the acres of conifer forest that have been gained from or lost to each vegetation/land cover type, as well as the acreage of conifer forest that has persisted over this time period (the y-axis has been scaled proportionally).

Habitat Type

- Developed/Disturbed
- Water
- Grassland
- Shrubland
- Hardwood Forest
- Conifer Forest
- Forest (Hardwood or Conifer)

Locator Map

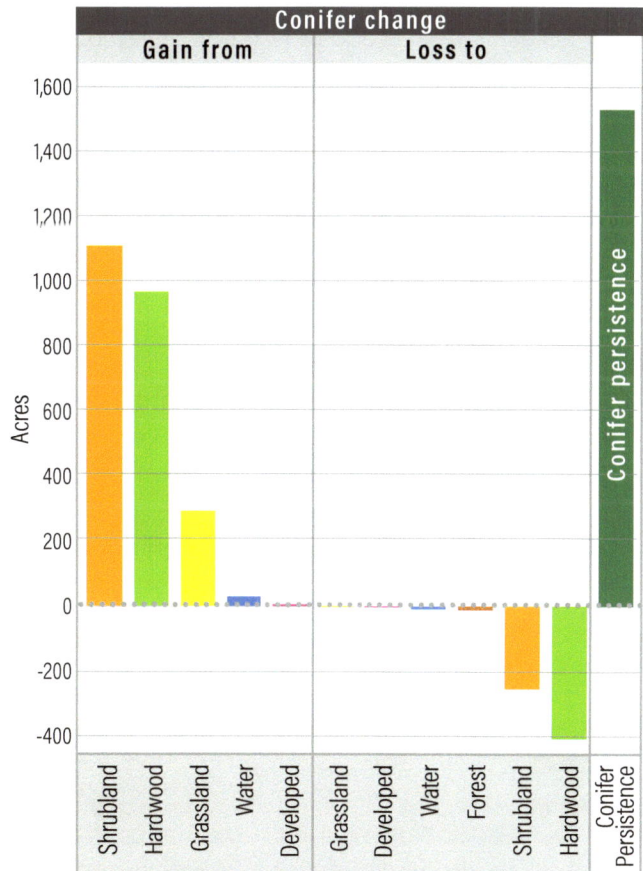

Analysis of General Land Office Survey Data

TIME PERIOD: 1852 - 2017

This analysis classified vegetation cover at 175 points throughout the watershed using GLO survey data from the mid-19th century (1852-64), and examined vegetation cover at these same locations in both historical (1946-48) and modern (2015-17) aerial imagery (see discussion of methods on pages 32-34).

Of the 175 points classified in this analysis, the GLO surveys indicate that, during the mid-19th century, 13-17% were grassland, 39-43% were shrubland, 30-32% were hardwood forest, 4-5% were conifer forest, 7% were water, and 1% were developed/disturbed (the ranges represent uncertainty in the classification [e.g., points classified as grassland/shrubland]; Fig. 4.10 on page 104 and Fig. 4.12 on page 107). Classification of vegetation cover at the same points using historical aerial imagery indicates that, by 1946-48, 14% were grassland, 31% were shrubland, 16% were hardwood forest, 10% were conifer forest, 26% were water, and 2% were developed/disturbed. Classification using the modern aerial imagery indicates that by 2015-17, 5% of the points were grassland, 21% were shrubland, 21% were hardwood forest, 19% were conifer forest, 26% were water, and 8% were developed/disturbed.

As discussed in Chapter 2, the fact that the GLO points are not evenly or randomly distributed throughout the study area raises the possibility that the dataset represents a biased sample of vegetation cover within the watershed. To assess the degree of bias within the GLO dataset, the relative proportion of each vegetation type across the 175 points, for both the 1946-48 and 2015-17 time periods, was compared with the corresponding proportion in the random sample of 5,000 points used in the aerial imagery point analysis (Table 4.3). The results show that there is fairly close agreement (generally within 10%) in the estimates of vegetation cover derived from the GLO and aerial imagery point analyses, suggesting that the degree of spatial bias in the GLO dataset is relatively minimal. The exceptions are in the estimates of contemporary shrubland cover (which differ by ~14%) and the estimates of both historical and contemporary water extent (which differ by ~20%); the latter finding suggests that the GLO data are likely somewhat overrepresented in the San Andreas Valley, in the areas subsequently inundated by the San Andreas and Crystal Springs reservoirs. Many of the GLO points in the San Andreas Valley documented the presence of hardwood or riparian forests or water during the mid-19th century, prior to reservoir construction. Thus, the GLO data likely somewhat overestimate the proportion of the watershed occupied by hardwood forest and water.

Examination of vegetation cover at each of the 175 points over time suggests that the trajectories of vegetation change were somewhat different between the mid-19th century

Table 4.3. Comparison of relative proportions of each vegetation/land cover class in 1946-48 and 2015-17 based on the GLO analysis and aerial imagery point analysis.

Time period: 1946-48					
Class	# points in GLO analysis	% in GLO analysis	# points in aerial imagery analysis	% in aerial imagery analysis	% Difference
Grassland	24	13.71%	937	18.74%	5.03%
Shrubland	55	31.43%	1958	39.16%	7.73%
Hardwood Forest	28	16.00%	1074	21.48%	5.48%
Conifer Forest	18	10.29%	466	9.32%	-0.97%
Water	46	26.29%	353	7.06%	-19.23%
Developed/Disturbed	4	2.29%	212	4.24%	1.95%
Total	**175**		**5000**		

Time period: 2015-17					
Class	# points in GLO analysis	% in GLO analysis	# points in aerial imagery analysis	% in aerial imagery analysis	% Difference
Grassland	9	5.14%	408	8.16%	3.02%
Shrubland	36	20.57%	1746	34.92%	14.35%
Hardwood Forest	37	21.14%	1260	25.20%	4.06%
Conifer Forest	33	18.86%	768	15.36%	-3.50%
Water	46	26.29%	335	6.70%	-19.59%
Developed/Disturbed	14	8.00%	483	9.66%	1.66%
Total	**175**		**5000**		

and the 1940s than they were between the 1940s and present (Fig. 4.11); these trajectories are consistent with documented patterns of land use change within the watershed (see Chapter 3). During the earlier era (mid-19th century through 1946-48), for example, one readily apparent trend is the conversion of areas formerly occupied by hardwood (or riparian) forest to water as a result of reservoir construction within San Andreas Valley. Conifer forest expanded into areas formerly occupied by shrubland and hardwood forest, and hardwood forest expanded into shrublands as well, likely due in part to changes in fire frequency and other disturbance regimes. Grassland cover increased somewhat in areas formerly occupied by shrubland and hardwood forest, likely reflecting efforts by settlers in the mid-19th century to clear woody vegetation for cropland and livestock grazing.

During the latter period (1946-48 to 2015-17), the trajectories of vegetation change suggested by the GLO analysis are generally consistent with the patterns evident from the aerial imagery point analysis and the vegetation mapping comparison. Grassland cover decreased as a result of development and conversion to woody vegetation types. Conifer extent increased as a result of encroachment into shrubland and hardwood forest. Hardwood forest expanded in some areas formerly occupied by grassland and shrubland, and decreased somewhat in other areas due to conifer forest encroachment. These patterns are further examined in the following chapters.

GLO Vegetation Change

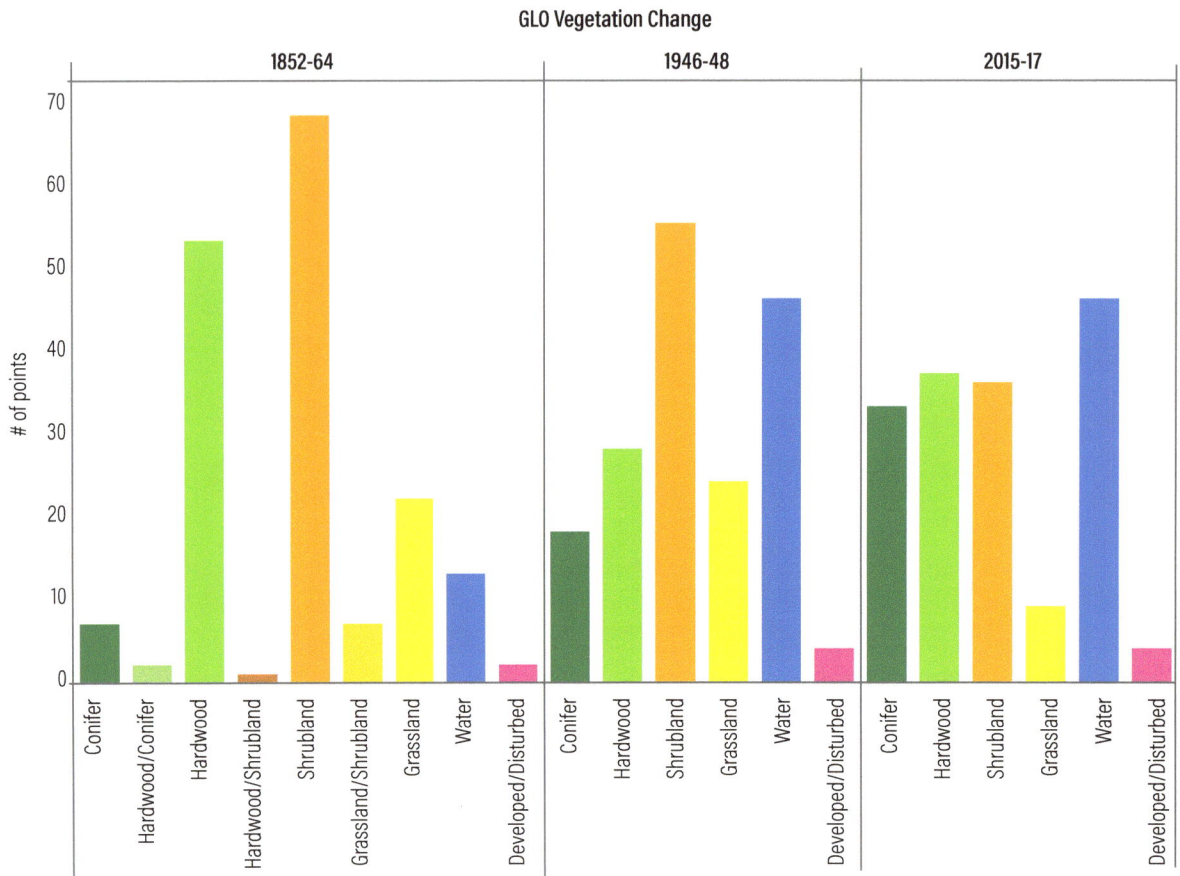

Figure 4.10. Bar chart showing the vegetation/land cover at the 175 points used in the analysis of GLO survey data for three time periods: 1852-64, 1946-48, and 2015-17. Point classifications for the 1852-64 time period were derived from descriptions in the GLO surveys, while point classifications for the 1946-48 and 2015-17 time periods were derived from historical and modern aerial imagery, respectively (see pages 32-34 for detailed methods).

Old growth Douglas-fir forest.
(Photo courtesy SFPUC)

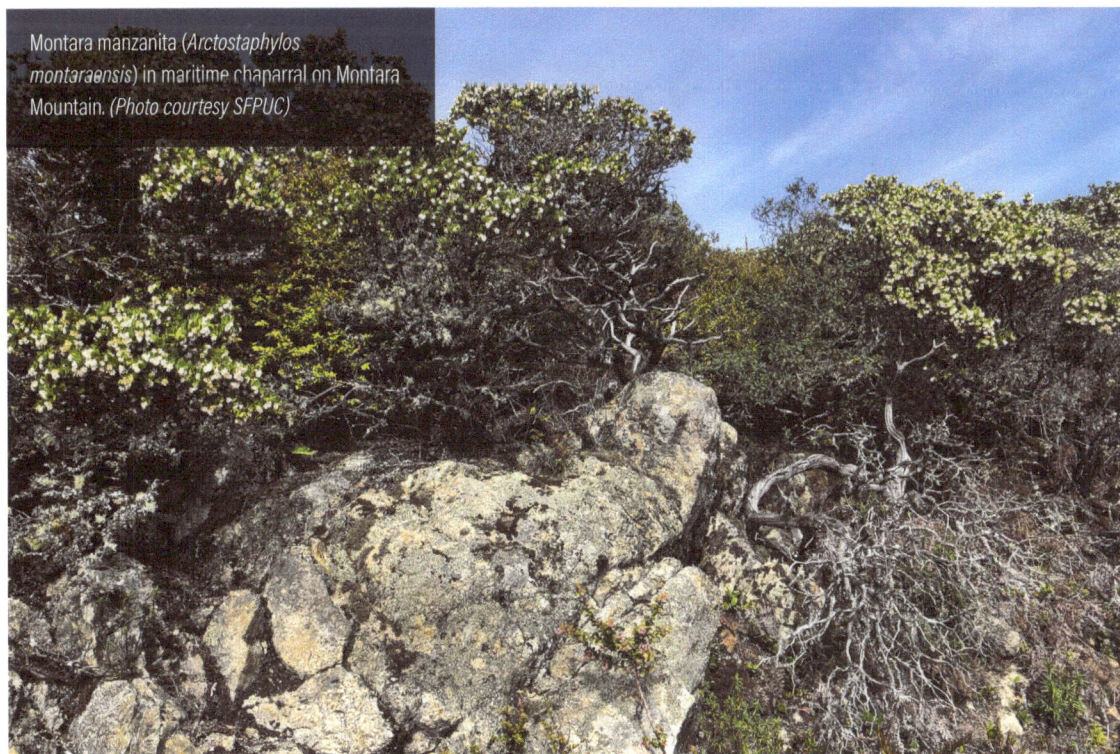

Montara manzanita (Arctostaphylos montaraensis) in maritime chaparral on Montara Mountain. (Photo courtesy SFPUC)

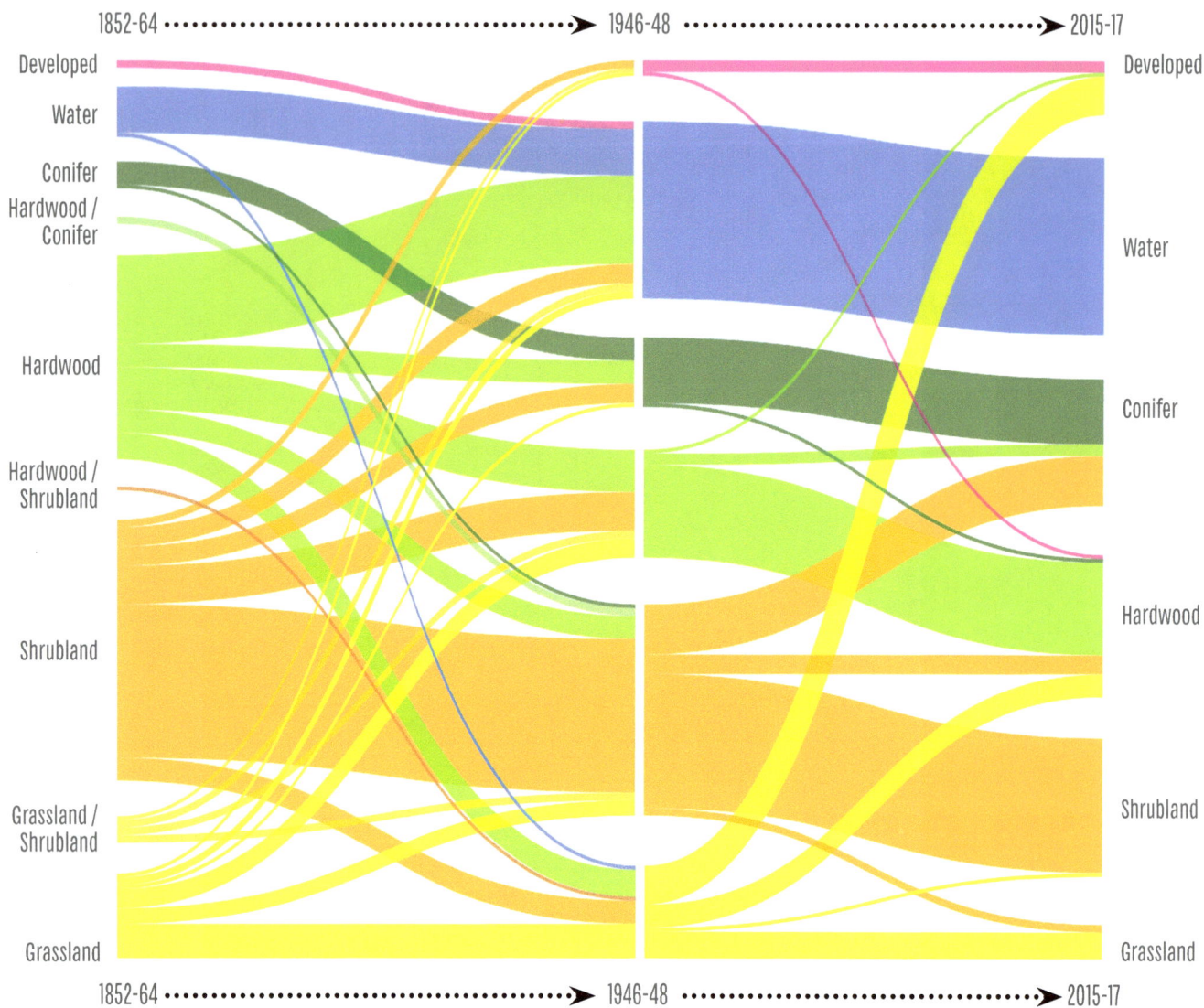

1852-64 ●●●●●●●●●●●●●●●●●●●●●●●●●●●●➤ 1946-48 ●●●●●●●●●●●●●●●●●●●●●●●●➤ 2015-17

Developed
Water
Conifer
Hardwood / Conifer
Hardwood
Hardwood / Shrubland
Shrubland
Grassland / Shrubland
Grassland

Developed
Water
Conifer
Hardwood
Shrubland
Grassland

1852-64 ●●●●●●●●●●●●●●●●●●●●●●●●●●●●➤ 1946-48 ●●●●●●●●●●●●●●●●●●●●●●●➤ 2015-17

Figure 4.11. Vegetation type conversion that occurred within the study area based on the GLO survey analysis. The bars on the left side represent the frequency of each vegetation type in 1852-64 among the 175 points analyzed; the bars in the middle represent the frequency of each vegetation type at the same points in 1946-48; and the bars on the right represent the frequency of each vegetation type at the same points in 2015-17. The lines connecting the two sides of the chart illustrate the conversion "pathways" that have occurred over these two time periods; the thickness of each line corresponds to the total area that has undergone a given type of conversion. Note that ambiguity in the GLO survey descriptions precluded definitive classification of some points, and thus a fraction of the points for the 1852-64 time period are classified as one of two possible vegetation types.

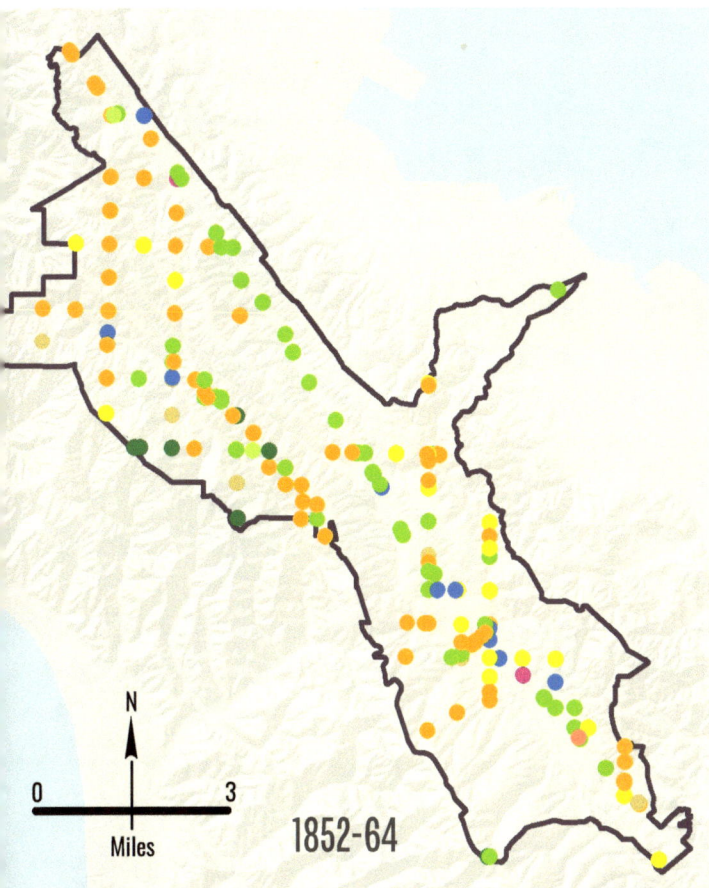

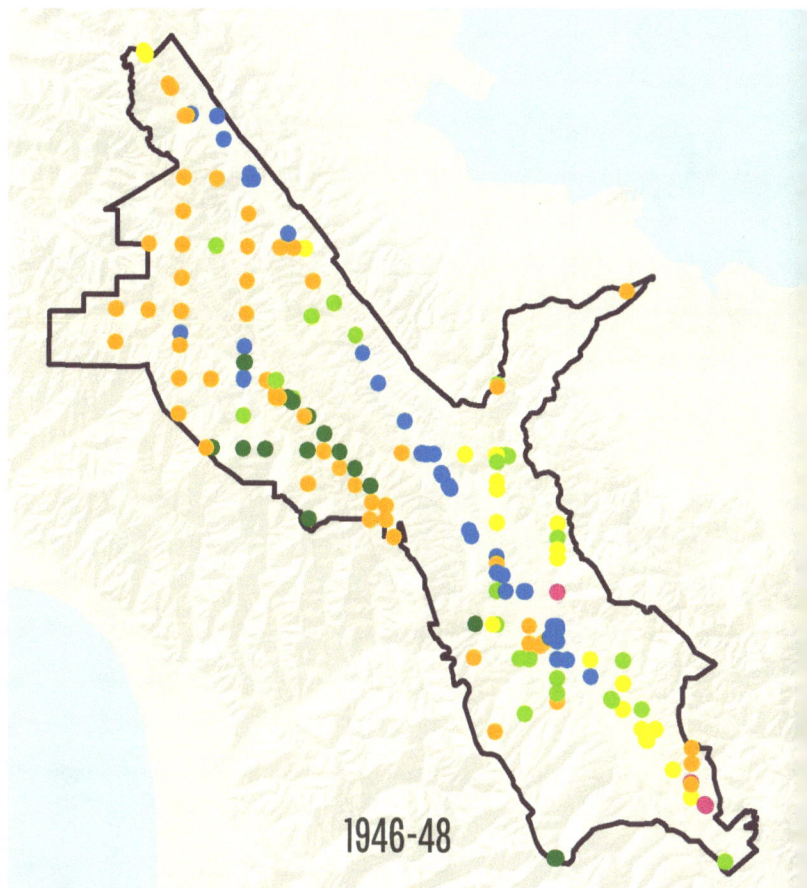

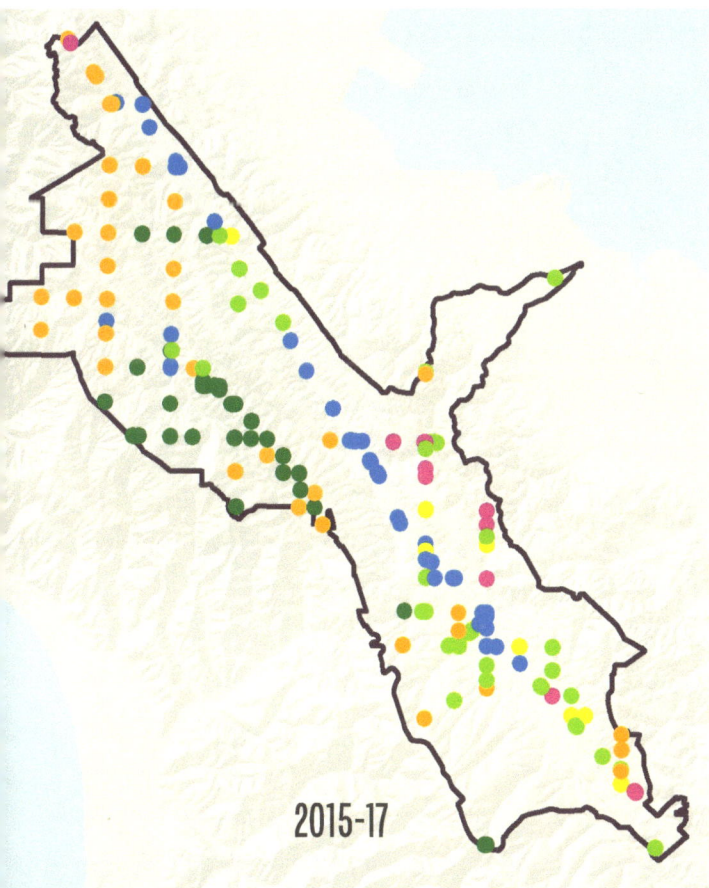

Habitat Type

- Developed/Disturbed
- Water
- Conifer Forest
- Hardwood/Conifer Forest
- Hardwood Forest
- Hardwood Forest/Shrubland
- Shrubland
- Grassland/Shrubland
- Grassland

Figure 4.12. Mid-19th century, mid-20th century, and contemporary vegetation cover at 175 points based on GLO survey field notes (above left) and re-analysis of the same points using historical (above right) and contemporary (left) aerial imagery.

Serpentine grassland in summer, with farewell to spring (*Clarkia rubicunda*), purple needle grass (*Stipa pulchra*), and California oatgrass (*Danthonia californica*). *(Photo courtesy SFPUC)*

CHAPTER 5 Grasslands

Overview

Grasslands historically occupied thousands of acres in the Peninsula Watershed (Fig. 5.1). They were the dominant vegetation type on the eastern side of San Andreas Valley, along Buri Buri and Pulgas ridges—often in association with serpentine bedrock and soils—and occupied large areas on the southwestern side of San Andreas Valley as well. They were also prevalent along many of the ridges throughout the watershed, including large areas along Sawyer, Fifield, Spring Valley, and Cahill ridges. Along with anthropogenic burning, extensive areas characterized by serpentine geology and soils were likely a major factor contributing to the prevalence and diversity of grasslands within the Peninsula Watershed historically (see pages 135-141).

The pre-modification composition of California grasslands is a matter of active debate, but evidence suggests that grasslands in many coastal settings were dominated by perennial bunchgrasses. In contrast, grasslands in the Central Valley and other more xeric sites further inland were likely dominated by forbs and annual grasses; today these inland grasslands are often dominated by introduced annual grasses such as soft brome (*Bromus hordeaceus*), red stemmed filaree (*Erodium cicutarium*), wild oat (*Avena fatua*), and Italian rye grass (*Festuca perennis*; Heady et al. 1992, Jackson and Bartolome 2002, Bartolome et al. 2007, Minnich 2008). Coastal grasslands in California often have a higher plant species richness than inland grasslands (Stromberg et al. 2001), and serpentine grasslands support an especially high plant diversity, including a number of rare and endemic species.

Early botanical records from the Peninsula Watershed include native grasses such as purple needle grass (*Stipa pulchra*), foothill needle grass (*Stipa lepida*), California oatgrass (*Danthonia californica*), blue fescue (*Festuca idahoensis*), tufted hairgrass (*Deschampsia cespitosa*), meadow barley (*Hordeum brachyantherum*), small leaf bentgrass (*Agrostis microphylla*), and pine bluegrass (*Poa secunda*), and a variety of native grassland-associated forbs such as giant mountain dandelion (*Agoseris grandiflora* var. *grandiflora*), Johnny jump-up (*Viola pedunculata*), coastal tarweed (*Deinandra corymbosa*), western blue eyed grass (*Sisyrinchium bellum*), blue dicks (*Dichelostemma capitatum*), wicker buckwheat (*Eriogonum luteolum*), clovers (*Trifolium* spp.), narrow leaved mule ears (*Wyethia angustifolia*), California buttercup (*Ranunculus californicus*), purple sanicle (*Sanicula bipinnatifida*), dwarf brodiaea (*Brodiaea terrestris* ssp. *terrestris*), California

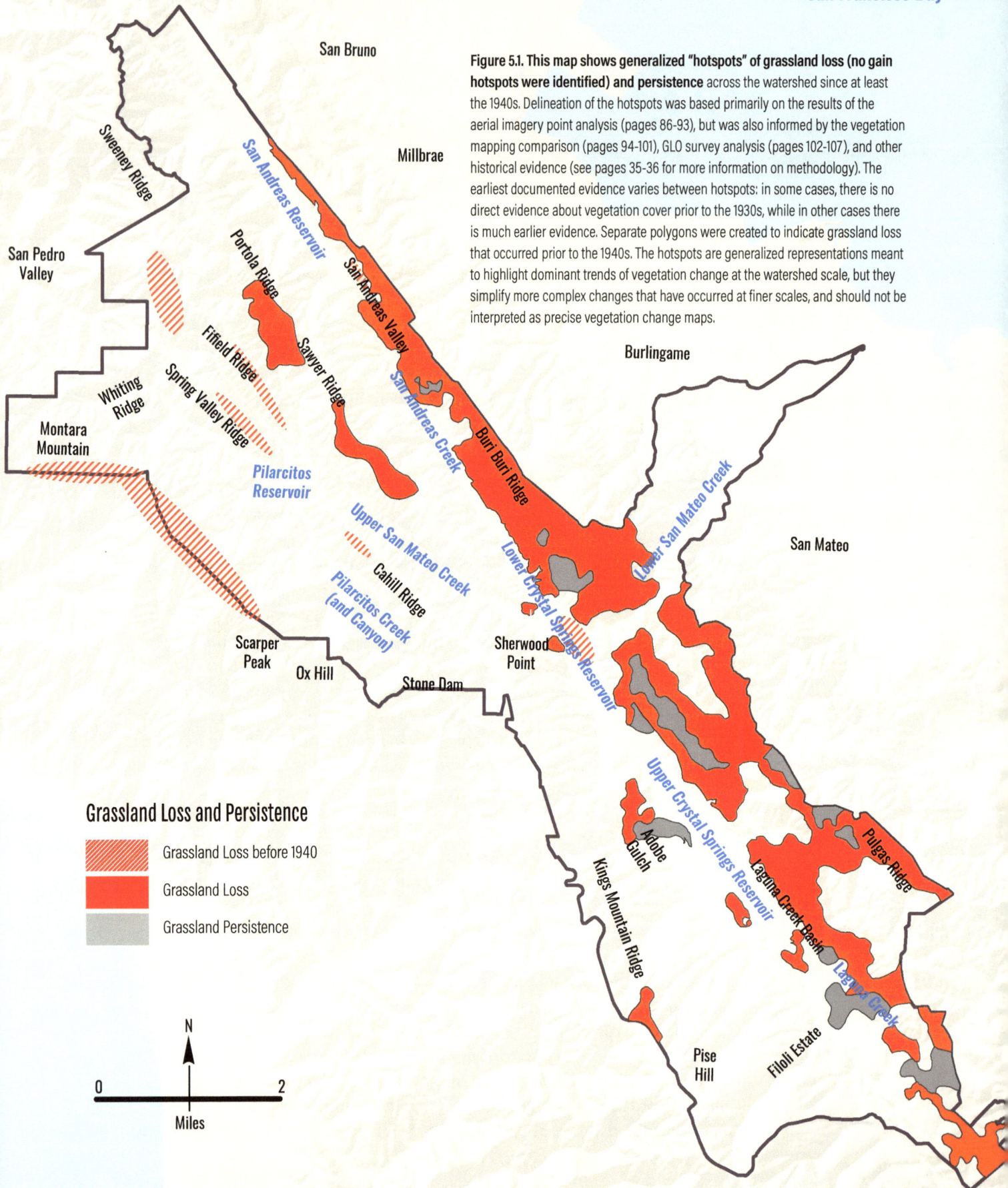

San Francisco Bay

San Bruno

Millbrae

Figure 5.1. This map shows generalized "hotspots" of grassland loss (no gain hotspots were identified) and persistence across the watershed since at least the 1940s. Delineation of the hotspots was based primarily on the results of the aerial imagery point analysis (pages 86-93), but was also informed by the vegetation mapping comparison (pages 94-101), GLO survey analysis (pages 102-107), and other historical evidence (see pages 35-36 for more information on methodology). The earliest documented evidence varies between hotspots: in some cases, there is no direct evidence about vegetation cover prior to the 1930s, while in other cases there is much earlier evidence. Separate polygons were created to indicate grassland loss that occurred prior to the 1940s. The hotspots are generalized representations meant to highlight dominant trends of vegetation change at the watershed scale, but they simplify more complex changes that have occurred at finer scales, and should not be interpreted as precise vegetation change maps.

Sweeney Ridge

San Andreas Reservoir

Portola Ridge

San Pedro Valley

Fifield Ridge

Sawyer Ridge

San Andreas Valley

San Andreas Creek

Burlingame

Buri Buri Ridge

Lower San Mateo Creek

Whiting Ridge

Spring Valley Ridge

Pilarcitos Reservoir

San Mateo

Montara Mountain

Upper San Mateo Creek

Pilarcitos Creek (and Canyon)

Cahill Ridge

Lower Crystal Springs Reservoir

Scarper Peak

Ox Hill

Stone Dam

Sherwood Point

Upper Crystal Springs Reservoir

Pulgas Ridge

Adobe Gulch

Laguna Creek Basin

Kings Mountain Ridge

Laguna Creek

Pise Hill

Filoli Estate

Grassland Loss and Persistence

Grassland Loss before 1940

Grassland Loss

Grassland Persistence

N

0   2

Miles

goldfields (*Lasthenia californica*), brownie thistle (*Cirsium quercetorum*), harlequin lotus (*Hosackia gracilis*),[1] and Choris' popcornflower (*Plagiobothrys chorisianus* var. *chorisianus*)[2] (data from Consortium of California Herbaria). The watershed also supported extensive serpentine grasslands, with a number of serpentine-obligate species (see page 135). As with nearly all grasslands throughout California, the introduction of non-native annual grasses in the 18th and 19th centuries significantly altered the composition of some of the grasslands within the Peninsula Watershed (see pages 74-77). Nevertheless, the watershed still supports extensive native perennial grasslands dominated by species such as purple needle grass, foothill needle grass, and California oatgrass.

Grasslands in and around the watershed historically supported a wide range of wildlife, including large herbivores such as mule deer (*Odocoileus hemionus*) and tule elk (*Cervus elaphus nannodes*); small mammals like California ground squirrel (*Otospermophilus beecheyi*), California vole (*Microtus californicus*), and Botta's pocket gopher (*Thomomys bottae*); carnivorous mammals like American badger (*Taxidea taxus*); birds such as Grasshopper Sparrow (*Ammodramus savannarum*), Savannah Sparrow (*Passerculus sandwichensis*), Horned Lark (*Eremophila alpestris*), Western Meadowlark (*Sturnella neglecta*), American Pipit (*Anthus rubescens*), Say's Phoebe (*Sayornis saya*), and Golden Eagle (*Aquila chrysaetos*); reptiles and amphibians like the state and federally endangered San Francisco garter snake (*Thamnophis sirtalis tetrataenia*) and the federally threatened California red-legged frog (*Rana draytonii*); insects such as the once-abundant western bumblebee (*Bombus occidentalis*); and myriad other invertebrates (Brace 1869, Burroughs 1928, Ford and Hayes 2007, Evans et al. 2008, Gifford-Gonzalez et al. 2013, data from Global Biodiversity Information Facility). Grasslands also likely provided foraging habitat for bat species such as small-footed myotis (*Myotis ciliolabrum*), Townsend's big eared bat (*Corynorhinus townsendii*), and pallid bat (*Antrozous pallidus*)[3] (San Francisco Planning Department 2008, records from Vertnet). Several special status, grassland-associated butterfly species currently occur or have the potential to occur within the Peninsula Watershed, including the Mission blue butterfly (*Plebejus icarioides missionensis*), Callippe silverspot (*Speyeria callippe callippe*), and Bay checkspot butterfly (*Euphydryas editha bayensis*; see page 136; EDAW Inc. 2002, San Francisco Planning Department 2008, USFWS 2010).

A considerable body of evidence suggests that the native Ohlone inhabitants of the Peninsula used fire as a land management tool to remove woody vegetation, promote growth of edible or useful plant parts, and enhance habitat for game animals (see pages 50-51). While there is no direct evidence of indigenous burning in the Peninsula Watershed, it is highly likely that portions of the watershed, like many other parts of the Peninsula, were burned at frequent intervals. If so, this practice would have been a major driver—in addition to grazing by tule elk and other native herbivores—in creating and maintaining

---

1 California Rare Plant Rank 4.2.

2 California Rare Plant Rank 1B.2.

3 A CDFW Species of Special Concern.

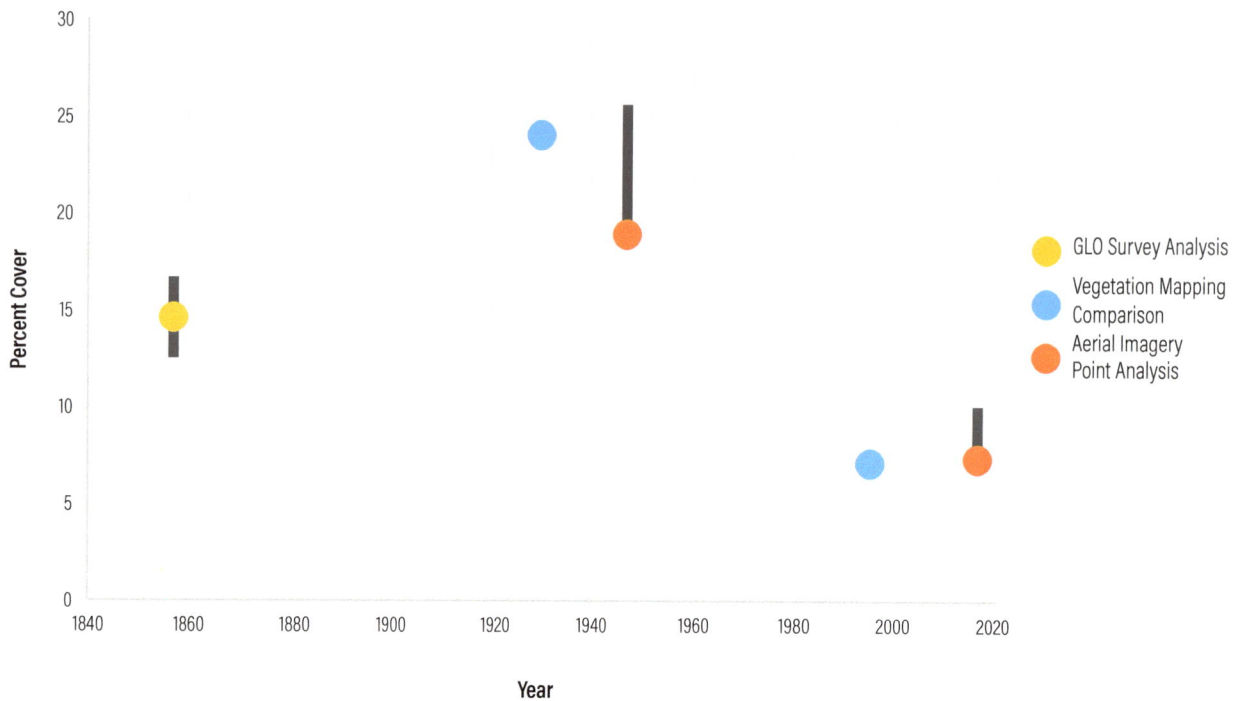

Figure 5.2. This diagram shows the change in estimated grassland extent over time throughout the study area, based on the three quantitative analyses described in Chapter 4: the aerial imagery point analysis (ca. 1947 and ca. 2016), vegetation mapping comparison (ca. 1930 and ca. 1995), and GLO survey analysis (ca. 1857). Gray boxes around the ca. 1947 and ca. 2016 points represent the range in the grassland frequency classified by four mappers for 200 overlapping points in the aerial imagery point analysis (see pages 27-29). The gray box around the ca. 1857 point represents the uncertainty associated with GLO points classified as grassland/shrubland (see pages 32-34).

grasslands within the watershed prior to the late 18th century. With the subjugation of native cultures and removal of native people from the land during the Mission era (1769-1833), the frequency of fire within these ecosystems probably declined considerably, likely resulting in a decrease in grassland extent between the late 18th century and mid-19th century. By the mid-19th century, GLO survey data suggest that grasslands occupied approximately 13-17% of the watershed (Fig. 5.2).

Following Spanish and Mexican colonization, the watershed's grasslands were among the first areas converted to grazing and and other agricultural land uses (see pages 56-63). Intensive grazing and agriculture continued, and in some areas expanded, with the influx of American settlers during the late 19th century. In the absense of regular burning, browsing of shrubs and saplings by livestock, as well as intentional clearing of woody vegetation for cultivation, were likely the major drivers in maintaining, and in some cases creating, grasslands during this period. Indeed, the results of the quantitative analyses presented in Chapter 4 suggest that grassland extent increased somewhat between the mid-19th and mid-20th centuries, from roughly 13-17% in the mid-19th century to 19% by the 1946-48 (see Fig. 5.2). (Grasslands created by clearing of woody vegetation for livestock grazing likely did not support the same species richness as older established grasslands.)

In contrast, grassland extent declined substantially during the latter part of the 20th century. The aerial imagery analysis suggests that grassland extent declined by 56% between the 1940s and present, while the Wieslander/modern vegetation mapping comparison indicates a decline of 70% between ca. 1930 and ca. 1995 (see Fig. 5.2). Urban development, including construction of Highway 280, was a major driver of grassland loss during this period, particularly along the eastern margin of the watershed and in the areas around Lower San Mateo Creek. Encroachment of woody vegetation was another major contributor: between ca. 1930 and ca. 1995, 21% of grassland area converted to shrubland—of which more than 90% is mapped as coyote brush alliance (Schirokauer et al. 2003)—and another 19% converted to hardwood forest. This encroachment of woody vegetation was likely driven in large part by the lack of regular disturbances such as fire and grazing during this period (see Chapter 3).

Though grassland extent has been greatly diminished, grasslands remain along the western slopes of Buri Buri and Pulgas ridges on the eastern side of the watershed (an area with extensive serpentine soils) and on the southwestern side of Upper Crystal Springs Reservoir (see Fig. 5.1). In general, grasslands in serpentine areas have tended to be more persistent than non-serpentine grasslands over the past 70-80 years, suggesting that these grasslands may be more resistant to woody vegetation encroachment (see page 138).

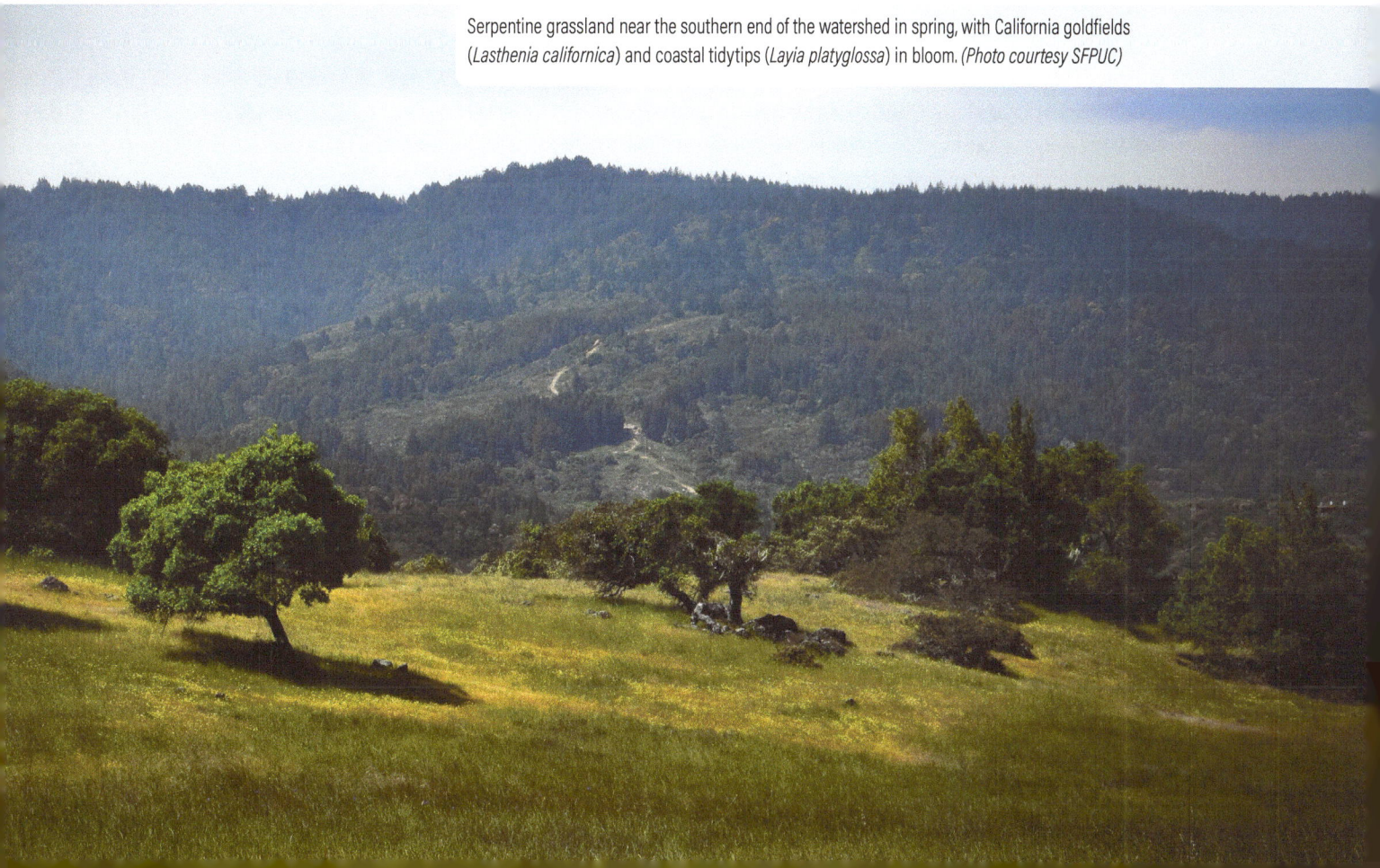

Serpentine grassland near the southern end of the watershed in spring, with California goldfields (*Lasthenia californica*) and coastal tidytips (*Layia platyglossa*) in bloom. *(Photo courtesy SFPUC)*

Grasslands dominated the eastern sides of San Andreas Valley and the Laguna Creek Basin

A large amount of grassland in this area has been lost due to development or conversion to shrubland or forest

Prior to European contact, grasslands were the dominant vegetation type on the eastern side of San Andreas Valley and the Laguna Creek Basin, and grasslands in this area represent the vast majority of grasslands documented within the study area historically. The earliest documentation of extensive grasslands on the eastern side of the valley comes from the Portolá Expedition of 1769. Traveling south through the valley along San Andreas Creek with the Portolá Expedition in November of 1769, Friar Juan Crespí observed "high ranges of knolls of sheer soil, very grass-grown" (Crespí and Brown 2001) on the west slope of Buri Buri, east and south of present-day San Andreas Reservoir. Further south Crespí likewise noted "high knoll ranges of sheer soil and grass" along Pulgas Ridge east of present-day Upper and Lower Crystal Springs reservoirs (Crespí and Brown 2001).

Grasslands remained the dominant vegetation type on the eastern side of San Andreas Valley and the Laguna Creek Basin throughout the 19th and early 20th centuries. GLO surveyors in the 1850s described the southwestern slope of Buri Buri Ridge (north of San Mateo Creek) as "barren rolling hills" with "good grazing" and "no timber," and similarly described portions of the northwestern slope of Pulgas Ridge as "good for grazing purposes" and "bare of timber and covered with wild oats" (Tracy 1852, Lewis 1860).[4] Charles Loring Brace, traveling through the valley in 1867, described the hills to the east as "covered with grain to the very tops... generally wild oats" and "bare of trees, except occasionally dark green clumps with flattened tops, of evergreen oaks" (Brace 1869). Grasslands are the dominant vegetation cover visible in landscape photographs of Buri Buri Ridge and Pulgas Ridge in the late 1800s and early 1900s (Fig. 5.3), though smaller patches oak woodland and other woody vegetation are also visible in some areas (see further discussion below). "Barren," "windswept," "bare of trees," and "meadow land" are typical descriptions of private parcels in this part of the watershed from the 1907-1914 Spring Valley rate case (e.g., parcels 17, 18.2, 48, 90, 218; see Appendix B), suggesting that grassland was the dominant vegetation type (SVWC vs. San Francisco 1916).

---

4 GLO surveys and other historical accounts were often focused on the potential economic value of the land rather than detailed botanical information, and thus in many cases grassland presence must inferred from comments like "good grazing."

Serpentine grassland in spring, with purple needle grass (*Stipa pulchra*) and California oatgrass (*Dantonia californica*). *(Photo courtesy SFPUC)*

Figure 5.3. Early landscape photos document the extensive grasslands on the eastern side of San Andreas Valley and the Laguna Creek Basin. A) An 1875-77 photo, looking east across Upper Crystal Springs Dam during its construction, shows grasslands covering the hills east and south of the reservoir. B) An 1881 photo of San Andreas Reservoir, showing grasslands on the east side of the valley and shrublands on the west side. C) This 1907-14 photo is described as showing the "general character of land just west of crest of Buri Buri Ridge." D) This 1907-14 photo, taken from Sawyer Ridge, shows grasslands along Buri Buri Ridge south of San Andreas Reservoir. Note the patches of oak woodland within gullies running perpendicular to the ridgeline. E) A 1907-14 photo, looking northwest, showing the grass-covered hills east of Upper Crystal Springs Reservoir. F) A 1907-14 photo taken near the future intersection of Cañada and Edgewood roads, looking east. Stephens (1856b) described this area as "the Pedrigal" (pedregal), a "low plateau thickly strewn with stones." Evidence suggests that hardwood forest was more prevalent in this area in the mid-19th century, prior to clearing for cultivation (see page 182).

(A: Photo M-1519_ca.1875-77, courtesy SFPUC; B: BANC PIC 1954.016 -- fALB, courtesy The Bancroft Library, UC Berkeley; C: Photo F3725_1180-13, courtesy California State Archives; D: Photo F3725_1180-22, courtesy California State Archives; E: Photo F3725_1180-5, courtesy California State Archives; F: Photo F3725_1180-2, courtesy California State Archives)

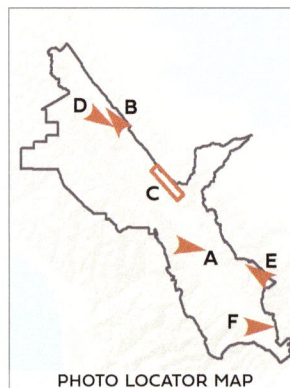

PHOTO LOCATOR MAP

B

D

E

F

REPHOTOGRAPHY

1907-14

2017

Figure 5.4. Rephotograph pair showing hardwood forest encroachment into grasslands just south of Lower Crystal Springs Dam. *(Top: Photo F3725_1180-10, courtesy California State Archives; Bottom: Photo by SFEI-ASC, August 2018)*

PHOTO LOCATOR MAP

Though grasslands were the dominant vegetation type in this region, early data sources suggest heterogeneity within these larger tracts of grasslands. In particular, smaller patches of oak woodlands were present along the canyons and gullies, which likely supported a more mesic microclimate with more groundwater availability. GLO surveyor John Addison, for instance, described Buri Buri Ridge as "oats hills with oak groves in the indentures in the ground" (Addison 1857). (It is unknown whether the reference to "oats hills" indicates native California oatgrass [*Danthonia californica*] or non-native *Avena* spp.). An account from the 1907-14 court case describes one parcel (48) east of Upper Crystal Springs Reservoir as "bare of timber along lake, except in ravines" (SVWC vs. San Francisco 1916). Similarly, a parcel (73) near the northern end of Lower Crystal Springs Reservoir was "indented by small gulches, in which are found comparatively dense growth of oak" (SVWC vs. San Francisco 1916). Landscape photos further illustrate these conditions (Fig. 5.3 and Fig. 5.4).

Though grasslands on the eastern side of San Andreas Valley and the Laguna Creek Basin largely persisted into the early 20th century, a large proportion of these grasslands (estimated at 56%) have been lost over the past 70-80 years as a result of urban development, shrub encroachment, and forest encroachment. Construction of low density residential housing on either side of Lower San Mateo Creek Canyon resulted in substantial grassland loss during the mid-20th century. A large amount of additional grassland loss resulted from construction of Highway 280 in the 1970s, and to a lesser degree from construction of Highway 92 in 1964.

Hardwood forest has replaced a large amount of grassland in parts of the valley, particularly along Buri Buri Ridge south and east of San Andreas Reservoir and east of Lower Crystal Springs Reservoir. Repeat photography illustrates the dramatic shifts from grassland to oak woodland, planted eucalyptus and acacia forest, and other forest types in these areas over the past century or so (Fig. 5.5). Planted conifer forest has replaced grasslands along portions of eastern San Andreas Valley as well. For instance, a stand of Monterey cypress planted by a group of Boy Scouts around 1950 (J. Avant, personal comm.) occupies former grasslands on the eastern side of San Andreas Reservoir. Encroachment of coastal scrub has also contributed to grassland loss, particularly on the southeast side of Upper Crystal Springs Reservoir. In many cases, former grassland areas currently occupied by forest likely transitioned through an intermediate shrubland stage (see page 40).

Grassland has also persisted in a number of locations along the eastern side of San Andreas Valley, most notably to the north of Crystal Springs Golf Course, along Buri Buri Ridge to the east of Sherwood Point, along Pulgas Ridge east of Upper Crystal Springs Dam, to the east and south of Cemetery Hill, and near Cañada Road in the southern part of the study area. Grassland persistence has been disproportionately high in serpentine areas—see page 138 for further discussion.

REPHOTOGRAPHY

Figure 5.5. These rephotographs show vegetation change around Lower Crystal Springs and San Andreas reservoirs. The pair of photographs on the left were taken about 1 km northwest of Lower Crystal Springs Dam, looking northwest along the reservoir at the eastern slope of Sawyer Ridge and the western slope of Buri Buri Ridge. Water levels were higher at the time of the historical photograph, as the 2018 rephoto shows additional lands exposed. Planted trees and hardwood forest encroachment into grasslands along Buri Buri Ridge are apparent. The rephotograph comparison on the right was taken from the western side of San Andreas Dam, facing east. In the historical photo (ca. 1900) grasslands dominate the hills on the eastern side of the dam; in the modern rephoto, the grasslands have been displaced by forest.

A - ca. 1900-29

A -2018

A) Top: BANC PIC 1985 .036 fALB, courtesy Bancroft Library, University of Cailfornia, Berkeley; Bottom: Photo by SFEI-ASC, August 2018

B) Top: Photo M-1721_ca1900, courtesy SFPUC; Bottom: Photo by SFEI-ASC, August 2018

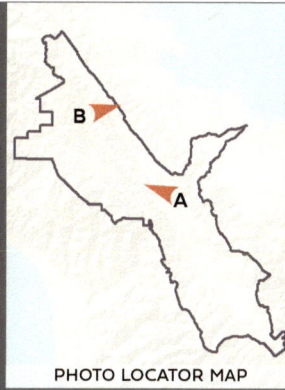

PHOTO LOCATOR MAP

B - ca. 1900

B - 2018

Grasslands occupied some valley floor portions of the Laguna Creek Basin and lower hillslopes to the west

Grasslands in this area have been almost entirely eliminated

Prior to agricultural development and reservoir construction, much of the valley floor within the Laguna Creek Basin was occupied by wetlands, riparian forests, and hardwood forests (see chapters 7 and 9), though evidence suggests that grasslands extended across the valley in some areas. For example, in November of 1774, while camped with the Anza Expedition near present-day Lower Crystal Springs Dam, Friar Francisco Palóu observed that there was "good grass for the animals," along with "a great deal of wood, and water both in the large creek of the hollow and in a little stream coming down from the hills" (see chapters 7 and 9; Palóu et al. 1969). Nineteenth century photographs of this area, taken just before construction of Lower Crystal Springs Dam, show patches of grassland extending across the valley (Fig. 5.6). Grasslands in many parts of the valley were used as pasture for livestock grazing in the early to mid-19th century, and woody vegetation was cleared in many parts of the valley to facilitate this land use. Grazing was likely a key driver in preventing encroachment by woody vegetation in the absence of fire, but likely also resulted in soil erosion, changes in species composition, and other impacts. By the late 1850s and 60s, wheat fields, orchards, and other cultivated crops had also been planted in a number of areas formerly occupied by grasslands (see pages 58-63). Construction of Upper and Lower Crystal Springs dams in the late 19th century submerged the remaining grasslands and other habitat types on the valley floor.

While woody vegetation dominated the hills to the west of the Laguna Creek Basin, patches of grasslands existed around Adobe Gulch, Sherwood Point, and other locations; some of these patches may have been contiguous with the larger expanse of grassland on Pulgas Ridge prior to construction of the Crystal Springs reservoirs. For instance, while traveling south through the valley with the Portolá Expedition in November of 1769, Friar Juan Crespí observed that the hills above present-day Upper Crystal Springs Reservoir formed a "green mountain range wooded with low trees and in spots clad in grass alone" (Crespí and Brown 2001). Like adjacent grasslands on the valley floor, many of these areas were likely subject to intensive grazing (and in some areas cultivation) prior to reservoir construction, though in many cases they persisted into the mid-20th century. Early photographs, for instance, show patches of grassland at Sherwood Point (Fig. 5.6b and Fig. 5.7), on the western shore of Lower Crystal Springs Reservoir across from the dam (Fig. 7.7 on page 183), and Adobe Gulch (Fig. 5.9a,c on pages 126-127). Witnesses in the 1907-14 Spring Valley court case described the area opposite Lower Crystal Springs Dam as "more or less open" (parcel 38), while portions of the parcel to the south were "comparatively

Figure 5.6. Photos of San Andreas Valley prior to completion of Lower Crystal Springs Dam. A) This ca. 1887 photo is looking south along the Laguna Creek Basin from the future dam site. Grasslands are visible on the right side of the photograph extending across the valley. B) This 1888 photo is looking north from the future dam site. Grasslands are visible in the valley adjacent to the dam site (on the left side of the image), and on Sherwood Point (in the center). Hardwood forests occupy the eastern slope of Cahill Ridge within Upper San Mateo Creek Canyon (see page 184). Riparian scrub borders San Andreas Creek in the center of the image. *(A: Photo M-1756_ca1887; B: Photo M-1766_7-26-1888; courtesy SFPUC)*

REPHOTOGRAPHY

1927

2018

PHOTO LOCATOR MAP

Figure 5.7. This rephotograph pair (1927-2018) shows loss of grasslands at Sherwood Point and on the hillslope across from Lower Crystal Springs Dam due tree planting (including a large number of eucalyptus) and forest encroachment. *(Top: Photo A-160_4-21-1927, courtesy SFPUC; Bottom: Photo by SFEI-ASC, August 2018)*

open but... fast reverting to brush-covered land by non-use" (parcel 68.2; see Appendix B). The Adobe Gulch area, just south of present-day Highway 92, was for the most part "bare rolling land rising from the lake" with the exception of "some oak timber on northeasterly corner" (parcel 46). Outside of areas dominated by oak woodlands, portions of the watershed near the southern end of Upper Crystal Springs Reservoir were "[comparatively] open" and "level meadow land" (parcel 49). The 1930s Wieslander VTM mapping shows a swath of grasslands between Upper Crystal Springs Reservoir and the shrublands and forests on Kings Mountain Ridge to the west.

Most of the grasslands in this part of the watershed have been lost over the past 70-90 years as a result of shrub and forest encroachment and intentional tree planting. Sherwood Point, for example, is today occupied by coast live oak forest and planted eucalyptus (see Fig. 5.7). The former grasslands across from Lower Crystal Springs Dam are occupied by coyote brush-dominated shrublands and Douglas-fir-dominated conifer forest (see Fig. 7.7). Grasslands within the Adobe Gulch area have been invaded by Monterey pine and Monterey cypress (which have expanded from planted stands; see page 210), as well as coyote brush and coast live oak; recent restoration efforts have restored grasslands to approximately 40 ac of Adobe Gulch (Winzler & Kelly 2010, AECOM 2020; Fig. 5.8).

Figure 5.8. Comparison of historical (1946) and modern (2015) aerial imagery of Adobe Gulch area showing grassland conversion to shrubland (coyote brush) and conifer forest (Monterey cypress and Monterey pine). Recent restoration efforts have restored grasslands in a portion of this area. *(Top: imagery courtesy historicaerials.com; Bottom: Pictometry, Inc. 2015)*

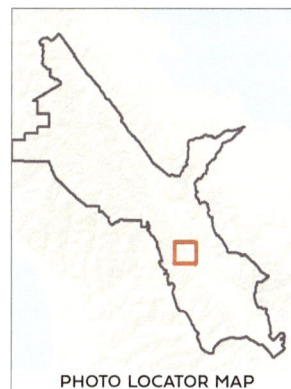

PHOTO LOCATOR MAP

grassland

PHOTO LOCATOR MAP

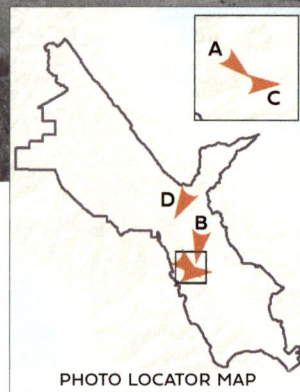

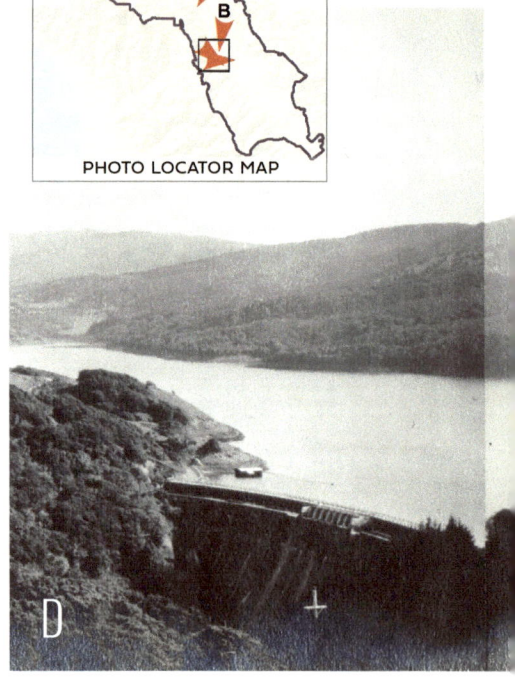

Figure 5.9. Early landscape photos showing grasslands on the west side of the Laguna Creek Basin. A) This 1907-14 photo, taken from northwest of Adobe Gulch and facing southeast, shows grasslands (along with scattered trees and shrubs) in the Adobe Gulch area. B) This 1875-77 photo, looking west across Upper Crystal Springs Dam, shows the vegetation on the shoreline of the reservoir. Hardwood forest occupies the area fringing the lake (see page 179), while a strip of shrubland occupies slightly higher elevation. A patch of grassland is visible on the upper right side of the image. C) This photo, taken above Lower Crystal Springs Dam, looking to the southwest, highlights open grassland, with some windswept oaks, in the area north of Adobe Gulch. D) This 1907-14 photo highlights a patch of grassland on the hillslope opposite Lower Crystal Springs Dam. *(A: Photo F3725_1180-7, courtesy California State Archives; B: Photo M-1518_ca.1875-77, courtesy SFPUC; C: Photo F3725_1180-6, courtesy California State Archives; D: Photo F3725_1180-12, courtesy California State Archives)*

Grasslands occupied extensive areas along ridges

Most of these grasslands had converted to shrubland by the 1940s

In addition to occupying extensive areas of Buri Buri and Pulgas ridges, and to a lesser extent portions of the Laguna Creek Basin valley floor and lower hillslopes to the west, grasslands were also present along ridges in many parts of the study area. Substantial evidence indicates that patches of grasslands occurred along Sawyer, Fifield, Spring Valley, Cahill, and Montara Mountain ridges, though they had largely been converted to shrubland by the 1940s. Because it was not possible to develop a reliable estimate for the extent of these grassland patches prior to landscape modification, their historical presence is indicated by generalized ovals in Fig. 5.1. The historical presence of grasslands in these settings is consistent with other research documenting ridgetop grasslands, or "bald hills," in other locations throughout coastal California (Stromberg et al. 2001).

Spanish expeditions in the 18th century did not travel close enough to ridges in the interior of the watershed to provide any account of vegetation cover in these areas, and thus the earliest evidence for grasslands on ridges comes from mid-19th century sources such as GLO surveys. In September of 1864, GLO surveyor A.W. Von Schmidt traversed many of the ridges in the northwestern portion of the watershed and noted the presence of grasslands along the ridges (in contrast to the hillslopes between ridges, which he generally described as covered in dense brush or chaparral (see page 149). For example, traveling west across Fifield and Spring Valley ridges approximately 0.5 mi north of present-day Pilarcitos Reservoir, he wrote that the area was "mostly good grazing land" though the "slopes of [the] ridge [are] covered in chaparral." The crest of Fifield Ridge further to the north, as well as the crest and much of the western slope of Sawyer Ridge in the vicinity of Fifield Dairy (west and southwest of present-day San Andreas Dam), were both characterized as "good grazing and farming land," in contrast to surrounding "brush" and "chaparral." The ridgeline south of Montara Mountain, which he crossed over multiple times, was also described as "good grazing" land in three separate locations (Von Schmidt 1864b). GLO accounts also indicate the presence of grassland on portions of Kings Mountain Ridge just north of Pise Lookout: Tracy (1853b) described the area as "bare hills" (in contrast to the "timber and chaparral in gulches") and Stephens (1856a) noted the lack of trees.

Serpentine grassland in summer, with purple needle grass (*Stipa pulchra*) and California oatgrass (*Danthonia californica*). *(Photo courtesy SFPUC)*

Late 19th and early 20th century observers also documented the presence of grasslands along many of these ridges. In 1895, botanist J. Burtt Davy observed that the ridge south of Montara Mountain had "clear spaces with grass, and abundance of... lupine." This observation was echoed by Brewer (1903), who described the same area as "grass-covered and bare of trees." Photographs and testimony from the 1907-14 SVWC court case also document the presence of grasslands along Fifield, Sawyer, Spring Valley, and Montara Mountain ridges (Fig. 5.10, next page). Fifield Ridge, for instance, was described as "pasture land" and "comparatively open" (SVWC vs. San Francisco 1916; Fig. 5.11, next page), and the ridge south of Montara Mountain was similarly described as having "large open tracts with patches of brush" (SVWC vs. San Francisco 1916; Fig. 5.12, next page).

Evidence suggests that grazing played a role in maintaining many of these ridgetop grasslands during the mid-19th through early 20th centuries, and in some cases grazing and other agricultural activities may have been responsible for the expansion or even creation of grasslands during this period. Unlike many other parts of the watershed, including the San Andreas Valley and the Laguna Creek Basin, the northwestern portion of the study area was not a part of a Mexican land grant (likely because of the lack of extensive grasslands), and thus was likely not subject to intensive grazing during the early 19th century (see page 59). However, the Fifield Dairy,

Sawyer Ridge

Fifield Ridge

Spring Valley Ridge

Figure 5.10. (above) In this 1907-14 photo, taken from the ridge between Montara Mountain and Scarper Peak and looking east across the watershed, patches of grassland are visible in the distance on Sawyer, Fifield, and Spring Valley ridges. (Photo F3725_1180-36, courtesy California State Archives)

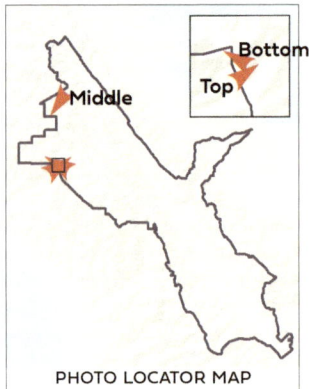

PHOTO LOCATOR MAP

Top
Bottom
Middle

Figure 5.11 (right). A 1907-14 photo showing patches of grassland along Fifield Ridge. (Photo F3725_1180-41, courtesy California State Archives)

Figure 5.12 (below). A 1907-14 photo showing patches of grassland along the ridge south of Montara Mountain. (Photo F3725_1180-37, courtesy California State Archives)

Pilarcitos Reservoir

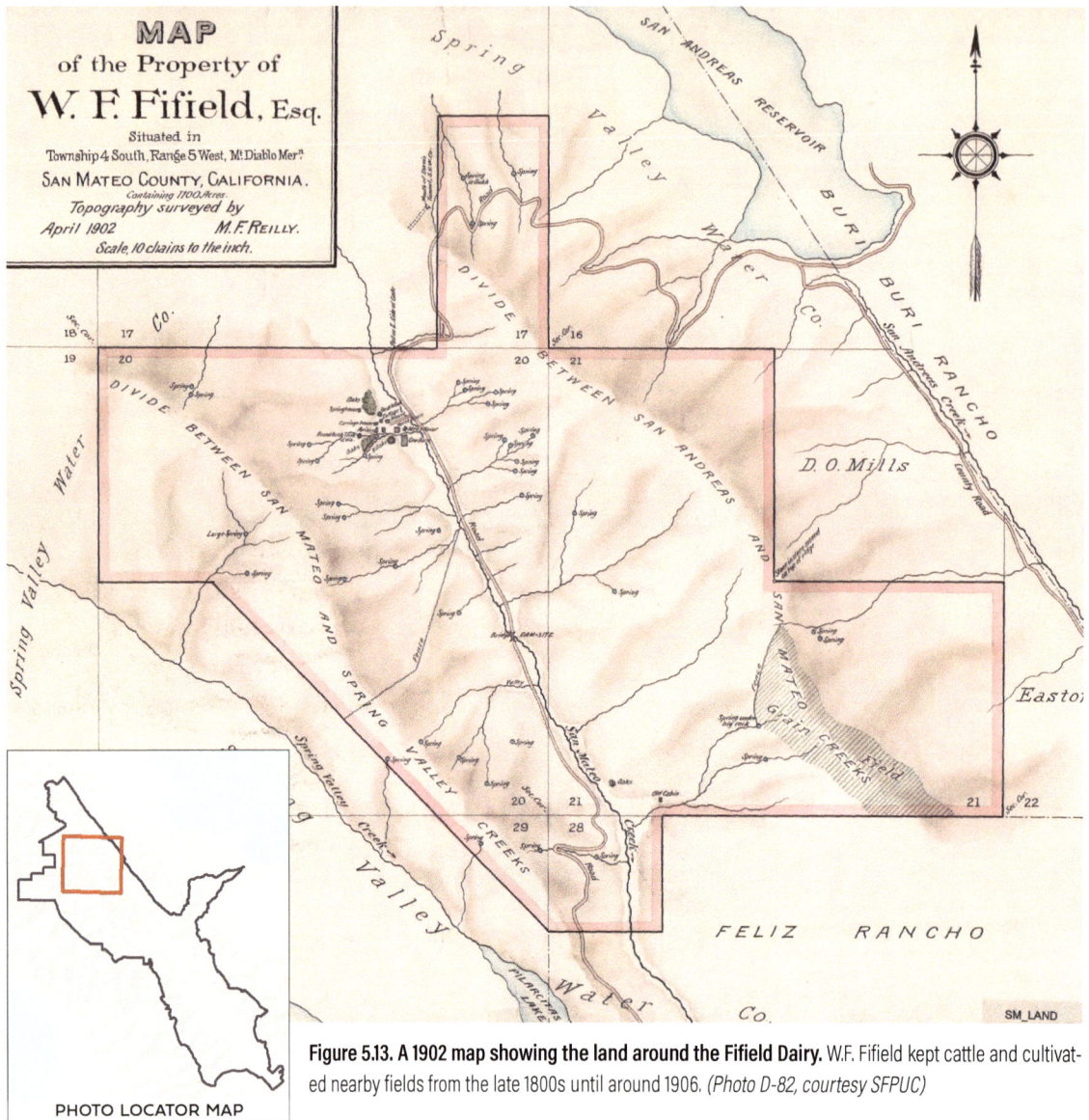

Figure 5.13. A 1902 map showing the land around the Fifield Dairy. W.F. Fifield kept cattle and cultivated nearby fields from the late 1800s until around 1906. (Photo D-82, courtesy SFPUC)

PHOTO LOCATOR MAP

Figure 5.14. A 1907-14 court case photo showing grassland, and encroaching shrubland, along Sawyer Ridge. (Photo F3725_1180-24, courtesy California State Archives)

PHOTO LOCATOR MAP

established in the 1860s, encompassed 1,066 ac in the northern part of watershed, including portions of Fifield and Sawyer Ridges (Fig. 5.13). Davy (1895) described the the western slope of Sawyer Ridge within the Fifield Dairy as "much cow bitten," indicating that grazing likely played a major role in maintaining grasslands within the Fifield Dairy, especially in the absence of regular fire (Davy 1895). Many portions of Kings Mountain Ridge were also grazed during the late 19th/early 20th centuries (Bromfield 1893, Unknown 1912; see Fig. 3.11 on page 63 and Fig. 8.4 on page 199).

As SVWC acquired land throughout the late 19th and early 20th centuries, the prohibition on grazing was extended to more parts of the watershed. With the reduced frequency of disturbance, grasslands once maintained or expanded by grazing or fire on ridgetops (and elsewhere) began to transition to woody-dominated habitat types. W. F. Fifield sold his lands to SVWC in 1906, and shrubs quickly began to expand into former grassland areas. Witnesses in the 1907-14 SVWC court case, for instance, stated that "the crest of Sawyer Ridge... is mostly open, and has been cleared, but is fast reverting to brush-covered land" (parcel 153; see Appendix B). Further to the south, the ridge was described as "comparatively bare land, with patches of brush which show indications of spreading" (parcel 39; Fig. 5.14). Further south, along Kings Mountain Ridge, witnesses reported that "portions... along the crest of the ridge... were cleared land but are fast becoming overgrown with brush" (parcel 199.2).

It is unknown whether grasslands preceded grazing in these ridgetop settings. The Fifield Dairy, for instance, was established at approximately the same time that GLO surveyor Von Schmidt reported the presence of "good grazing land." On the one hand, the fact that "bald hills" have been commonly observed in numerous locations throughout California suggests that they may have been part of the pre-agricultural landscape in the Peninsula Watershed as well, maintained by frequent indigenous burning. On the other hand, references to "cleared" lands suggest that to some extent the lack of woody

vegetation in these areas was the result of intentional removal. It is also quite plausible that grasslands maintained by indigenous burning existed in these areas prior to Spanish colonization, experienced scrub encroachment following the prohibition on burning, and were subsequently cleared by Euro-American settlers.

Most of the ridgetop grasslands documented in the 19th and early 20th centuries appear to have been lost prior to the 1930s and 40s, given their relative absence in Wieslander VTM mapping and 1946-48 aerial imagery. Today, most of these areas are dominated by coastal scrub, though contemporary vegetation mapping and aerial photos do reveal some persistent patches of grassland, especially on the upper western slope of Sawyer Ridge.

Serpentine grassland on Fifield Ridge. *(Photo courtesy SFPUC)*

Serpentine grasslands supported (and still support) a specialized plant community

These grasslands have been more resistant to encroachment of woody vegetation than non-serpentine grasslands

Areas with serpentine geology and soils currently occupy approximately 2,040 ac within the study area, particularly along the western slopes of Buri Buri and Pulgas ridges, and to a lesser extent along Fifield Ridge, in areas historically dominated by grasslands (Fig. 5.15). The full historical extent of serpentine grasslands within San Andreas Valley is unknown, as the earliest geologic mapping post-dated reservoir construction. Lawson (1895) speculated that the main bodies of serpentine rocks on the eastern side of San Andreas Valley extended westward as far as the San Andreas Fault (beneath the current reservoirs) in some areas.

Serpentine rocks and soils are generally characterized by low calcium and high magnesium concentrations, nutrient deficiency (i.e., low levels of nitrogen, phosphosus, and potassium), high levels of heavy metals such as nickel and chromium, and in some cases seasonally low water availability. These properties make serpentine soils a relatively harsh substrate, and as a result serpentine areas often give rise to a short-statured and sparse plant community; many plant species are excluded from serpentine soils altogether. Some species, however, have evolved adaptations that enable them to tolerate serpentine conditions, giving them a competitive advantage over other native taxa and making serpentine areas relatively resistant to invasion by non-native plant species. In addition, over 200 plant species in California—a large number of them forbs—are considered serpentine endemics or obligates (restricted to serpentine soils), and many of these species are rare, threatened, or endangered (Kruckeberg 1984, Safford et al. 2005, Harrison and Viers 2007, Safford and Miller 2020).

Early botanical records from the Peninsula Watershed include a number of species restricted to serpentine grasslands and other serpentine habitats (including wetlands, shrublands, and woodlands), as well as species associated with (but not restricted to) serpentine habitats (Table 5.1). Most of these species were first reported in the areas around the Crystal Springs reservoirs during the early part of the 20th century (data from Consortium of California Herbaria). Many of these species are now rare or have a protected status, including local or regional endemics such as San Mateo thorn mint (*Acanthomintha duttonii*), Crystal Springs fountain thistle (*Cirsium fontinale* var. *fontinale*), San Francisco collinsia (*Collinsia multicolor*), Marin dwarf flax (*Hesperolinon congestum*), Serpentine leptosiphon (*Leptosiphon ambiguus*), and Crystal Springs lessingia (*Lessingia arachnoidea*).

Serpentine Grasslands

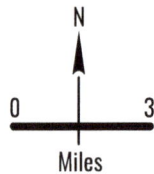

☐ Study Area

🟩 Serpentine Areas

N

0 — 3

Miles

Figure 5.15. Map showing serpentine areas within the watershed.
(data from SSURGO and Brabb et al. 1998)

In addition to a diverse flora, serpentine grasslands provide habitat for a number of rare animal species which are largely restricted to serpentine areas. The Edgewood blind harvestman (*Calicina minor*), for instance, is an arachnid measuring just 1 mm long that only occurs in serpentine grasslands, and is found in moist areas under serpentine rocks. Originally collected around the Crystal Springs reservoirs in 1966, the only known extant populations are in Edgewood Park and Natural Preserve, on the southeast side of the Peninsula Watershed (Shanks n.d.).

The federally threatened Bay checkerspot butterfly (*Euphydryas editha bayensis*) is also restricted to serpentine grasslands around San Francisco Bay. Historically distributed throughout the San Francisco Peninsula and elsewhere around the Bay Area, it was first described from a specimen collected in 1933 in Hillsborough, just to the east of the Peninsula Watershed (Sternitsky 1937; USFWS 1998). Its main larval host plant, dwarf plantain (*Plantago erecta*), is most abundant on serpentine soils, though not restricted to them (Harrison and Viers 2007). Loss of grasslands resulting from development, spread of non-native species such as Italian ryegrass (*Festuca perennis*), and other factors have extirpated the Bay checkerspot butterfly from most of its historical range (USFWS 1998). One of the only remaining populations exists in Edgewood Park and Natural Preserve, where habitat restoration efforts have focused on the use of grazing, mowing,

Serpentine grasslands. (Photos courtesy SFPUC)

prescribed fire, and other means to preserve native serpentine grassland flora and limit the spread of non-native grasses (Weiss 2002).

Based on the results of the aerial imagery point analysis, serpentine grasslands have been proportionally more persistent than non-serpentine grasslands within the Peninsula Watershed since the 1940s. Among points classified as grasslands in the 1940s within serpentine areas, 52% converted to non-grassland types while 48% remained grassland. Conversely, among points classified as grasslands in the 1940s in non-serpentine areas, 71% converted to non-grassland types and only 29% remained grassland (Fig. 5.16). Twenty-four percent of all serpentine grasslands and 47% of non-serpentine grasslands were lost due to tree or shrub encroachment. An additional 28% of serpentine grasslands and 24% of non-serpentine grasslands were lost due to development and transportation corridors (including a very small number of points that converted to water).

For example, in the 1940s Fifield Ridge supported both serpentine and non-serpentine grasslands. Between 1946 and 2015, coyote brush encroached into the non-serpentine grasslands, virtually eliminating them, while the serpentine grasslands have remained relatively stable (Fig. 5.17). This differential persistence suggests that serpentine grasslands in the watershed may be more resilient than non-serpentine grasslands to native shrub encroachment resulting from altered disturbance regimes and other environmental changes. The future resilience of serpentine grasslands is uncertain, however, particularly in light of the spread of invasive plant species and other anthropogenic modifications. For instance, research suggests that increased levels of atmospheric nitrogen deposition from anthropogenic activities may be decreasing the natural resistance of serpentine grasslands to non-native species invasions (Dukes and Shaw 2007, Vallano et al. 2012).

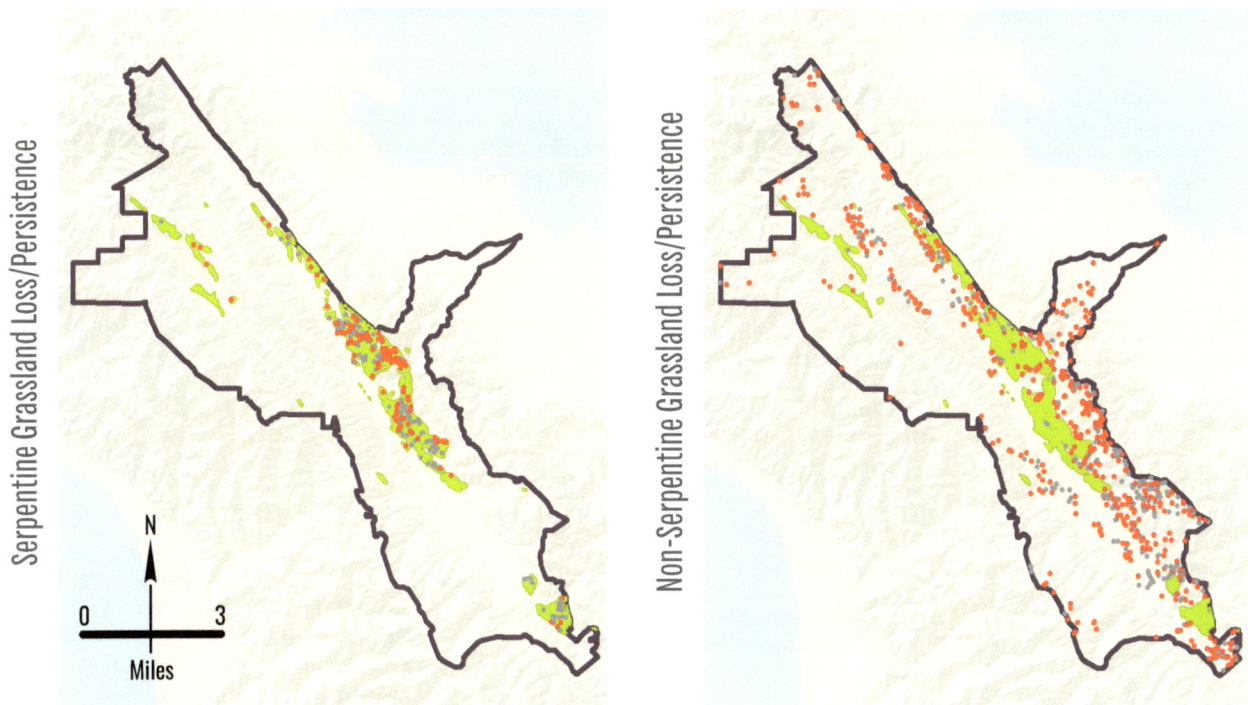

Figure 5.16. Maps showing loss (red) and persistence (gray) of grasslands present in the 1940s in serpentine (left) and non-serpentine (right) areas. Among serpentine grasslands present in the 1940s, 48% remain and 52% were lost. Among non-serpentine grassland present in the 1940s, 29% remain and 71% were lost.

PHOTO LOCATOR MAP

Gain/Loss

- Study Area
- Serpentine Areas
- Loss
- Persistence

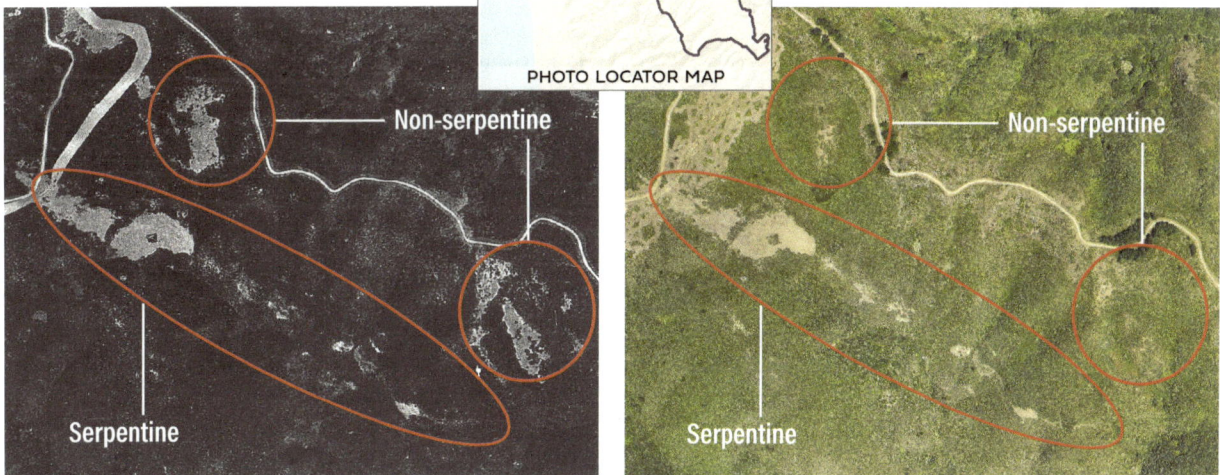

Figure 5.17. This figure highlights the differential persistence of serpentine grasslands over non-serpentine grasslands on Fifield Ridge. In the 1940s (left), Fifield Ridge supported a mix of serpentine and non-serpentine grasslands. By 2015 (right), most of the non-serpentine grasslands had been invaded by coyote brush, while the serpentine grasslands had remained largely stable. (Left: imagery coutesy historicaerials.com; Right: Pictometry Inc., 2015)

Table 5.1. Selected list of serpentine associated species observed in the Peninsula Watershed historically. Serpentine affinities are from Safford and Miller (2020). *(data from Consortium of California Herbaria)*

Common Name	Latin Name	Serpentine Affinity	CA Rare Plant Rank	Locality (year)
San Mateo thorn mint	*Acanthomintha duttonii*	Strict endemic	1B.1	"Crystal Springs Lake" (1903)
Sickle leaf onion	*Allium falcifolium*	Broad endemic/ strong indicator	None	"Crystal Springs Lake" (1903)
Serpentine columbine	*Aquilegia eximia*	Broad endemic/ strong indicator	None	"Crystal Springs [Lake]" (1901, 1903, 1912)
Bifid sedge	*Carex serratodens*	Broad endemic	None	"Crystal Springs Lake" (1902); "In moist places on the open hillsides" (1902)
Fountain thistle	*Cirsium fontinale var. fontinale*	Strict endemic	1B.1	"Crystal Springs" (1897); "Spring Valley" (1901); "Crystal Springs Lake... around spring on serpentine" (1932); "Crystal Springs Lake... open hills, springs" (1936); "Crystal Springs Lake... in moist gully, fed from spring seepage. Dry, grassy hills east of lake" (1937); "Crystal Springs Lake... along brooklet on serpentine" (1939)
San Francisco collinsia	*Collinsia multicolor*	Weak indicator/ indifferent	1B.2	"On the Half Moon Bay road" (1907); "Crystal Springs [Lake]" (1902); "On the road to Crystal Springs" (1896); "Pilarcitos Lake and Canyon" (1893); "San Mateo Creek below Crystal Springs Dam" (1941)
Hairy bird's beak	*Cordylanthus pilosus subsp. pilosus*	Strong indicator	None	"Near Pulgas Tunnel" (1923)
Beaked cryptantha	*Cryptantha flaccida*	Weak indicator	None	"Crystal Springs Lake... along banks and roadsides" (1902)
Franciscan wallflower	*Erysimum franciscanum*	Strong indicator	4.2	"Lake San Andreas" (1895); "Crystal Springs Lake" (1895, 1916); "Crystal Springs Lakes... on open grassy hillsides" (1902); "Spring Valley" (1903); "Spring Valley Lakes" (1930); Meadow near Spring Valley Lakes" (1937); East side Crystal Springs Lake... exposed, somewhat rocky knolls and roadbank" (1938); "E side of Crystal Springs Lake... soil containing much disintegrated serpentine rock" (1938); "Fifield Ridge, just above San Mateo Creek and north of Pilarcitos Lake... Among sparse bushes and loose rock on sunny knoll" (1939); "Above southeast shore of Crystal Springs Lake, soutwest base Pulgas Ridge... With small herbage, rock-strewn knoll above shore" (1939); "Below Crystal Springs Lake... Shady, damp, rich slope of ravine with Rhus under Quercus agrifolia" (1939)[1]
Marin dwarf flax	*Hesperolinon congestum*	Strict endemic	1B.1	"Above Crystal Springs on the Half Moon Bay road" (1907); "Crystal Springs Lake" (1903, 1910); "E side Upper Crystal Springs Lake... on open serpentine slope" (1940)
Serpentine leptosiphon	*Leptosiphon ambiguus*	Strict endemic	4.2	"Crystal Springs Lake" (1903, 1906)

---

1   Some localities omitted.

Table 5.1. Continued.

Common Name	Latin Name	Serpentine Affinity	CA Rare Plant Rank	Locality (year)
Crystal springs lessingia	*Lessingia arachnoidea*	Strict endemic	1B.2	"Crystal Springs Lake" (1902); "Crystal Springs Lake, east side, in adobe soil where locally common" (1931); "Between Lake San Andreas and Crystal Springs Lake" (1926); "East side of Crystal Springs Reservoir" (1909); "Hills near Crystal Springs Lake" (1926); "Crystal Springs Reservoir" (1896, 1902); "San Mateo Canyon, open grassy slopes" (1941); "Near San Andreas Lake... grassland" (1949); "Crystal Springs Lake... serpentine soil" (1934)
California fairypoppy	*Meconella californica*	Only seen on serpentine in the watershed (H. Bartosh pers. comm.)	None	"By road from San Mateo to Crystal Springs Reservoir" (1894); "Crystal Springs Lake... along stony banks" (1902); "Crystal Springs Lake" (1903); "San Mateo Creek" (1895); "Buri-Buri Ridge... rocky slope in full sun" (1934)
Divaricate phacelia	*Phacelia divaricata*	Weak indicator	None	"Near Crystal Springs Lake" (1896); "Crystal Springs Lake... small colonies along stony banks and roadsides" (1902); "Near Spring Valley Lakes... grassy meadow" (1937)
Leather oak	*Quercus durata [recognized as var. durata; H. Bartosh pers. comm.]*	Strict endemic	None	"Near shore... of Crystal Springs... rocky ground" (1908); "Crystal Springs Reservoir" (1908); San Francisco Watershed Reserve, southern boundary" (1949)

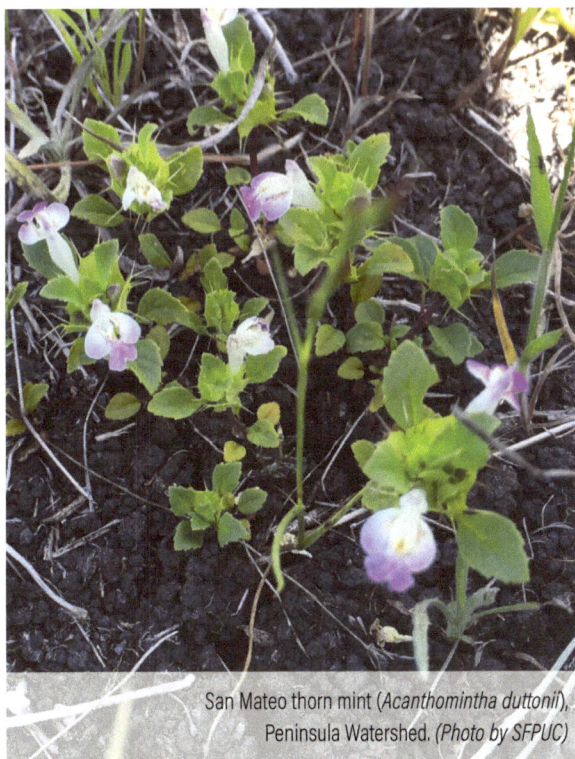

San Mateo thorn mint (*Acanthomintha duttonii*), Peninsula Watershed. (*Photo by SFPUC*)

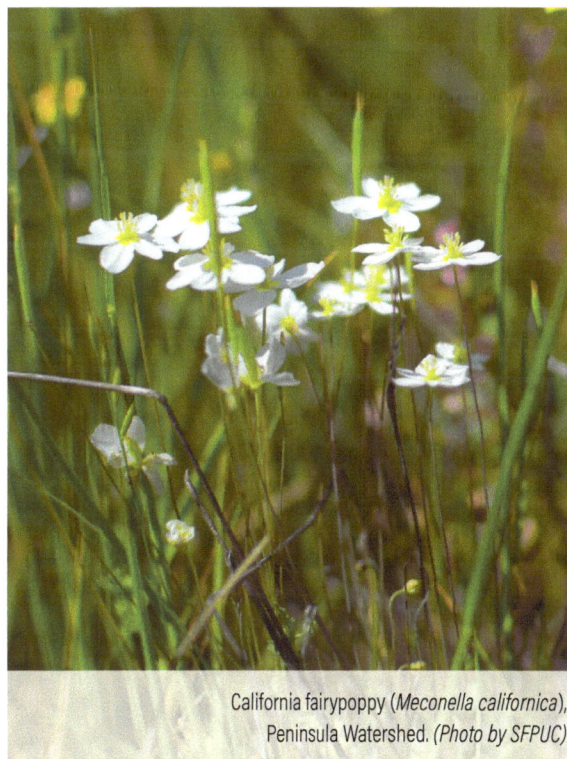

California fairypoppy (*Meconella californica*), Peninsula Watershed. (*Photo by SFPUC*)

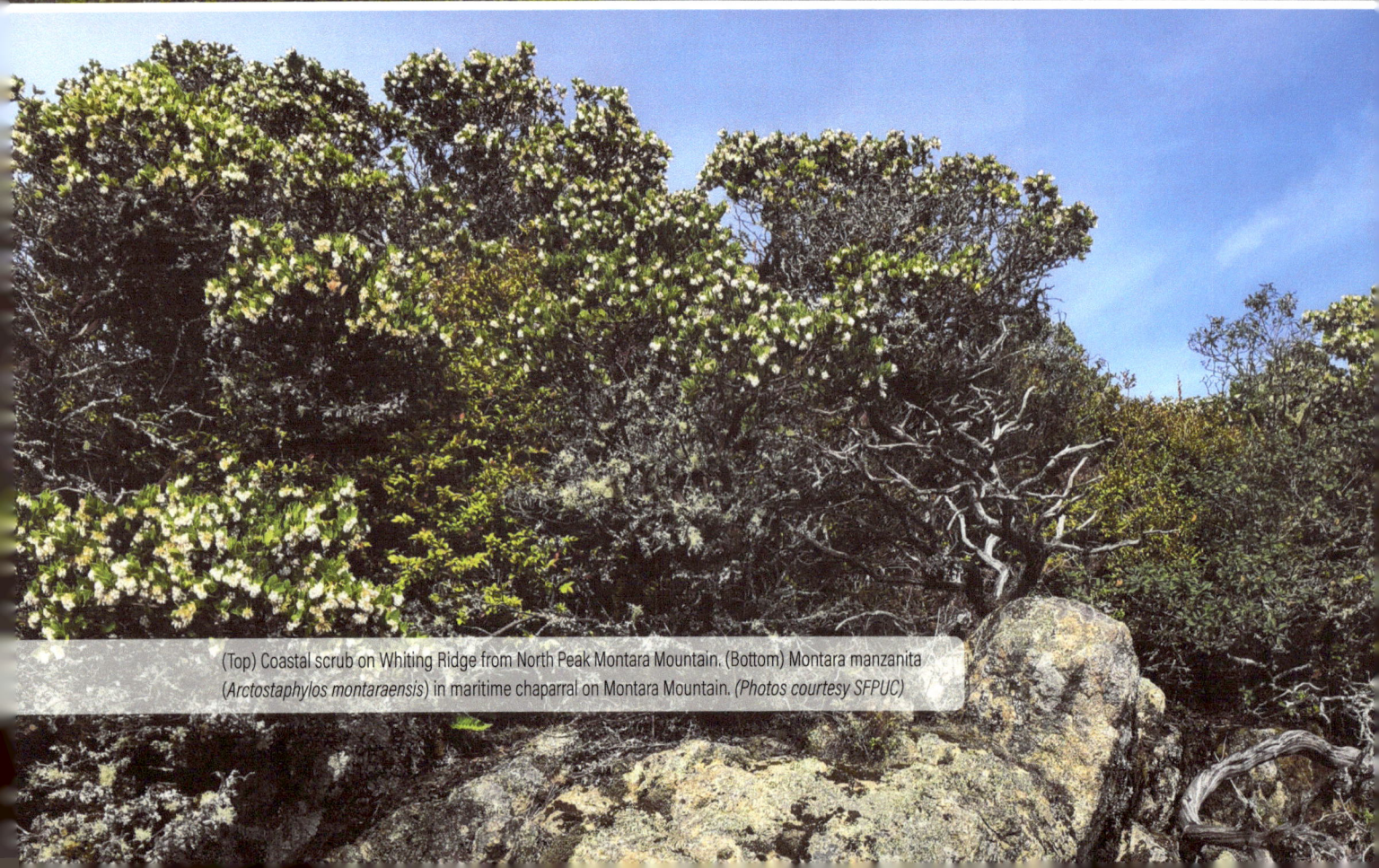

(Top) Coastal scrub on Whiting Ridge from North Peak Montara Mountain. (Bottom) Montara manzanita (*Arctostaphylos montaraensis*) in maritime chaparral on Montara Mountain. *(Photos courtesy SFPUC)*

CHAPTER 6

Shrublands

Overview

The most extensive vegetation type within the study area historically, shrublands blanketed much of the northwest portion of the watershed—including much of the slopes of Sweeney, Portola, Sawyer, Fifield, Whiting, and Spring Valley ridges and the ridge south of Montara Mountain—as well as the upper slopes of Kings Mountain Ridge on the southwestern side of the watershed. The watershed supported a diversity of shrubland plant communities, which can be broadly categorized as either coastal scrub (often dominated by coyote brush [*Baccharis pilularis*], coffeeberry [*Frangula californica*], or California sagebrush [*Artemisia californica*]) or chaparral (often dominated by chamise [*Adenostoma fasciculatum*], manzanita [*Arctostaphylos* spp.], or ceanothus [*Ceanothus* spp.]). Most historical data sources do not differentiate between shrubland types—they are often referred to using general terms such as "brush," "scrub," "chaparral" (used generically in many historical sources), or "chamisal"—and thus by necessity these communities have been combined into the larger "shrubland" grouping. However, coastal scrub and chaparral encompass distinct plant communities that vary widely in terms of distribution, successional dynamics, adaptation to fire, drought tolerance, wildlife support, and other ecological factors (see page 164), and evidence suggests that they have experienced different trajectories within the watershed over (at least) the past century (see pages 162-167). Thus, wherever possible, this chapter differentiates the unique historical patterns and trajectories of these vegetation communities.

Shrub species historically documented within coastal scrub and/or chaparral communities in the watershed include buck brush (*Ceanothus cuneatus*), Jim brush (*Ceanothus sorediatus*), giant chinquapin (*Chrysolepis chrysophylla* var. *minor*), California barberry (*Berberis pinnata*), peak rushrose (*Crocanthemum scoparium*), coast silk tassel (*Garrya elliptica*), pitcher sage (*Lepechinia calycina*), silver bush lupine (*Lupinus albifrons* var. *collinus*), blue witch nightshade (*Solanum umbelliferum*), arcuate bush mallow (*Malacothamnus arcuatus*),[1] and many others (see pages 162-163). In addition, shrublands supported a diverse array of herbaceous plants, such as purple woodland monolopia (*Monolopia gracilens*),[1] western morning glory (*Calystegia purpurata* ssp. *purpurata*), Indian paintbrush (*Castilleja* spp.), cobwebby thistle (*Cirsium occidentale* var. *venustum*), chaparral clematis (*Clematis lasiantha*), yerba santa (*Eriodictyon californicum*),

---

1 California Rare Plant Rank 1B.2.

coast buckwheat (*Eriogonum latifolium*), woodland tidytips (*Layia gaillardioides*), tall tidytips (*Layia hieracioides*), San Francisco coyote mint (*Monardella villosa* ssp. *franciscana*), warrior's plume (*Pedicularis densiflora*), branching phacelia (*Phacelia ramosissima*), stinging phacelia (*Phacelia malvifolia*), California milkwort (*Polygala californica*), California cudweed (*Pseudognaphalium californicum*), California chicory (*Rafinesquia californica*), Fremont's death camas (*Toxicoscordion fremontii*), and chia sage (*Salvia columbariae*) (data from Consortium of California Herbaria). Shrublands in the watershed currently support a number of sensitive plant communities, such as golden chinquapin thickets and coin leaf manzanita (*Arctostaphylos sensitiva*) stands, and rare plants such as Montara manzanita (*A. montaraensis*), Kings Mountain manzanita (*A. regismontana*), western leatherwood (*Dirca occidentalis*), and arcuate bush mallow (*Malacothamnus arcuatus;* all California Rare Plant Rank 1B.2; Nomad Ecology 2020; T. Corelli pers. comm.).

Shrubland communities in the watershed historically provided habitat and cover for a range of wildlife, including birds such as Western Scrub-Jay (*Aphelocoma californica*), Bewick's Wren (*Thryomanes bewickii*), White-crowned Sparrow (*Zonotrichia leucophrys*), Song Sparrow (*Melospiza melodia*), California Quail (*Callipepla californica*), Mountain Quail (*Oreortyx pictus*), Greater Roadrunner (*Geococcyx californianus*), Wrentit (*Chamaea fasciata*), Allen's Hummingbird (*Selasphorus sasin*), and Common Poorwill (*Phalaenoptilus nuttallii*); mammals such as brush rabbits (*Sylvilagus bachmani*) and numerous rodents; and reptiles such as striped racer (*Coluber lateralis*) and western rattlesnake (*Crotalus oreganus*; Evans 1873, Alexander and Hamm 1916, Burroughs 1928, CalPIF 2004, data from Global Biodiversity Innovation Facility). The federally endangered San Bruno elfin butterfly (*Callophrys mossii bayensis*), which utilizes coastal scrub and grassland habitats that support broadleaf stonecrop (*Sedum spathulifolium*; its larval hostplant), has been documented within the Peninsula Watershed along Whiting Ridge near Montara Mountain (USFWS 2010).

The trajectory of shrublands within the watershed prior to the mid-19th century is unknown. As discussed previously, indigenous burning was likely a key factor in maintaining extensive grasslands in the watershed prior to Spanish contact, and would likely have resulted in a shifting mosaic of grasslands and coastal scrub in parts of the watershed. In general, removal of indigenous burning would thus be expected to lead to expansion of shrublands into grasslands. However, the simultaneous introduction of grazing and other agricultural land uses to parts of the watershed in the late 18th or early 19th centuries would have had an opposing effect, tending to favor open, herbaceous habitats, and the ultimate effect on shrubland extent and distribution would have been determined in large part by the relative impacts of these two factors. Further research using phytolith analysis could help to resolve outstanding questions about shrubland/grassland extent prior to the existence of reliable documentary evidence (see page 242).

The analysis of GLO survey data (see page 102) suggests that, by the mid-19th century, shrublands occupied approximately 39-43% of the watershed (Fig. 6.1). Shrubland extent decreased slightly during the late 19th and early 20th centuries, to an estimated 37% of the watershed by 1928-32 and 39% by 1946-48. Livestock grazing and active clearing of woody

Mature northern coastal scrub on North Peak Montara Mountain, with huckleberry (*Vaccinium ovatum*), coast barberry (*Berberis pinnata*), gooseberry (*Ribes* sp.), California coffeeberry (*Frangula californica*), thimbleberry (*Rubus parviflorus*), coyote brush (*Baccharis pilularis*), and poison oak (*Toxicodendron diversilobum*). *(Photo courtesy SFPUC)*

vegetation for agriculture were likely factors contributing to shrubland decline during this period, particularly around major operations such as the Fifield Dairy and Jersey Farm Dairy. Fire suppression may have contributed to the decline: while the lack of fire would be expected to favor shrubland encroachment into grasslands, it would also be expected to favor encroachment of trees into shrublands, in some cases leading to type conversion from shrubland to hardwood or conifer forest.

Over the past 80-90 years, shrublands have continued to decline gradually, and today occupy approximately 35% of the watershed (see Fig. 6.1). The areas of the greatest shrubland loss have been around Sawyer Ridge, western Cahill Ridge in Pilarcitos Canyon, the area between Montara Mountain and Scarper Peak west of Pilarcitos Reservoir, and Kings Mountain Ridge on the southwestern side of the watershed (Fig. 6.2). Shrubland loss has likely been driven by a combination of factors, including urban development, tree planting, natural successional processes and facilitation of tree establishment by shrubs, climatic changes, and fire suppression. A dynamic equilibrium between shrubland and hardwood or conifer forest—driven by natural succesional processes and periodic non-anthropogenic fires—may have existed historically in parts of the watershed not subject to regular indigenous burning, though fire suppression efforts may have shifted this equilibrium toward forest-dominated communities. In some areas, such as Kings Mountain Ridge, planted stands of Monterey cypress or other trees have directly displaced

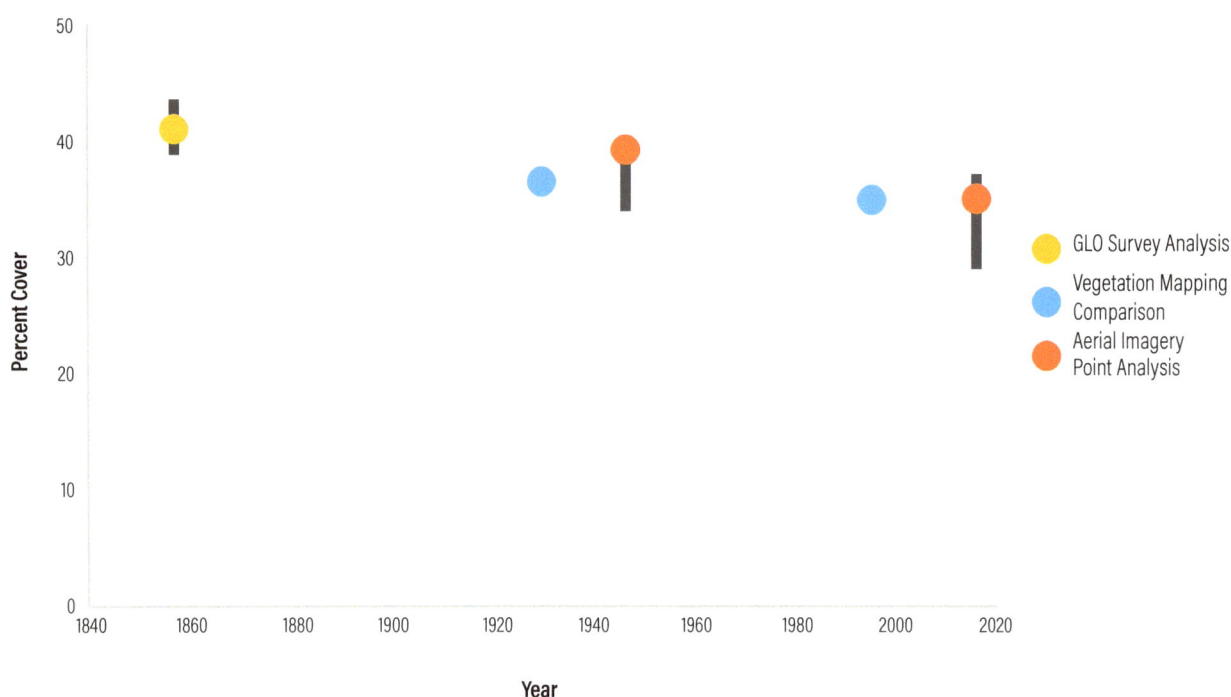

Figure 6.1. This diagram shows the change in estimated shrubland extent over time throughout the study area, based on the three quantitative analyses described in Chapter 4: the aerial imagery point analysis (ca. 1947 and ca. 2016), vegetation mapping comparison (ca. 1930 and ca. 1995), and GLO survey analysis (ca. 1857). Gray boxes around the ca. 1947 and ca. 2016 points represent the range in the shrubland frequency classified by four mappers for 200 overlapping points in the aerial imagery point analysis (see pages 27-29). The gray box around the ca. 1857 point represents the uncertainty associated with GLO points classified as hardwood forest/shrubland or grassland/shrubland (see pages 32-34).

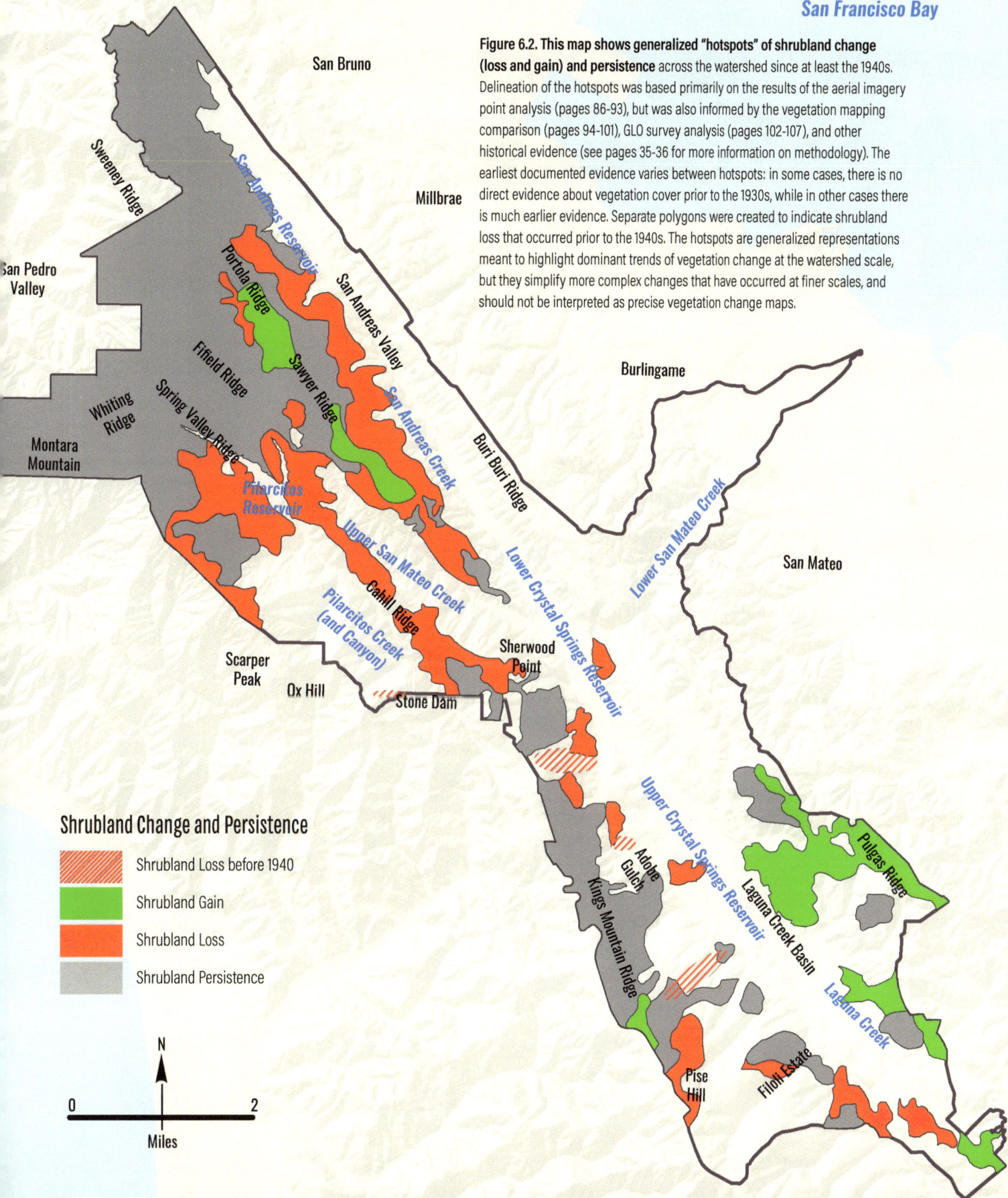

San Francisco Bay

Figure 6.2. This map shows generalized "hotspots" of shrubland change (loss and gain) and persistence across the watershed since at least the 1940s. Delineation of the hotspots was based primarily on the results of the aerial imagery point analysis (pages 86-93), but was also informed by the vegetation mapping comparison (pages 94-101), GLO survey analysis (pages 102-107), and other historical evidence (see pages 35-36 for more information on methodology). The earliest documented evidence varies between hotspots: in some cases, there is no direct evidence about vegetation cover prior to the 1930s, while in other cases there is much earlier evidence. Separate polygons were created to indicate shrubland loss that occurred prior to the 1940s. The hotspots are generalized representations meant to highlight dominant trends of vegetation change at the watershed scale, but they simplify more complex changes that have occurred at finer scales, and should not be interpreted as precise vegetation change maps.

Shrubland Change and Persistence

Shrubland Loss before 1940

Shrubland Gain

Shrubland Loss

Shrubland Persistence

N

0   2

Miles

> Fresh green chaparral and tall, full-foliaged trees stretch out on every side, and we ride down a road embowered with shrubbery... Coveys of tufted quail rise and whirr away as we gallop on, and rabbits creep into the bushes at every turn in the road.
>
> —EVANS 1873, DESCRIBING RIDE THROUGH SAN ANDREAS VALLEY

shrublands, while urban development has displaced shrublands in Lower San Mateo Creek Canyon. There has also been a small amount of recent shrubland loss caused by mortality from *Phytophthora* spp. infection (M. Ingolia pers. comm.).

Despite this gradual loss, shrublands still represent the single most extensive vegetation type in the Peninsula Watershed (as they did historically). More than 70% of the shrublands present in the 1940s remain today, as do a large fraction of the shrublands present in the mid-19th century. In addition, substantial shrubland gain has occurred in some portions of the watershed, likely driven by a combination of fire suppression and removal of grazing pressure. Much of this gain has occurred on the western side of Pulgas Ridge near the southern end of Upper Crystal Springs Reservoir and along the crest of Sawyer Ridge, where coyote brush has encroached into areas formerly dominated by grassland (see Fig. 6.2). Thus, the net decline in shrubland extent overall belies a more complex trajectory, with shrubland loss in some areas and scrub encroachment in other areas.

Buck brush (*Ceanothus cuneatus*) in chaparral on Sawyer Ridge. *(Photo courtesy SFPUC)*

Shrublands dominated much of the northwestern portion of the watershed

Forest encroachment has replaced shrubland along the eastern and southwestern slopes of Sawyer Ridge, around Upper San Mateo Creek, and around Pilarcitos Reservoir and Canyon

Shrublands dominated a large swath of the northwestern portion of the watershed historically, extending from Montara Mountain on the west to the San Andreas Valley on the east. In 1769, after descending Sweeney Ridge and travelling south through the San Andreas Valley, Friar Juan Crespí observed that the hills along the eastern slope of Sweeney and Sawyer ridges were "very green with low woods" (Crespí and Brown 2001). Engineer Miguel Costansó, travelling with the same expedition, commented similarly, stating that along the west side of the hollow the hills were topped with "[handsome] savins, with scrub-oaks and other lesser trees" (Costansó et al. 1969). Though these descriptions are somewhat ambiguous, and likely encompass multiple vegetation types (potentially including scattered redwoods, or "savins" [*Sequoia sempervirens*]), in light of later evidence it is reasonable to interpret them as references to shrubland intermixed with patches of forest or woodland. Oberlander (1953) concurs, stating, "The description indicating low green cover applies well to the present chaparral, primarily the Prunus and Baccharis vegetation types... which now occupy Sweeney Ridge."

The earliest unambiguous documentation of widespread shrubland cover in the northwestern portion of the watershed dates from a series of GLO surveys conducted in the mid-19th century. The northernmost tip of the watershed, northwest of present-day San Andreas Reservoir, was described as "chaparral" by both Robert Matthewson (1858) and William Lewis (1859b). To the west, A. W. Von Schmidt (1864b) described the eastern slope of Sweeney Ridge and parts of Fifield Ridge as "covered with brush," and the eastern slopes of Portola and Sawyer ridges as "dense chaparral." Further south, along San Mateo Creek near present-day Lower Crystal Springs Reservoir, Lewis (1860) reported "dense brush... impossible to continue." The eastern slope of Montara Mountain west of present-day Pilacitos Reservoir was "covered with chaparral" (Von Schmidt 1864b), while the upper portion of the ridge south of Montara Mountain was described as "chamisal" (a term commonly used by early surveyors to describe chaparral; Lewis 1860). Shrublands are dominant in this portion of the watershed in numerous landscape photographs dating back to the 1860s (Figs. 6.3 and 6.4).

Figure 6.3. Early landscape photos showing extensive shrublands in the northwest portion of the watershed. A) A ca. 1896 photo showing the eastern slope of Sawyer Ridge downstream of San Andreas Dam. B) A 1907-14 photo showing the western slope of Cahill Ridge between Pilarcitos and Stone dams, mostly covered in shrubland and scattered conifers (especially in the gulches). C) A 1907-14 photo taken from above San Andreas Reservoir facing roughly northwest, showing the "character of the east slope of Sweeney Ridge along the west shore of San Andreas Lake." D) A 1907-14 photo taken from Fifield Ridge looking to the northeast, showing the extensive shrublands between Fifield Ridge and San Andreas Reservoir (in the top left of the photo). E) A 1907-14 photo, taken from Fifield Ridge, showing the east fork of Pilarcitos Reservoir, Upper San Mateo Creek Canyon, Cahill Ridge, and the southern part of Spring Valley Ridge. F) A 1907-14 photo taken from the Montara Mountain area looking east, showing "at left Spring Valley Ridge, Fifield Ridge and Sweeney Ridge, with Cahill Ridge and Sawyer Ridge on right." G) A 1907-14 photo showing shrublands on either side of the northern portion of Fifield Ridge.

A) *Photo M-2042_ca1896, courtesy SFPUC*
B) *Photo F3725_1180-27, courtesy California State Archives*
C) *Photo F3725_1180-23, courtesy California State Archives*
D) *Photo F3725_1180-25, courtesy California State Archives*
E) *Photo F3725_1180-29, courtesy California State Archives*
F) *Photo F3725_1180-36, courtesy California State Archives*
G) *Photo F3725_1180-41, courtesy California State Archives*

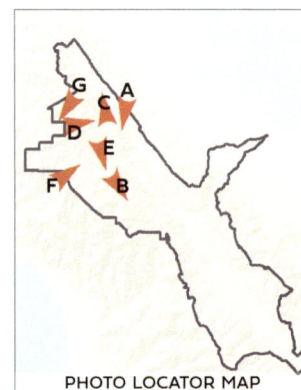

PHOTO LOCATOR MAP

Figure 6.4. These early photos, taken in 1866-67 during the construction of Pilarcitos Dam, show the shrublands to the north and west of the dam. *(Left: photo D-3105, courtesy SFPUC. Right: image of the stereograph by Carleton E. Watkins (American, 1829–1916), The Dam, Spring Valley Water Works, ca. 1866, from Pacific Coast, albumen silver prints, image (each): 2 15/16 × 3 1/16 in. (7.4 × 7.7 cm); mount: 3 1/4 × 6 3/4 in. (8.2 × 17.1 cm). George Eastman Museum, purchase)*

Field notes written by the botanist Joseph Burtt Davy during an April 1895 collecting trip through the northern part of the watershed provide insight into the composition of these shrublands. Climbing the eastern slope of Sawyer Ridge west of San Andreas Reservoir, Davy (1895) observed a number of chaparral and hardwood species, writing, "Thick brush clothes this slope, consisting mainly of *Arbutus* [Pacific madrone, *Arbutus menziesii*], *Cerasus ilicifolius* [holly-leaf cherry, *Prunus ilicifolia*], *Quercus densiflora* [tanoak, *Notholithocarpus densiflorus*], *Rubus vitifolius* [California blackberry, *Rubus ursinus*], Blue Ceanothi [blue blossom, *Ceanothus thyrsiflorus*], *Umbellularia* [California bay, *Umbellularia californica*], *Rhamnus californica* [California coffeeberry, *Frangula californica*], and higher up *Arctostaphylos* [manzanita, *Arctostaphylos* spp.], *Vaccinium* [likely huckleberry, *Vaccinium ovatum*], *Helianthemum scoparium* [peak rush-rose, *Crocanthemum scoparium*]." Continuing on, Davy noted "scrub at base of slope" on the western side of Upper San Mateo Creek (the eastern slope of Fifield Ridge), and "brush thickish in places... [with] *Ribes Menziesii* [canyon gooseberry] in good form" on the western slope of Spring Valley Ridge near Pilarcitos Reservoir. Ascending the eastern slope of the ridge between Montara Mountain and Scarper Peak to the west of Pilarcitos Reservoir, Davy observed "*Ribes Menziesii, R. divaricatum* [spreading gooseberry]; higher up scrub is largely *Arctostaphylos andersonii* [heartleaf manzanita, California Rare Plant Rank 1B.2]."[2]

One scenario that should be considered is whether the extensive shrublands documented by GLO survyeors and others in the northwestern part of the watershed in the mid-19th century represented relatively recent encroachment into areas formerly dominated by grasslands as a result of the cessation of indigenous burning. Several factors indicate that this

2 Davy's record may have been a misidentification of Kings Mountain manzanita (*Arctostaphylos regismontana*, also California Rare Plant Rank 1B.2; T. Parker pers. comm.), though heartleaf manzanita (*A. andersonii*) was recently documented in the watershed (Nomad Ecology 2020).

scenario is unlikely. In addition to the direct evidence from Crespí and Costansó of chaparral vegetation along the eastern slopes of Sweeney and Sawyer ridges (see above), the fact that this portion of the watershed was never included in a petition for a private land grant in the 1830s or 40s (unlike many other parts of the watershed) suggests that the land was relatively undesireable for use as pasturage for livestock. However, the possibility that scrub encroachment during the late 18th and early 19th centuries resulted in substantial conversion of grassland to shrubland in this portion of the watershed cannot be completely ruled out. Future research using phytolith analysis to assess the likelihood of grassland presence in this area could help to resolve this question (see Chapter 10).

Shrublands persisted throughout much of the northwestern portion of the watershed during the 20th century, particularly along the eastern slope of Sweeney Ridge, the northwestern slope of Sawyer Ridge, the slopes of Fifield and Spring Valley ridges, and the eastern slope of Montara Mountain west of Pilarcitos Reservoir. For example, testimony from the 1907-1914 SVWC court case describes the eastern slope of Sweeney Ridge as "dense wind beaten brush" (parcel 43), "steep brush-covered hills" (parcel 12), and "covered in chaparral" (parcels 13.1 and 144; see Appendix B). Similarly, the slopes of Fifield Ridge were "heavily overgrown with chaparral" (parcel 153), "bare low brush" (parcel 196), and "heavily brush-covered" (parcel 104). The slopes of Spring Valley Ridge were "steep and brush-covered" (parcel 5-2), while the eastern slope of Montara Mountain was "covered with low brush, grease-wood, and bear-berry" (parcel 5-2). The ca. 1930 Wieslander VTM mapping, the 1946-48 and 2015-17 aerial photos, and the modern vegetation mapping indicate that shrubland has remained the dominant vegetation type in these areas throughout the 20th century.

While shrubland persisted across much of the northwestern portion of the watershed during the 19th and 20th centuries, hardwood and conifer encroachment has resulted in shrubland-to-forest conversion in some areas. Rephotography, aerial imagery analysis, and other evidence indicates that, since the early 20th century, hardwood encroachment has replaced shrublands along portions of eastern and southwestern Sawyer Ridge, the lower eastern slope of Fifield Ridge near Upper San Mateo Creek, and the lower Montara Mountain area northwest of Pilarcitos Reservoir; conifer encroachment has replaced shrublands along portions of western Cahill Ridge (east of Pilarcitos Canyon) and the Scarper Peak area west of Pilarcitos Reservoir (see Fig. 8.2). Repeat photographs illustrate the dramatic expansion of forest that has occurred in these areas over the past century (Figs. 6.5 and 6.6; see also Fig. 8.9 on page 207). Because indigenous burning was likely relatively infrequent in this part of the watershed, this afforestation may simply represent natural succession following the last non-anthropogenic fire, though fire suppression has likely contributed to this shift. (See chapters 7 and 8 for more information on forest gain in these areas.)

In some areas—such as Cahill Ridge and eastern Sawyer Ridge—19th and early 20th century evidence indicates that scattered trees were already intermixed with shrubs, creating a mosaic of shrubland and hardwood or conifer forest. Scattered conifer trees, or small patches of conifers, were present within the shrubland matrix along Cahill Ridge east

REPHOTOGRAPHY

Figure 6.5. This rephotograph pair shows the eastern slope of Sawyer Ridge west of San Andreas Dam, ca. 1928 (top) and 2018 (bottom). In the ca. 1928 photo, hardwood forest occupies the lower hillslope, while shrubland occupies the upper hillslope; the contemporary photo shows expansion of hardwood forest into former shrubland areas. *(Top: Photo D-1428, courtesy SFPUC; Bottom: Photo by SFEI-ASC, August 2018)*

ca. 1910

2018

Figure 6.6. This rephotograph pair, looking south from the southern part of Fifield Ridge, highlights the encroachment of conifers into former shrubland areas around Pilarcitos Reservoir between 1907-14 and 2018. (Top: Photo F3725_1180-29, courtesy California State Archives; Bottom: Photo by SFEI-ASC, August 2018)

PHOTO LOCATOR MAP

Left

Right

1907-14

2018

and south of Pilarcitos Reservoir at least as early as the 1860s. A drawing by the famed artist Edward Vischer, for instance, showing the area around Upper San Mateo Creek Dam just east of Pilarcitos Reservoir, depicts the vegetation on the hillslopes as a mixture of shrublands and sparse conifer cover (Fig. 6.7). (The area depicted in the drawing corresponds to the area on the far-left hand side of the images in Fig. 6.6, where scattered conifers are visible in the early 20th century.) This depiction is consistent with GLO surveys of the area from the same period: while travelling along Cahill Ridge south of present-day Pilarcitos Reservoir in 1861, GLO surveyor Robert Matthewson (1861) observed "chaparral... [with] wild plum... lilac... fir... oak." Vegetation on the western slope of Cahill Ridge (on the east side of Pilarcitos Canyon) in the early 20th century was dominated by a "dense, wind-beaten growth of chaparral extending to the crest of the Cahill Ridge, with scattering fir in most of the gulches" (parcel 5-2; Fig. 6.8; see Appendix B). Today, all of these areas are dominated by Douglas-fir forest, with smaller patches of coyote brush and coffeeberry shrubland (see Fig. 2.1 on page 23).

Eastern Portola and Sawyer ridges likewise supported a mix of shrubs and trees in some areas—the descriptions by Crespí and Costansó (see above) provide evidence for this heterogeneity as early as 1769. While shrubland was the dominant vegetation type, scattered trees were present in some areas, particularly on the lower hillslopes (see Fig. 6.5). For example, describing the road ascending Sawyer Ridge west of San Andreas Dam, the *Daily Alta California* (1887) states that it "leads at once into underbrush... passes through wild lilac, buckeye and any variety of smaller trees [and] on nearing the top it gets above the growth of trees." Early 20th century descriptions of eastern Sawyer Ridge from the 1907-14 SVWC court case also indicate heterogeneous vegetation cover: upper portions of the slope were "covered with chaparral" (parcel 153) or "densely overgrown with chaparral and some hardwood timber" (parcel 39), while lower portions of the slope were "heavily overgrown with chaparral and oak" (parcel 89), and "densely overgrown with hardwood and chaparral" (parcel 39; see Appendix B).

SPRING VALLEY WATER-WORKS.

OUTLET OF THE MAIN TUNNEL, AND COMMENCEMENT OF THE FLUME to the CITY.

Scenery 4 miles above CRYSTAL — SPRINGS; 27 miles to SAN FRANCISCO.

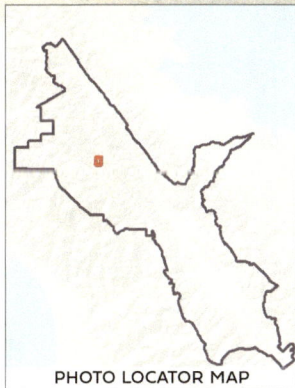

PHOTO LOCATOR MAP

Figure 6.7. (above) This 1860s drawing by artist Edward Vischer shows shrublands and sparse conifer cover on the hills around San Mateo Creek Dam #1 just east of Pilarcitos Reservoir. *(Photo vdp00142_0016, courtesy Claremont Colleges Digital Collections)*

Figure 6.8. (right) This 1932 photo, looking south down Pilarcitos Canyon, shows the shrublands on the west side of the canyon (with scattered Douglas-fir in the gulches). *(Photo number 286968, courtesy Wieslander Vegetation Type Mapping Collection, Marian Koshland Bioscience and Natural Resources Library, University of California, Berkeley, www.lib.berkeley.edu/BIOS/vtm/)*

286968

Mature northern coastal scrub with gooseberry (*Ribes* sp.), California coffeeberry (*Frangula californica*), thimbleberry (*Rubus parviflorus*), coyote brush (*Baccharis pilularis*), and poison oak (*Toxicodendron diversilobum*). (*Photo courtesy SFPUC*)

Shrublands dominated the upper eastern slopes of Kings Mountain Ridge

Conifer forest encroachment and planted trees have replaced shrublands on some parts of Kings Mountain Ridge

In addition to the large expanses of shrublands covering much of the northwestern part of the watershed, shrublands were also the dominant vegetation type on the upper eastern slopes of Kings Mountain Ridge in the southwestern part of the watershed. Though somewhat distant from the routes traveled by early Spanish expeditions through the San Andreas Valley and Laguna Creek Basin, a journal entry by Friar Juan Crespí may provide some early evidence for shrublands in this part of the watershed. As previously noted (see discussion of grasslands on the southwest side of San Andreas Valley, page 122), while travelling south through the Laguna Creek Basin with the Portolá Expedition in November 1769, Crespí described Kings Mountain Ridge as a "green mountain range wooded with low trees and in spots clad in grass alone" (Crespí and Brown 2001). As with his account of "low woods" on the eastern flanks of Sweeney and Sawyer ridges further north (see page 149), the description of "low trees" here is somewhat ambiguous, but could reasonably be interpreted as a reference to mixed chaparral (which later sources show was dominant on the upper hillslope) and hardwood forest (which dominated the lower hillslope; see Chapter 7).

GLO surveyors recorded shrublands throughout this portion of the watershed during the mid-19th century; as on Cahill Ridge and eastern Sawyer Ridge (see above), scattered trees or patches of forest were often intermixed with the shrublands, particularly on the lower portions of the slope. For instance, ascending the slope of Kings Mountain Ridge to the southwest of present-day Upper Crystal Springs Reservoir in 1852, surveyor C.C. Tracy wrote that he was traveling up the "summit of [the] lateral spur from the mountain range among very dense chaparral." He continued, "The spurs become so precipitous and covered with timber, chaparral and matted vines it was impossible to carry this line farther... very dense chaparral" (Tracy 1853b). Further north, surveyors C. C. Tracy (1853b), Thomas Stephens (1856a), and Robert Matthewson (1860) again reported chaparral in a number of areas.

Testimony and photographs from the 1907-1914 SVWC court case document extensive shrublands throughout the region, and in some areas indicate that shrublands had formerly been cleared for cultivation (Fig. 6.9). For instance, the hillslope north of present-day highway 92 was "densely covered with brush" (parcel 68.3) and "heavily overgrown with chaparral and brush, excepting near the lake" (parcel 68.2); lower reaches of the slope had a "dense growth of hardwood timber" (parcel 68.2) or were "covered with dense chaparral, interspersed with oaks" (parcel 122; see Appendix B). The upper hillslopes around present-day highway 92 consisted of "rough and brush-covered" land, some of which had been cleared but was "reverting to brush-covered hills" (parcel 132.1). Parcels further south were likewise "covered with chaparral" (parcel 68.4), "covered with brush and wind-swept" (parcel 68.5), "heavily overgrown with chaparral" (parcel 134), and "covered with dense brush and scrub oak" (parcel 132.3). In the southwestern corner of the watershed, shrubland gave way to redwood-dominated conifer forests along the upper slopes of Kings Mountain Ridge; however, shrubland was the dominant vegetation type on the lower hillslopes in this area (parcel 194).

Shrubland appears to have been somewhat more extensive on Kings Mountain Ridge during the 19th century than it was by the mid-20th century: some of the areas described as chaparral by GLO surveyors or other early observers had become forested by the 1940s (based on historical aerial imagery). Some of this early conversion was the result of active planting of Monterey cypress or other species by local landowners during the late 19th or early 20th centuries (see page 210). Much of the shrubland that was present in the early to mid-20th century on Kings Mountain Ridge has persisted, though encroachment of hardwood and conifer forest (including expansion of planted stands of Monterey cypress) has displaced shrubland in some areas. See chapters 7 and 8 for further discussion of forest expansion in this area.

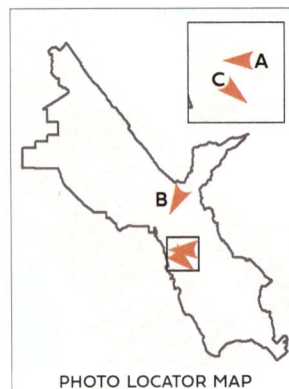

PHOTO LOCATOR MAP

Figure 6.9. Landscape photos of shrublands on Kings Mountain Ridge. A) Shrublands are visible along much of Kings Mountain Ridge in this 1907-14 photo, taken from the west side of Upper Crystal Springs Reservoir. B) This 1907-14 photo shows the eastern slope of Kings Mountain Ridge west of Lower Crystal Springs Reservoir dominated by hardwood forest at lower elevations and shrubland at higher elevation. C) This 1907-14 photo shows shrubland covering much of the upper eastern slope of Kings Mountain Ridge, and transitioning to hardwood forest and grassland at lower elevations. *(A: Photo F3725_1180-16; B: Photo F3725_1180-12; C: Photo F3725_1180-7; all courtesy California State Archives)*

Shrublands

Early shrublands included a range of coastal scrub and chaparral communities

Diverse shrublands west of the San Andreas Fault have been lost due to afforestation and tree planting, while low diversity shrublands have been gained on the eastern side of the watershed as a result of coyote brush encroachment

While it was not possible to systematically map changes among specific shrubland communities across the watershed, the Wieslander VTM mapping and other early records do provide useful clues about shrubland diversity within different parts of the watershed, and reveal differences in the trajectories experienced by coastal scrub and chaparral communities over the past 80-90 years. Unfortunately, the species identifications and alliance-level classifications in the Wieslander VTM mapping are in some cases highly questionable (or clearly wrong), and thus it was determined that the mapping (at least within the Peninsula Watershed) could not be used to confidently assess changes at the alliance-level. It was possible, however, to categorize many (though not all) of the shrubland species assemblages or vegetation alliances as either coastal scrub or chaparral; vegetation alliances in the modern vegetation mapping (MRLCC 2001, Schirokauer et al. 2003) were likewise categorized as coastal scrub or chaparral where possible (Fig. 6.10; Appendix A).

The Wieslander VTM mapping shows that coyote brush-dominated coastal scrub, often with substantial cover of California coffeeberry as well (an association also noted by Oberlander [1953]), dominated much of the northwestern portion of the watershed, including much of the eastern slope of Sweeney Ridge, the slopes of Fifield and Spring Valley ridges, the eastern slope of Montara Mountain, and portions of the western slope of Cahill Ridge south of Pilarcitos Dam. According to Wieslander VTM mapping and plot data from this area, other associated species included toyon (*Heteromeles arbutifolia*), holly-leaf cherry (*Prunus ilicifolia*), poison oak (*Toxicodendron diversilobum*), blueblossom ceanothus (*Ceanothus thyrsiflorus*), California blackberry (*Rubus ursinus*), blue elderberry (*Sambucus nigra* ssp. *cerulea*), oso berry (*Oemleria cerasiformis*), bracken fern (*Pteridium aquilinum* var. *pubescens*), yerba santa (*Eriodictyon californicum*), ocean spray (*Holodiscus discolor*), California huckleberry (*Vaccinium ovatum*), and pitchersage (*Lepechinia calycina*). Coyote brush-dominated coastal scrub also occupied large areas along the upper eastern slope of Kings Mountain Ridge, and in some areas was associated with sticky monkeyflower (*Diplacus aurantiacus*), poison oak, coffeeberry, ocean spray, and various grasses.[3]

In addition to coyote brush-dominated coastal scrub, the Wieslander VTM mapping shows extensive stands of chaparral or coastal scrub dominated by holly-leaf cherry and toyon (an association also noted by Oberlander [1953]), which occupied large areas along the eastern slope of Sweeney Ridge, the western slope of Sawyer Ridge (around San Mateo Creek Dam No. 1), and the western slope of Cahill Ridge within Pilarcitos Canyon. Chamise (*Adenostoma fasciculatum*)-dominated chaparral was prevalent along the southern tip of Sawyer Ridge, and to a lesser extent in patches along Kings Mountain Ridge, Pulgas Ridge, and Lower San Mateo Creek Canyon. Several stands of chaparral dominated by manzanita (*Arctostaphylos* spp.) were recorded along the eastern slopes of Portola and Sawyer ridges (to the north and south of San Andreas Dam) and in parts of the southwestern portion of the watershed north and west of the Filoli Estate.[4]

Comparison of Wieslander VTM and modern vegetation mapping shows that, watershed-wide, areas categorized as coastal scrub have decreased by approximately 8% (499 ac), areas categorized as chaparral have decreased by 40% (712 ac), and uncategorized areas (i.e., vegetation alliances that could not be definitely categorized as either coastal scrub or chaparral) have increased by approximately 61% (781 ac). Visual comparison of the maps (see Fig. 6.10) shows that much of the uncategorized area in the modern vegetation mapping falls within areas categorized as coastal scrub in the Wieslander VTM mapping, suggesting that the shrubland communities in these areas have likely persisted over time. Because of the lack of spatial accuracy in the Wieslander VTM mapping, the vegetation mapping comparison should not be used in isolation to draw conclusions about vegetation changes at fine spatial scales, though it is useful for visualizing and quantifying broad patterns of coastal scrub and chaparral persistence and change across the watershed.

---

3 As noted previously, the Wieslander VTM mapping and plot data also include a number of misidentifications representing species outside of their natural range, such as fragrant sumac ("Rhus trilobata" [*R. aromatica*], white sage (*Salvia apiana*), and wild mockorange (*Philadelphus lewisii*).

4 These manzanita species are misidentified as Hoary manzanita (*Arctostaphylos canescens*) and Woolly leaf manzanita (*A. tomentosa*). *A. tomentosa* likely represents brittle leaf manzanita (*A. crustacea* spp. *crustacea*; T. Parker and T. Corelli pers. comm.), which had not been identified as a separate taxon at the time the Wieslander surveys were conducted. It is unknown which species *A. canescens* refers to.

COASTAL SCRUB AND CHAPARRAL

Shrublands within the Peninsula Watershed comprise a complex and heterogeneous assemblage of plant associations (ABI et al. 2003, Nomad Ecology 2020). Though vegetation classification systems differ somewhat in how these associations are grouped, at the coarsest level they can be divided into two types of plant communities: coastal scrub and chaparral. Coastal scrub is limited to areas where climate is moderated by marine influences, and is characterized by relatively soft-stemmed low shrubs, a prominent herbaceous understory, and frequent intermixing with coastal prairie (Ford and Hayes 2007). Northern coastal scrub, which extends from central/northern California into southern Oregon and is dominated by evergreen shrubs, is generally distinguished from more xeric southern coastal scrub, which extends from central/southern California into Baja California and is dominated by drought-deciduous taxa (Axelrod 1978, Holland 1986, Ford and Hayes 2007, Rundel 2007, Sawyer et al. 2009). In the San Francisco Bay Area, northern coastal scrub is typically dominated by coyote brush, and to a lesser extent coffeeberry, California sagebrush, blue blossom *(Ceanothus thyrsiflorus)*, poison oak *(Toxicodendron diversilobum)*, and other shrubs (Baxter and Parker 1999, ABI et al. 2003, Wrubel and Parker 2018, Nomad Ecology 2020).

Chaparral, a shrubland type dominated by evergreen sclerophyllous shrubs, occurs from Oregon to Baja California, and extends inland to the Sierra Nevada foothills and further east within mid-elevation zones (Keeley and Davis 2007). While chaparral is most prevalent at elevations between 300 and 1,500 m, in coastal settings characterized by marine climate influence (e.g., summer fog or cloud cover) a subtype known as maritime chaparral can extend to lower elevations (Griffin 1978, Keeley and Davis 2007). Maritime chaparral is typically dominated by narrowly distributed or locally endemic species of manzanita or ceanothus, along with more widespread species like chamise, and is recognized for its high shrub diversity (Holland 1986, Keeley and Davis 2007, Vasey et al. 2012, Vasey et al. 2014, T. Parker pers. comm.). Within the Peninsula Watershed, maritime chaparral co-occurs with more xeric chaparral communities dominated by chamise, holly-leaf cherry *(Prunus ilicifolia)*, and other shrubs (ABI et al. 2003, Nomad Ecology 2020).

While coastal scrub and chaparral are both characteristic of California's Mediterranean type climate, plant species within these vegetation communities often exhibit marked differences in their adaptations to summer drought and fire. Coastal scrub species are highly drought tolerant, and in general respond to low summer water availability by dramatically limiting their physiological activity. Chaparral species, which tend to be more deeply rooted, are able to access soil moisture at depth and thus maintain more physiological activity year-round (Mooney and Dunn 1970, Parker 2020, T. Parker pers. comm.). Many chaparral species have "refractory" seeds which require fire (either direct exposure to heat or exposure to charred wood) in order to germinate. These include shrub species such as chamise and numerous manzanita and ceanothus species, as well as subshrubs, suffrutescents, and annuals such as California sagebrush, chaparral clematis *(Clematis lasiantha)*, and mock parsley *(Apiastrum angustifolium;* Keeley 1991, Keeley and Davis 2007). Coastal scrub, while adapted to periodic stand-replacing fires, does not require fire for regeneration, and is typically dominated by species that readily resprout following fire such as coyote brush, California coffeeberry, and poison oak (Ford and Hayes 2007, T. Parker pers. comm.). In the absence of fire or other disturbance, coyote brush-dominated coastal scrub readily invades adjacent grasslands (McBride and Heady 1968, Hobbs and Mooney 1986, Callaway and Davis 1993, Ford and Hayes 2007; see pages 40-41 for more discussion). §

1928-32

1995-2001

N

0   3

Miles

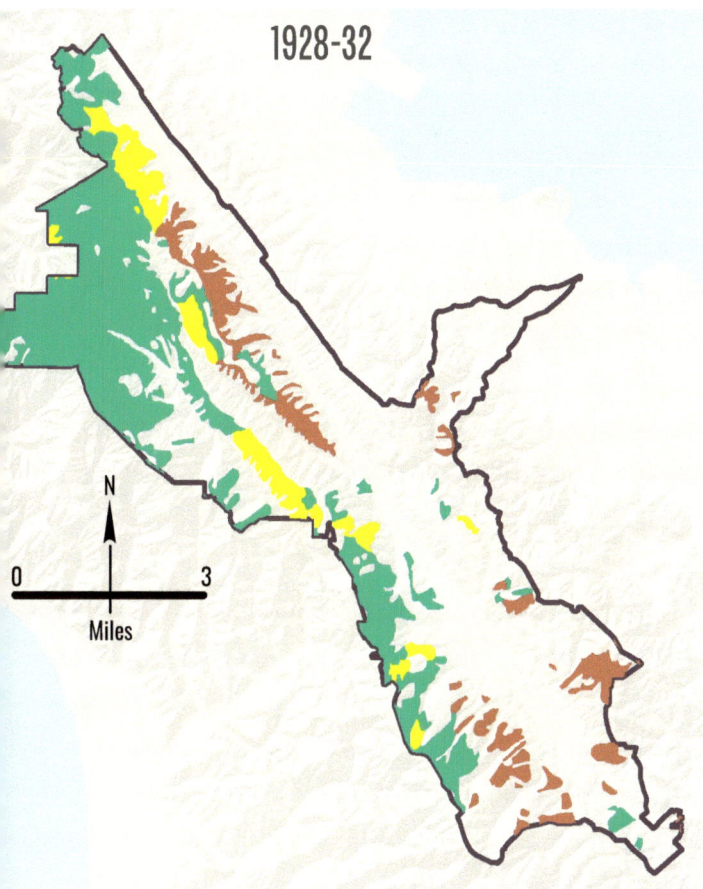

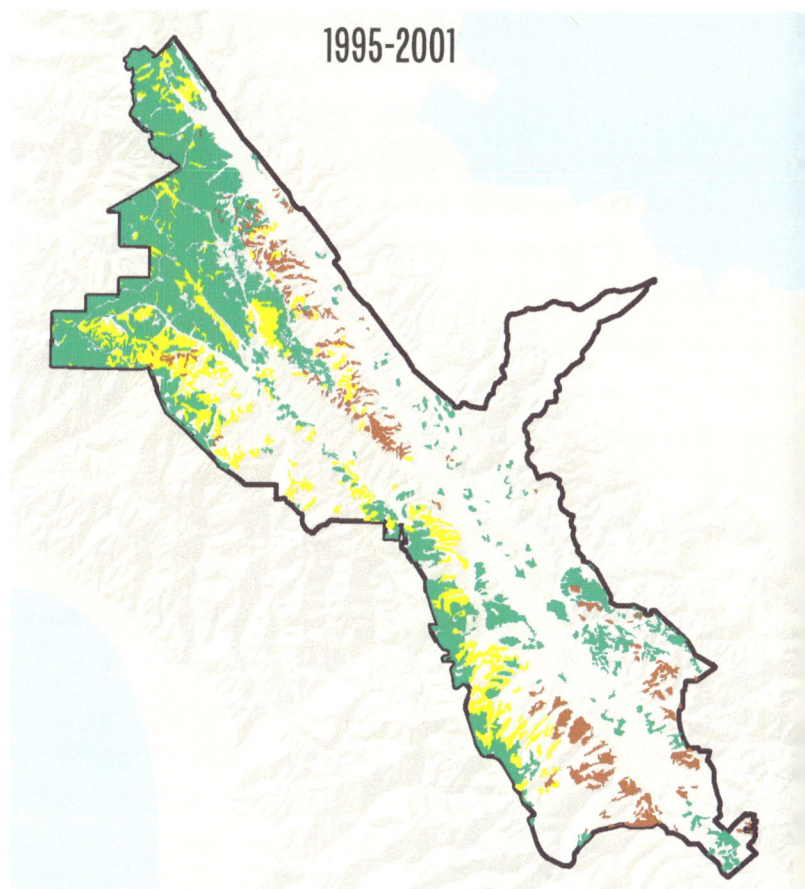

Figure 6.10. Shrubland alliances categorized as coastal scrub or chaparral (or uncategorized) in the Wieslander VTM mapping (1928-32, Kelly et al. 2005) and modern vegetation mapping (1995-2001; Schirokauer et al. 2003). See Appendix A for classification crosswalks. While the maps are useful for visualizing broad patterns of coastal scrub and chaparral persistence and change across the watershed, because of the lack of spatial accuracy in the Wieslander VTM mapping they should not be used in isolation to draw conclusions about vegetation changes as fine spatial scales.

Distribution of coastal scrub and chaparral, 1928-32 and 1995-2001

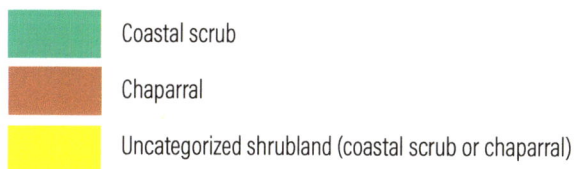

Coastal scrub

Chaparral

Uncategorized shrubland (coastal scrub or chaparral)

Diverse coastal scrub has persisted throughout much of the northern portion of the watershed, as well as along Kings Mountain Ridge. The main exceptions are in areas to the west of Pilarcitos Dam and along the western slope of Cahill Ridge in Pilarcitos Canyon, which have converted to Douglas-fir-dominated conifer forest, and along portions of Kings Mountain Ridge that have been planted to Monterey cypress and Monterey pine or converted to Douglas-fir or coast live oak woodland. Coyote brush has expanded considerably in a number of areas, most notably into areas formerly dominated by grasslands along Pulgas Ridge and Buri Buri Ridge, the southern part of the Laguna Creek Basin, the ridgeline and western slope of Sawyer Ridge, the Adobe Gulch area (some of which has recently been restored to grassland; see page 234), and the northern end of San Andreas Valley (Fig. 6.11; see Chapter 5 for further discussion of grassland loss). In addition, a large number of points showing grassland-to-shrubland conversion in the aerial imagery point analysis fall within areas currently mapped as non-shrubland vegetation types (e.g., grassland, hardwood forest), indicating that shrub encroachment may be more widespread than suggested solely by the vegetation mapping comparison (in other words, shrub cover in these non-shrubland vegetation types has increased but has not yet resulted in complete type conversion in some areas).

Among chaparral communities, large areas identified in the Wieslander VTM mapping as manzanita- or chamise-dominated chaparral have converted to hardwood forest or other non-shrubland vegetation types, though chaparral has persisted in many areas as well (see Fig. 6.10). Along southern Sawyer Ridge, for instance, encroachment of coast live oak-dominated hardwood forest has displaced large areas of chamise-dominated chaparral. Likewise, along Pulgas Ridge and Kings Mountain Ridge, some areas formerly dominated by chaparral have been displaced by coast live oak and California bay woodlands. Chaparral loss has occurred in Lower San Mateo Creek Canyon due to both urban development and hardwood forest encroachment. Within Pilarcitos Canyon, chaparral or coastal scrub formerly dominated by holly-leaf cherry (according to the Wieslander VTM mapping) has been displaced by conifer forest (Douglas-fir) encroachment.

PHOTO LOCATOR MAP

Figure 6.11. Comparison of 1946-48 (top) and 2015-17 (bottom) aerial photos showing scrub encroachment into grassland on the western slope of Pulgas Ridge, near the southeastern end of Upper Crystal Springs Reservoir. (Top: imagery coutesy historicaerials.com; Bottom: Pictometry, Inc. 2015)

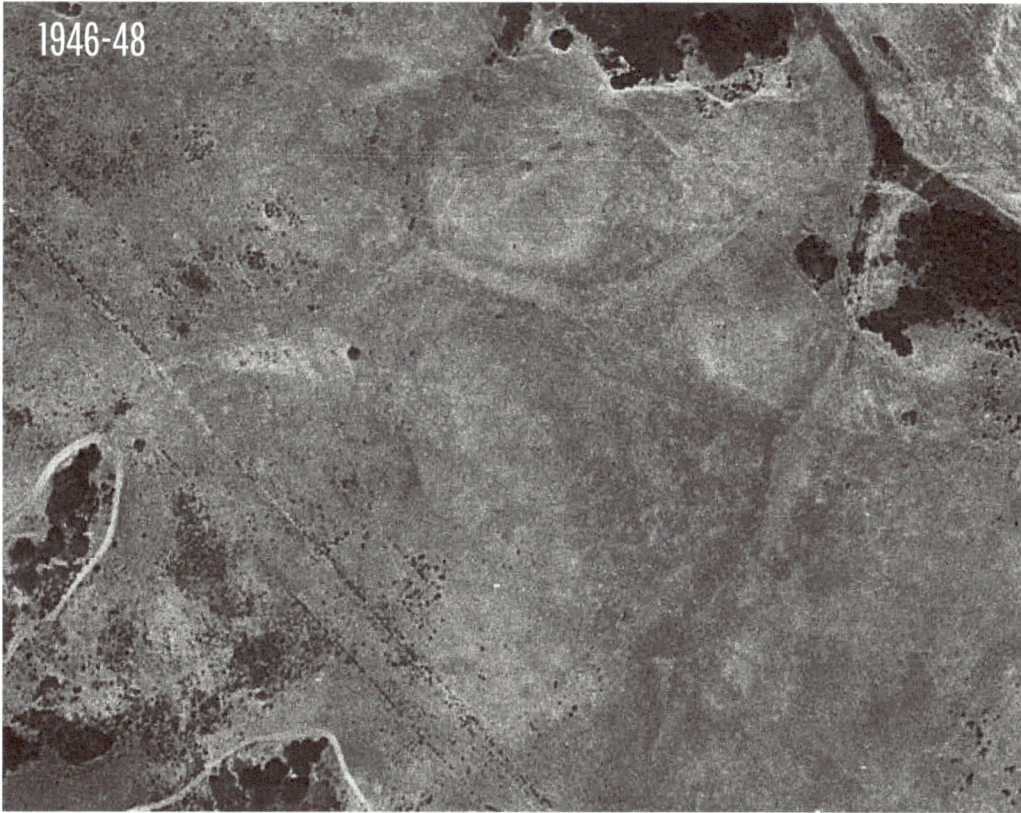

1946-48

2015-17

A heritage coast live oak (*Quercus agrifolia*) in a coast live oak woodland with a poison oak (*Toxicodendron diversilobum*) understory. (Photo courtesy SFPUC)

Hardwood Forests

Overview

Hardwood forests were widely distributed throughout the Peninsula Watershed historically, and were the second-most extensive vegetation type in the mid-19th century after shrublands (Fig. 7.1). This diverse vegetation type occupied large portions of San Andreas Valley, lower hillslopes along eastern Sawyer Ridge, ravines along lower Buri Buri Ridge, lower canyons and hillslopes around the Laguna Creek Basin, hillslopes within Lower San Mateo Creek Canyon, northeast-facing slopes of Cahill Ridge in the Upper San Mateo Creek Canyon, and areas around Sherwood Point (Fig. 7.2). Due to the difficulty in differentiating riparian and non-riparian forests in many historical data sources, the discussion of hardwood forests in this chapter (and the discussion of conifer forests in Chapter 8) unavoidably includes some riparian areas as well. A separate discussion of streams, riparian habitats, and wetlands is included in Chapter 9.

Coast live oak (*Quercus agrifolia*) was (and still is) the dominant hardwood tree in many parts of the watershed. Other hardwood forest trees included California bay (*Umbellularia californica*), interior live oak (*Quercus wislizeni*), valley oak (*Quercus lobata*), California buckeye (*Aesculus californica*), Pacific madrone (*Arbutus menziesii*), and tanoak (*Notholithocarpus densiflorus*; Palóu et al. 1969; Font and Brown 2011; Stephens 1856a,b; Matthewson 1861; Brace 1869; SVWC vs. San Francisco 1916; Oberlander 1953; Kelly et al. 2005; data from Consortium of California Herbaria). Understory vegetation in these hardwood forests included grasses, ferns such as California wood fern (*Dryopteris arguta*) and coffee fern (*Pellaea andromedifolia*); shrubs such as toyon (*Heteromeles arbutifolia*), western leatherwood (*Dirca occidentalis*),[1] California coffeeberry (*Frangula californica*), ocean spray (*Holodiscus discolor*), redberry buckthorn (*Rhamnus crocea*), poison oak (*Toxicodendron diversilobum*), currants and gooseberries (*Ribes* spp.), bitter cherry (*Prunus emarginata*), oso berry (*Oemleria cerasiformis*), beaked hazelnut (*Corylus cornuta* ssp. *californica*), and wood rose (*Rosa gymnocarpa*); and a wide variety of wildflowers such as checker lily (*Fritillaria affinis*), shooting star (*Primula hendersonii*), *Collinsia heterophylla* var. *heterophylla*, common bedstraw (*Galium aparine*), pink honeysuckle (*Lonicera hispidula*), western columbine (*Aquilegia formosa*), purple sanicle (*Sanicula bipinnatifida*), canyon larkspur (*Delphinium nudicaule*), small flowered nemophila (*Nemophila parviflora*), bent-flowered fiddleneck (*Amsinckia lunaris*),[1] Franciscan wallflower (*Erysimum franciscanum*),[2] common

---

1 California Rare Plant Rank 1B.2.

2 California Rare Plant Rank 4.2.

rupertia (*Rupertia physodes*), striped coral root (*Corallorrhiza striata*), and San Mateo woolly sunflower (*Eriophyllum latilobum*)[3] (data from Consortium of California Herbaria).

Oak woodlands in California support a large number of wildlife species, including roughly 5,000 species of arthropods (including 200 species of gall wasps), 60 species of amphibians and reptiles, 120 species of mammals, and nearly 150 species of birds (Pavlik et al. 2002, Tietje et al. 2005, Swiecki and Bernhardt 2006). Oak woodlands and other hardwood forests in the Peninsula Watershed historically provided food, shelter, and other resources for a diverse array of wildlife, including birds such as Acorn Woodpecker (*Melanerpes formicivorus*), Oak Titmouse (*Baeolophus inornatus*), Band-tailed Pigeon (*Patagioenas fasciata*), Western Bluebird (*Sialia mexicana*), and Hutton's Vireo (*Vireo huttoni*); mammals such as dusky-footed woodrat (*Neotoma fuscipes*) and western gray squirrel (*Sciurus griseus*); and reptiles and amphibians such as arboreal salamander (*Aneides lugubris*), California slender salamander (*Batrachoseps attenuatus*), and western skink (*Plestiodon skiltonianus*; Brace 1869, Burroughs 1928, Gifford-Gonzalez et al. 2013, data from Global Biodiversity Information Facility). The large nests of dusky-footed woodrats made an impression on early observer Charles Loring Brace, who noted in 1867 that "enormous rats' nests, some three feet high" were "a peculiar feature of the woods" around Lower San Mateo Creek Canyon (Brace 1869).

GLO survey data suggest that hardwood forests occupied approximately 30-32% of the watershed in the mid-19th century (see Fig. 7.1). Construction of the San Andreas and Crystal Springs reservoirs in the late 19th century eliminated many of the hardwood and riparian forests in San Andreas Valley, while urban development in San Mateo decreased hardwood forest extent within Lower San Mateo Creek Canyon. The combined impacts of these losses, along with browsing of seedlings and saplings by livestock, small scale clearing of trees by homesteaders, and several large fires during the late 19th century (see page 52), resulted in an overall decline in hardwood forest extent in the 19th and early 20th centuries. By the 1930s and 40s, hardwood forest cover had decreased to approximately 21% of the study area (see Fig. 7.1).

The acquisition of watershed land by SVWC, which was largely complete by 1930 (see page 72), was a major factor in the slowing and reversal of this trajectory during the latter portion of the 20th century. Ownership by SVWC (and later SFPUC) entailed exclusion of grazing, protection from development, and active fire suppression, all of which contributed to the expansion of hardwood forest into areas formerly dominated by grasslands or shrublands. The same drivers, however, also favored the expansion of conifer forests (see Chapter 8), in some cases resulting in the replacement of hardwood-dominated forests. Active planting of species like eucalpytus and acacia in the early to mid-20th century contributed substantially to the increase in hardwood forest cover during this period. For instance, areas mapped as "Eucalyptus spp. Alliance" in the modern vegetation mapping (Schirokauer et al. 2003) currently occupy at least 115 ac within the watershed. Overall, hardwood forest cover expanded to approximately 22-23% of the watershed by 1995-2001 and 25% by 2015-17 (see Fig. 7.1).

3 Federally and state endangered; California Rare Plant Rank 1B.1.

Clockwise, from top left: Western Bluebird (*Sialia mexicana; Photo by Becky Matsubara, courtesy CC BY 2.0*); Acorn Woodpecker (*Melanerpes formicivorus; Photo by Becky Matsubara, courtesy CC BY 2.0*); dusky-footed woodrat (*Neotoma fuscipes*) nest (*Photo by Martin Jambon, courtesy CC BY 2.0*); dusky-footed woodrat (*Photo by Mbmceach, courtesy CC BY 2.0*); California slender salamander (*Batrachoseps attenuatus; Photo by TJ Gehling, courtesy of CC BY 2.0*).

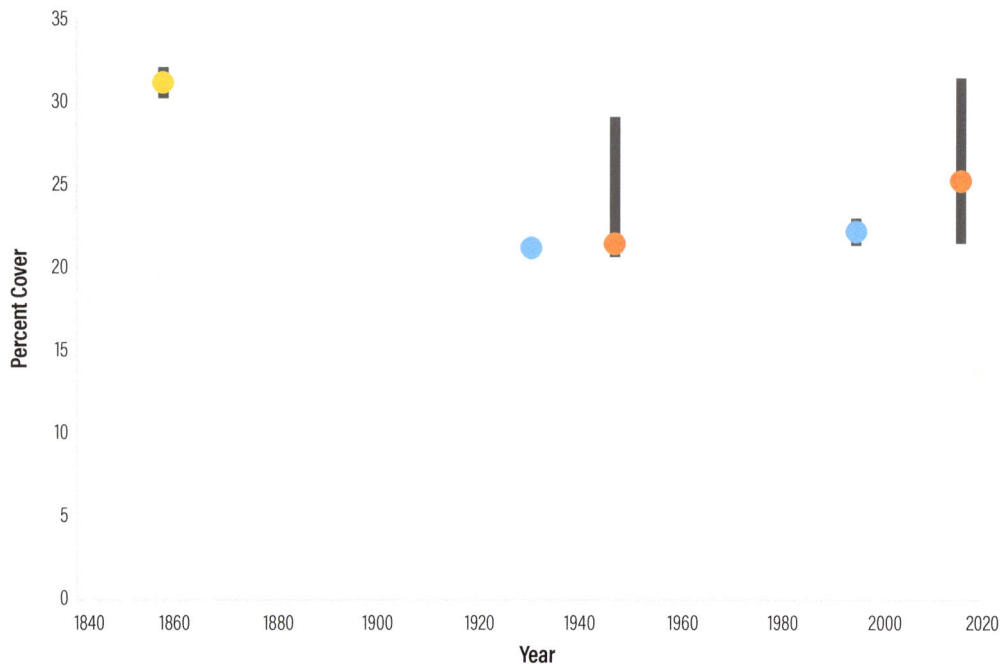

Figure 7.1. This diagram shows the change in estimated hardwood forest extent over time throughout the study area, based on the three quantitative analyses described in Chapter 4: the aerial imagery point analysis (ca. 1947 and ca. 2016), vegetation mapping comparison (ca. 1930 and ca. 1995), and GLO survey analysis (ca. 1857). Gray boxes around the ca. 1947 and ca. 2016 points represent the range in the hardwood forest frequency classified by four mappers for 200 overlapping points in the aerial imagery point analysis (see pages 27-29). The gray box around the ca. 1995 point represents the uncertainty associated with the forest (hardwood or conifer) class in the modern vegetation mapping (see pages 30-31). The gray box around the ca. 1857 point represents the uncertainty associated with GLO points classified as hardwood/conifer forest or hardwood forest/shrubland (see pages 32-34).

- GLO Survey Analysis
- Vegetation Mapping Comparison
- Aerial Imagery Point Analysis

In recent decades, Sudden Oak Death (SOD) has resulted in substantial mortality among oak populations in the watershed. First observed in the Peninsula Watershed in 2001, ongoing monitoring efforts have documented a 20% loss of coast live oaks and 50% loss of tanoaks due to SOD in plots throughout the watershed (Phytosphere Research 2013, Garbelotto 2017, M. Ingolia pers. comm.).

Patterns of change for hardwood forest are more spatially heterogeneous than for other vegetation types (Fig. 7.2). Hardwood forests have largely persisted in the southern part of the watershed, though there has been some conversion to conifer forest on the upper portions of southern Kings Mountain Ridge. In the northern part of the watershed, outside of areas lost due to reservoir construction, hardwood forest has persisted on the valley floor between San Andreas and Lower Crystal Springs reservoirs and on the eastern slope of Sawyer Ridge. Hardwood forest, including large stands of planted eucalpytus and associated naturalized populations, has expanded into areas formerly dominated by grasslands on Buri Buri Ridge, and into areas formerly dominated by shrublands on the upper eastern slope of Sawyer Ridge, but has been displaced by conifer forest along eastern Cahill Ridge in the Upper San Mateo Creek Canyon. Continued urban development in Lower San Mateo Creek Canyon (which is not under SFPUC ownership) has also resulted in continuing loss of hardwood forest.

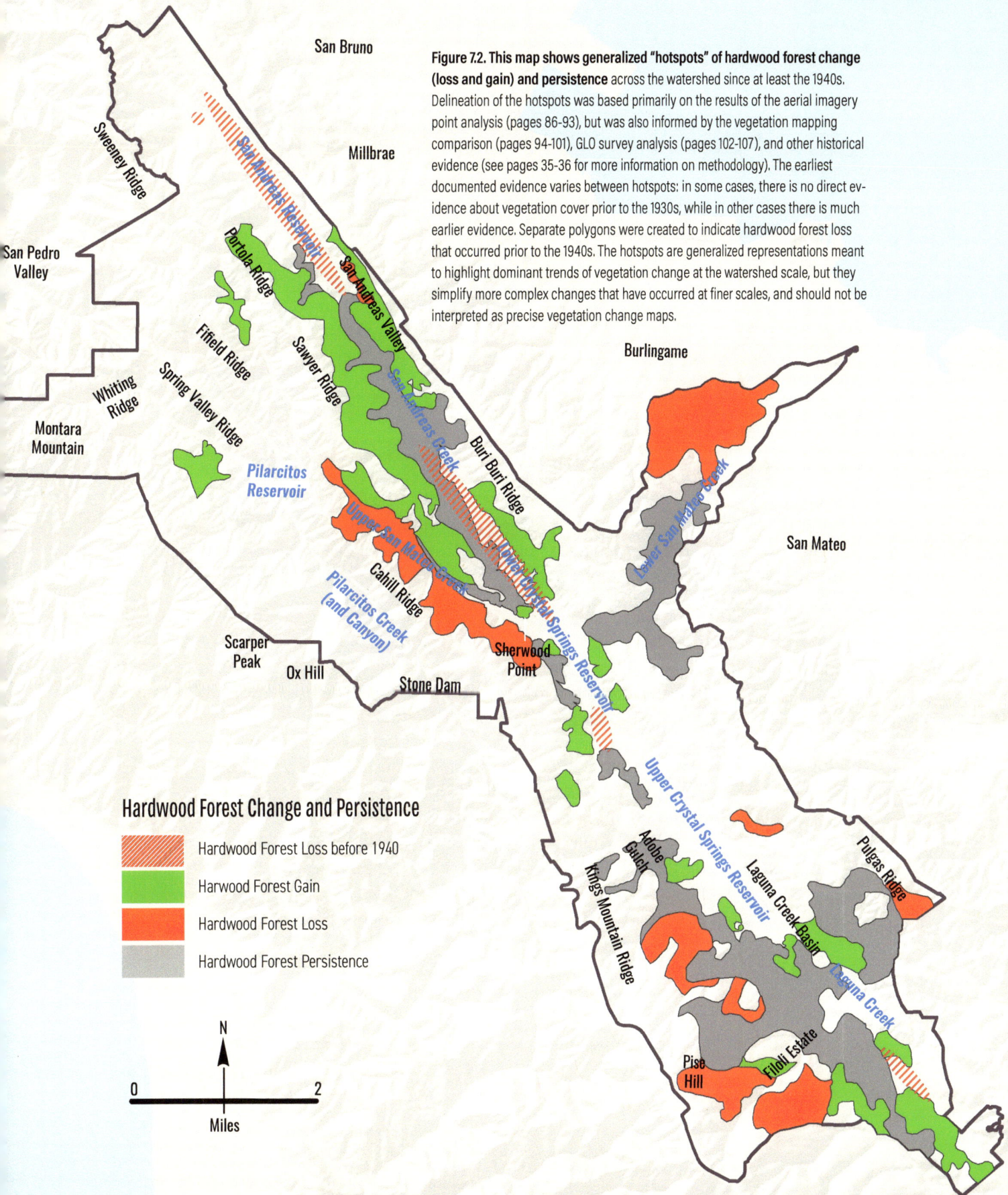

San Francisco Bay

San Bruno

Millbrae

Sweeney Ridge

San Pedro Valley

Whiting Ridge

Montara Mountain

Portola Ridge

Fifield Ridge

Spring Valley Ridge

San Andreas Reservoir

San Andreas Valley

Sawyer Ridge

San Andreas Creek

Pilarcitos Reservoir

Pilarcitos Creek (and Canyon)

Cahill Ridge

Upper San Mateo Creek

Buri Buri Ridge

Lower San Mateo Creek

Burlingame

San Mateo

Scarper Peak

Ox Hill

Stone Dam

Sherwood Point

Upper Crystal Springs Reservoir

Lower San Mateo Creek

Kings Mountain Ridge

Adobe Gulch

Laguna Creek Basin

Pulgas Ridge

Laguna Creek

Pise Hill

Filoli Estate

Figure 7.2. This map shows generalized "hotspots" of hardwood forest change (loss and gain) and persistence across the watershed since at least the 1940s. Delineation of the hotspots was based primarily on the results of the aerial imagery point analysis (pages 86-93), but was also informed by the vegetation mapping comparison (pages 94-101), GLO survey analysis (pages 102-107), and other historical evidence (see pages 35-36 for more information on methodology). The earliest documented evidence varies between hotspots: in some cases, there is no direct evidence about vegetation cover prior to the 1930s, while in other cases there is much earlier evidence. Separate polygons were created to indicate hardwood forest loss that occurred prior to the 1940s. The hotspots are generalized representations meant to highlight dominant trends of vegetation change at the watershed scale, but they simplify more complex changes that have occurred at finer scales, and should not be interpreted as precise vegetation change maps.

Hardwood Forest Change and Persistence

- Hardwood Forest Loss before 1940
- Harwood Forest Gain
- Hardwood Forest Loss
- Hardwood Forest Persistence

N

0   2

Miles

Hardwood and riparian forests occupied much of San Andreas Valley and lower portions of surrounding hillslopes

Though reservoir construction and other land use changes eliminated much of the hardwood forest from San Andreas Valley, this habitat type has expanded along the surrounding Buri Buri and Sawyer ridges

Hardwood and riparian forests were the dominant vegetation types historically in the San Andreas Valley, in the northeastern portion of the watershed. Hardwood forests also occupied ravines and gulches along Buri Buri Ridge to the east and the lower slopes of southern Sawyer Ridge to the west.

Traveling through this area in the late 18th century, Spanish explorers remarked on the extent of woody vegetation. As discussed previously (see page 149), descriptions of "low woods" (Crespí and Brown 2001) and "savins... scrub-oaks and other lesser trees" (Costansó et al. 1969) on the northwestern side of the valley by members of the 1769 Portolá Expedition were likely references to chaparral primarily, but may also have encompassed areas of hardwood forest as well as scattered redwoods. Descriptions by members of later Spanish expeditions provide additional insights. For instance, Captain Francisco Rivera, while traveling northwards through San Andreas Valley in November of 1774, encountered "rough and thickly tangled woods [on our left], pinning us up against the hill slope facing them" (Rivera et al. 1969). The "facing" hillslope that they were pinned against was Buri Buri Ridge on the eastern side of the valley, implying that relatively impenetrable "woods" occupied much of the valley floor. Similarly, traveling south through San Andreas Valley with the Anza Expedition in March of 1776, Friar Pedro Font observed "a great amount of woods in the hollow, [and] the thicker it got the farther we rode along, with a great deal of different sorts of good timber trees such as live oaks, madrons [sic], redwoods, cottonwoods, and other kinds as well" (Font and Brown 2011). The "woods" described by Rivera and Font are again somewhat ambiguous, and likely referred both to forested portions of the riparian corridors along San Andreas Creek and its tributaries, as well as the surrounding forests and woodlands outside of the riparian corridors. The Spanish explorers and other early sources also described "brush" along San Andreas Creek as well as scattered "lakes" (sag ponds) and emergent marshes in this part of the valley; see Chapter 9 for a discussion of these wetland and riparian habitats.

By the mid-19th century, farms, ranches, and homesteads had proliferated within San Andreas Valley, and hardwood forests and other habitat types were cleared in many areas to make room for agriculture (see page 58). Nevertheless, records from this period document the continued presence of oak groves and other hardwood forests in numerous locations throughout the northern portion of San Andreas Valley and adjacent areas. Colonel Albert Evans, traveling through San Andreas Valley in the mid-19th century,

PHOTO LOCATOR MAP

Fig. 7.3. This 1875 photograph by Carleton Watkins shows the landscape around the Crystal Springs Hotel (just west of present-day Lower Crystal Springs Dam) dotted with coast live oak and other trees. (BANC PIC 1974.006:19--ffALB, courtesy The Bancroft Library, UC Berkeley)

described "broad, spreading oaks" as well as "great spreading buckeyes and California laurels" (Evans 1873). The *Daily Alta California*, describing the area around the former Crystal Springs Hotel (to the west of present-day Lower Crystal Springs Dam, where San Andreas, San Mateo, and Laguna creeks joined) wrote in 1860:

> The house is embowered in trees. In front of it stand perhaps twenty tall and straight live oaks, whose ample foliage... springs lightly and gracefully up... The valley, formed by high hills—spurs of the Coast Range—is perfectly crowded with woods—not a mere tangled undergrowth of chaparral, but the most graceful foliage. (*Daily Alta California* 1860)

Other observers echoed this account, describing the area around the hotel as "thickly studded with the clustering live oaks" (*San Mateo Gazette* 1859) and a "dense forest of live oaks and other evergreen trees and shrubbery" (*San Mateo Gazette* in Lawrence 1922a; Fig. 7.3). Surveyors recorded numerous live oaks, as well as madrone, in the same vicinity between the 1850s and 80s (Addison 1857; Lewis 1860; Neumann 1874a,b,c; Neumann 1887). Further to the north, GLO surveyors documented live oaks and madrones in the valley to the south of present-day San Andreas Dam (Von Schmidt 1864b), and documented an "oak grove" (Matthewson 1858) and "timber in the ravines" (Tracy 1853a) near the northern end of present-day San Andreas Reservoir.

The construction of San Andreas Dam (1868) and Lower Crystal Springs Dam (1888) eliminated much of the hardwood and riparian forest on the floor of San Andreas Valley, though forests on the lower portions of adjacent hillslopes and in the surrounding ravines remained. Along the eastern slope of southern Sawyer Ridge, near the northwestern shore of Lower Crystal Springs, witnesses in the 1907-14 SVWC court case described the hillslope as "densely overgrown with chaparral, oak and madrone" (parcel 97); the lower hillslope was "covered" and "well wooded" with hardwood timber (parcels 110, 94), while the upper hillslope was dominated by chaparral (see Chapter 6) and "scattering oaks" (parcel 182; see Appendix B). Landscape photos highlight the prevalence of these forests along the lower eastern slope of Sawyer Ridge above Lower Crystal Springs Reservoir, and illustrate how hardwood forest gave way to chaparral higher up on the slope (Fig. 7.4). The portion of the valley between San Andreas and Lower Crystal Springs reservoirs, which had not been inundated, was described as "quite heavily wooded with oaks" (parcel 72). Further north, on the southwest side of San Andreas Reservoir, the area "along the lake and near the dam" was "quite well wooded" (in contrast to areas higher up the slope, which were described as "chaparral covered hills"; parcel 14). Further north along Portola Ridge, the hardwood forests on the lower hillslope gave way to chaparral (see Chapter 6).

PHOTO LOCATOR MAP

Figure 7.4. This 1907-1914 photo shows the dense hardwood forest lining the lower eastern slope of Sawyer Ridge south of San Andreas Reservoir. *(Photo F3725_1180-18, courtesy California State Archives)*

PHOTO LOCATOR MAP

Figure 7.5. This 1907-14 photo, taken from Sawyer Ridge and looking southeast across San Andreas Valley, shows patches of hardwood forest occupying gullies within the western slope of Buri Buri Ridge. See Chapter 5 for more detail regarding the heterogeneity of vegetation types present in this area. *(Photo F3725_1180-15, courtesy California State Archives)*

The woods are truly vocal with the songs of birds, and in an hour's ramble more than twenty varieties of wild flowers can be gathered... In whatever direction the eye wanders it falls upon new, bright foliage and gnarled trunks.

—*DAILY ALTA CALIFORNIA* 1860, DESCRIBING THE FORESTS AROUND THE CRYSTAL SPRINGS HOTEL

On the eastern side of San Andreas Valley, though grasslands dominated much of Buri Buri Ridge (see Chapter 5), oak woodlands were prevalent within the gullies and ravines on the lower portion of the ridge (Fig. 7.5; see also Fig. 5.3b in Chapter 5). The area along the eastern shore of Lower Crystal Springs Reservoir, to the north of the dam, was described by witnesses in the 1907-14 SVWC court case as "indented with gullies in which are found dense growths of oak" (parcel 90) or a "comparatively dense growth of oak" (parcel 73). The "growth of trees" in these gullies was described as "increasing towards the [south]," though parcels further north (around San Andreas Dam) also supported "scattering oak trees" (parcel 19) and "oak and laurel on [the] slope" (parcel 42).

Wieslander VTM mapping from ca. 1930 is consistent with these accounts, showing both the lower slopes of eastern Sawyer Ridge and the lower slopes and ravines of Buri Buri Ridge dominated by coast live oak; California bay and Pacific madrone are listed as subdominant species along lower Sawyer Ridge.

By the early to mid-20th century, numerous stands of non-indigenous trees (both hardwood and conifer) had been planted in the area, particularly on the eastern side of San Andreas Valley in areas formerly dominated by grasslands (see Chapter 5). At the Crystal Springs Golf Club, located on the perimeter of the watershed northeast of Lower Crystal Springs Reservoir, a planting plan in 1921 called for a large quantity of non-native trees to be planted, including 350 *Eucalyptus globulus*, 150 *Eucalyptus rostrata* (likely *Eucalyptus camaldulensis*), 300 *Acacia latifolia* (likely *Acacia longifolia* or *A. melanoxylon*), and several hundred individuals of non-indigenous conifer species (see Chapter 8; Roeding 1921). The eastern side of San Andreas Reservoir to the north was also planted to eucalyptus and other non-native trees. For instance, witnesses in the 1907-14 SVWC court case described planted "pine and eucalyptus trees" surrounding the watershed keeper's cottage on the northeast side of the reservoir (parcel 15; see Appendix B). The Wieslander VTM mapping shows seven stands of eucalyptus in the vicinity of San Andreas Dam, and another large stand at the Crystal Springs Golf Club to the south.

Hardwood forest has largely persisted in this part of the watershed since the early 20th century, and its distribution has expanded considerably on the upper portion of eastern Sawyer Ridge and the western slope of Buri Buri Ridge. Along Sawyer Ridge, the aerial imagery point analysis and vegetation mapping comparison indicate that hardwood forest (dominated by coast live oak woodland, according to Schirokauer et al. 2003) has expanded into areas formerly dominated by chaparral communities (see Fig. 6.5, Chapter 6). On Buri Buri Ridge, hardwood forest has encroached into areas formerly dominated by grasslands (see Fig. 5.5, Chapter 5). Much of this increase on Buri Buri Ridge appears to be due to the expansion of coast live oak woodland, though the proliferation of eucalyptus, acacia, and other non-native trees has also contributed to the shift.

Hardwood forest occupied canyons and lower hillslopes of the Laguna Creek Basin

Hardwood forests have largely persisted in these areas

To the south of San Andreas Valley, hardwood forest was the dominant vegetation cover on the lower slopes of Kings Mountain Ridge, to the west and south of present-day Upper Crystal Springs Reservoir. These forests extended onto the valley floor portions of the Laguna Creek Basin south of the reservoir, though the valley floor further north was largely occupied by riparian forests, willow groves, emergent marshes, and sag ponds (see Chapter 9). While grasslands dominated the slopes of western Pulgas Ridge on the east side of the Laguna Creek Basin, patches of hardwood forests existed within ravines and gulches.

As with San Andreas Valley further north, 18th century Spanish explorers described the hardwood forests present in this part of the watershed; as discussed earlier, these descriptions are often somewhat ambiguous, and encompass multiple vegetation types. For instance, while camped with the Portolá Expedition near present-day Upper Crystal Springs Dam on November 5, 1769, after traveling south through San Andreas Valley, Friar Juan Crespí reported, "A great many madronos [sic], small and large, have been met with during these two days' march." Continuing south through the Laguna Creek Basin the following day, he wrote, "To the right keeps on the green mountain range wooded with low trees and in spots clad in grass alone; a low place with a great deal of willows, madronos, and other, unknown sorts of trees ran at the foot of the mountains" (Crespí and Brown 2001). Friar Francisco Palóu, in November of 1774, similarly described this central part of the valley as having "a great deal of wood" as well as "good grass for the animals" (Palóu et al. 1969). As discussed in Chapter 6 (see page 159) , the reference to "low trees" may have included both hardwood forests on the lower slopes of Kings Mountain Ridge as well as chaparral on the upper hillslope. On the southern end of the valley, near the head of present-day Upper Crystal Springs Reservoir, Palóu reported "high mountains upon the south all grown over with red-wood, white and live-oaks, and other trees." Though Palóu does not elaborate on the distribution of these species, and there was likely a substantial amount of heterogeneity between hardwood and conifer forests, later sources generally agree that oak woodlands were dominant on the lower slopes of Kings Mountain Ridge, while redwood was more common at higher elevations on the southern portion of the ridge (see Chapter 8).

GLO surveyors in the mid-19th century documented oak-dominated hardwood forests intermixed with chaparral along much of the lower portion of Kings Mountain Ridge. For example, while approaching the peninsula that extends into present-day Lower Crystal Springs Lake near the west side of Upper Crystal Springs Dam in November of 1856, surveyor Thomas Stephens described a "spur of hills... thickly overgrown with Live Oaks" (Stephens 1856b). Climbing the slope south of present-day Upper Crystal Springs Dam in December of 1859, surveyor William Lewis reported "thick brush and timber," and further along noted "timber, on last ½ mile, chiefly oak" (Lewis 1859a). In the hills surrounding the Adobe Gulch grasslands (see page 122), surveyors documented live oak, madrone, buckeye, scrub oak, and chaparral (Lewis 1859a, Matthewson 1861). Though oak woodlands largely gave way to chaparral on the upper portion of the ridge, oaks and other hardwood trees occupied canyons and ravines higher up the slope as well. For example, in the canyon south of Adobe Gulch, surveyor A. W. Von Schmidt encountered an "impassable canyon" with "live oaks" approximately 500 m from the summit (Von Schmidt 1864c). Further south, the hills on the southwest side of present-day Upper Crystal Springs Reservoir were similarly covered in "very dense brush and timber" (Tracy 1853b).

By the early 20th century, clearing by settlers and farmers had removed hardwood forests from some portions of Kings Mountain Ridge (Fig. 7.6), while planted stands of Monterey cypress, eucalyptus, and other non-indigenous trees had displaced both native shrublands and hardwood forests in some areas (see page 210). Nevertheless, hardwood forests dominated

Figure 7.6. Hardwood forests are visible along the lower portion of Kings Mountain Ridge near the head of Upper Crystal Springs Reservoir in this 1907-14 photo. An abrupt division between hardwood forest and grassland at the center of the photo marks the location where forest had previously been cleared for agriculture. (Photo F3725_1180-17, courtesy California State Archives)

by oaks, California bay, and Pacific madrone persisted in many areas along the lower slopes of the ridge. The 1930s Wieslander VTM mapping, for instance, shows coast live oak-dominanted hardwood forest, intermixed with groves of coast redwoods and Monterey cypress, throughout much of this area. Additional evidence comes from witness testimony in the 1907-14 SVWC court case, which documents the presence of oak woodlands along much of the western shores of Lower and Upper Crystal Springs reservoirs. The southwestern shore of Lower Crystal Springs Reservoir, for instance, supported a "dense growth of hardwood timber... near the lake" (parcel 68.2; see Appendix B). Near Upper Crystal Springs Dam, witnesses described "a fringe of oak along the lake shore" (parcels 46, 47.1), while areas further south were characterized as having a "dense growth of oak and other natural trees and brush" (parcel 45). Still further south, the lower hillslopes around the southwestern corner of the reservoir were described as "heavily wooded with oak" (parcel 50), "covered with very fine, heavy timber" consisting of "fine oaks and laurels" (parcel 55), or covered with a "very dense growth of oaks and madrones" (parcel 49). Some of these trees attained a massive size, particularly on the rich alluvial soils of the valley floor: observers described "splendid, large oak trees" (most likely valley oaks [*Quercus lobata*]), for instance, just to the south of Upper Crystal Springs Reservoir. Hardwood forest cover in this area was heaviest on the east side of Old Cañada Road (which ran along the base of the hills on the western side of Upper Crystal Springs Reservoir; USGS 1902; parcels 49, 50, 194).

Hardwood forest has persisted throughout much of Kings Mountain Ridge over the past 70-90 years, based on aerial imagery point analysis and vegetation mapping comparison. Gain of hardwood forest has occurred in

PHOTO LOCATOR MAP

some areas, particularly where coast live oak and California bay woodland have expanded into areas formerly occupied by grasslands (near the reservoirs) and shrublands (higher up the slope). Hardwood forest loss has occurred in other areas, due in large part to encroachment by Douglas-fir, planted Monterey cypress, and other coniferous trees (Fig. 7.7; see Chapter 8 for further discussion).

The historical distribution of hardwood forest was much more limited on the eastern side of the Laguna Creek Basin, which was dominated by grasslands (see Chapter 5), though patches of oak woodland did occur within ravines along western Pulgas Ridge. For example, witnesses in the 1907-14 SVWC court case described the area east of the present-day Pulgas Water Temple as "indented with swales, separated by knolls, and... wooded with scattered large oaks" (Phelps parcel). Further north, witnesses described the area north of Upper Crystal Springs Dam as "wooded with oak in the gulches" (parcel 37; see Appendix B). The Wieslander VTM mapping identifies these areas as dominated by coast live oak. Hardwood forests in this part of the watershed have generally persisted, and in some cases expanded, over the past century (see Fig. 5.4 on page 118).

Compared with Pulgas Ridge and Kings Mountain Ridge, the impacts of early farming and settlement on forest cover were likely more pronounced in the bottomlands of the Laguna Creek Basin (see page 58). Surveyors in the mid-19th century noted a number of cultivated fields throughout the valley, though oak woodlands were present as well. Traveling south through the valley in 1856, for instance, surveyor Thomas Stephens reported entering "thick oak timber," dominated by "white" (valley) oaks, at the edge of a field just to the north of present-day Edgewood Road. He again reported "white oaks" bordered by a field to the south of Edgewood Road, and described the area as "covered with Oak and Chaparrel [sic]" (Stephens 1856b). The presence of fields on either side of these oak groves suggests that the forested area may have been more extensive prior to cultivation of these areas. Indeed, in 1860 the *San Mateo County Gazette* observed that "the 'Canada' is being fast settled up... Clearings, where needed, are also being made, the open or less wooded lands being mostly occupied" (Postel 2010). Much of this area was planted to grapes in the late 19th and early 20th centuries, though these ventures eventually failed (Peninou 2000, Postel 2010). Reports of the area from the 1907-14 SVWC court case describe "bare... abandoned vineyard[s]" (parcel 205) or "[former] vineyard land; now nearly bare" (parcel 210), while in some areas the "old vineyard[s]" supported a "new growth of timber" (parcel 211) and "a few scattering white oaks" (parcel 208). Over the past 70-90 years, hardwood forest has expanded into grasslands, shrublands, and formerly cultivated areas in this part of the watershed.

Hardwood Forests

REPHOTOGRAPHY

1907-14

2018

Figure 7.7. This rephotograph pair documents the encroachment of conifers into hardwood forest along the eastern slope of Kings Mountain Ridge west of Lower Crystal Springs Reservoir. In the 1907-14 photo (top), hardwood forest dominates much of the hillslope, and a patch of grassland occupies the shore opposite Sherwood Point in the center of the photograph. In the 2018 rephotograph (bottom), encroachment of conifers (more conical growth form) can be seen in the area formerly dominated by hardwood forest and grassland. *(Top: photo F3725_1180-12, courtesy California State Archives; Bottom: Photo by SFEI-ASC, August 2018)*

PHOTO LOCATOR MAP

Hardwood forests occupied lower hillslopes of Upper San Mateo Creek Canyon

Conifer forest has largely replaced hardwood forest on the eastern slope of Cahill Ridge, while hardwood forest has displaced shrublands on the western slope of Sawyer Ridge

Hardwood forests historically occupied extensive areas within Upper San Mateo Creek Canyon, including much of the eastern slope of Cahill Ridge (on the west side of the canyon), and lower slopes of western Sawyer Ridge (on the east side of the canyon) (Fig. 7.8; see also Fig. 5.6b on page 123). The ca. 1930 Wieslander VTM mapping shows much of the area dominated by coast live oak, along with substantial cover of California bay and Pacific madrone. The Wieslander mapping transitions to conifer forest, dominated by Douglas-fir, on the upper portions of the slope.

A witness in the 1907–14 SVWC court case reported that the eastern slope of Cahill Ridge "from the San Mateo Creek Dam No. 1, south to the east portal of Tunnel No. 1 Stone Dam Aqueduct, is extremely precipituous and gulchy, with quite a dense growth of fir, oaks, madrones, bay, and other natural trees and chaparral" (parcel 39; see Appendix B). The southeastern portion of the hillslope, across from Sherwood Point, was similarly "wooded with fir, oak, madrone, and chaparral," (parcel 92) while the knoll to the east was "well wooded with oaks" (parcel 38). Across the canyon, Sherwood Point supported "a number of very large oak trees" (parcel 92). The western slope of Sawyer Ridge, while mostly "densely overgrown with shrubs and chaparral" (parcel 39), supported hardwood forest on the lower hillslope (Wieslander VTM).

Encroachment of conifer forest has resulted in substantial loss of hardwood forest on the eastern slope of Cahill Ridge (on the west side of the canyon), based on the aerial imagery point analysis and vegetation mapping comparison. In contrast, hardwood forest has expanded upslope on the east side of the canyon, along western Sawyer Ridge, which has resulted in substantial loss of the former shrubland habitats in this area.

Hardwood Forests

PHOTO LOCATOR MAP

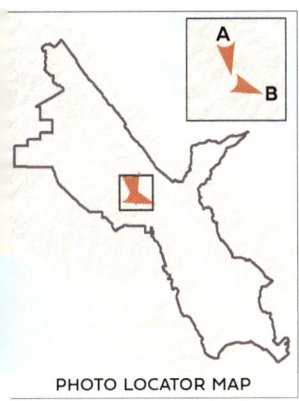

*Fig. 7.8c could not be accurately placed

Figure 7.8. Landscape photos show the hillslopes around Upper San Mateo Creek dominated by hardwood forest. A) Hardwood forest is visible in this 1907-14 photo on the slopes of Sawyer Ridge and Cahill Ridge; the east slope of Cahill Ridge is visible in the background and the west slope of Sawyer Ridge in the foreground. B) A mix of vegetation types is visible in this 1907-14 photo, with substantial conifer cover along with hardwood forest and shrubland. The photo's caption reads, "The west arm of Crystal Springs Lake showing the steep easterly slope of Cahill Ridge and the dense brush covered west slope of Sawyer Ridge." C) In this 1906 photo looking downstream in Upper San Mateo Creek Canyon, the Stone Dam Aqueduct, damaged in the 1906 earthquake, is visible along the eastern slope of Cahill Ridge (Schussler 1906). *(A: Photo F3725_1180-20, courtesy California State Archives; B: Photo F3725_1180-19, courtesy California State Archives; C: Photo Y-84, courtesy SFPUC)*

Lower San Mateo Creek Canyon supported a mosaic of hardwood forest and shrubland, while oak savanna occupied the plains to the east

Development has resulted in substantial loss of hardwood forests and savannas along Lower San Mateo Creek

Below Lower Crystal Springs Dam, San Mateo Creek flows through a steep canyon as it approaches the Bay. Beyond the riparian forests immediately adjacent to San Mateo Creek (see Chapter 9), the surrounding hillsides historically supported a mosaic of hardwood forest and shrublands (Fig. 7.9). Traveler Charles Loring Brace, who spent a summer living in a cottage near the canyon in 1867, described the scenery in the area as follows:

> The Crystal Springs Cañon... was dark even at noon-day... [the] road... winding... along the banks of a dashing crystal-clear stream, and beneath such weird trees. They were evergreen oaks (*Quercus crassipocula*) – heavy, moss-grown trunks, and great gray branches reaching out fantastically... and the leaves a roof of small black green leaves, giving an impervious shade. These trees were often growing amid wild gray rocks, tossed about in great confusion, and covering the sides of the hills above...

> Occasionally, as we walk through the forest, we are startled by coming suddenly on a bloody trunk, the Manzanita (*Arctostaphylos glauca*)... or the Madroña (*Arbutus Menziesii*) with its trunk of a bright red, where the bark is stripped off. The woods are now sprinkled like snow with the white flowers of the buckeye, a horse-chestnut (*Cornus Nuttalii*). (Brace 1869)

Oberlander (1953) interprets Brace's reference to "*Quercus crassipocula*" as coast live oak (*Quercus agrifolia*) and his reference to "*Cornus Nuttalii*" as buckeye (*Aesculus californica*). "*Arctostaphylos glauca*" may have been a misidentification of brittle leaf manzanita (*A. crustacea* ssp. *crustacea*; T. Parker and S. Simono pers. comm.).

Other 19th century observers echoed Brace's description of the canyon, noting a mix of chaparral and hardwood forest. Surveying just downstream of present-day Lower Crystal Springs Dam in 1859, for instance, William Lewis described some areas of the canyon as a "brushy ravine" covered in "chaparral," and others as covered in "brush and timber" (Lewis 1860). Later observers described the canyon as "deep steep + wooded" (Davy 1895), and reported that "large oak trees overhang the [entire length of the] road" through the canyon

Figure 7.9. This late 19th century photo by Carleton Watkins, titled "along Crytal Springs road," shows the mix of hardwood forest and shrubland on the slopes surrounding Lower San Mateo Creek Canyon. *(BANC PIC 1974.006:18--ffALB, courtesy The Bancroft Library, UC Berkeley)*

from San Mateo (*Daily Alta California* 1887). Witnesses in the 1907-14 SVWC court case described the area downstream of the dam, on the south side of the canyon, as "heavily covered with oak, madrone and buckeye" (parcel 36), "heavily wooded" (parcels 91, 62), and covered with "wind-swept oak and chaparral" (parcel 91; see Appendix B). The area further downstream, to the north of present-day Polhemus Road, was "beautifully wooded" with a combination of "clearings and wooded spaces' (Wilder 1925). The ca. 1930 Wieslander VTM mapping shows lower elevation portions of the canyon as dominated by coast live oak, which transitions to shrubland and grassland at higher elevations.

To the east of the canyon, San Mateo Creek flows across an alluvial plain to its outlet at the Bay. It was along this portion of San Mateo Creek that a Spanish mission outpost was established in 1793 (see page 56), and where the City of San Mateo was later established. Early observers described this plain, between the foothills to west and the tidal marshes to the east (see page 223), as an open, park-like savanna with extensive grasslands dotted with large oaks and other trees (Fig. 7.10). Beechey (1827), for instance, describes the landscape around the San Mateo mission outpost in 1826 as "a wide country of

meadow land, with clusters of fine oak free from underwood. It strongly resembled a nobleman's park: herds of cattle and horses were grazing upon the rich pasture, and numerous fallow-deer."

The landscape around the lowest reaches of San Mateo Creek was used heavily for grazing and agriculture during the Mission and Rancho eras (late 18th through mid-19th century; see page 56). With the arrival of American settlers and the growth of the City of San Mateo in the late 19th century, much of the original oak savanna and grassland habitat on the plain was eliminated. Further upstream, much of the hardwood forest and chaparral in Lower San Mateo Creek Canyon persisted into the 20th century: the Wieslander VTM mapping, for instance, shows the canyon downstream of the Hillsborough Park neighborhood as developed, while the canyon further upstream is largely undeveloped. Since the mid-20th century, however, urban development has extended throughout much of the upstream portion of the canyon as well (e.g., the Lakeview and Highlands neighborhoods), resulting in substantial loss of hardwood forest and other habitats in this portion of the study area.

Figure 7.10. Large oaks can be seen surrounding the town of San Mateo in this 1874 photo by Carleton Watkins. (Photo 2008-2312, courtesy California State Archives)

Redwoods at nearby Purisima Creek. *(Photo by Tanaka Purisima, courtesy CC 2.9)*

Conifer Forests

Overview

Conifer forests had the most limited spatial extent of any of the four major terrestrial vegetation types within the Peninsula Watershed historically, only occupying an estimated 4-5% of the study area in the mid-19th century (Fig. 8.1). There were two areas where conifer forests formed the principal vegetation type: a coast redwood (*Sequoia sempervirens*)-dominated forest occupied the southwestern corner of the watershed, on the upper portions of Kings Mountain Ridge, while a Douglas-fir (*Pseudotsuga menziesii*)-dominated forest occupied the eastern slope of Scarper Peak, on the west side of Pilarcitos Canyon (Fig. 8.2).

In addition to coast redwood and Douglas-fir, conifer forests included a number of associated understory plants. Understory trees, shrubs and ferns included species such as tanoak (*Notholithocarpus densiflorus*), currants and gooseberries (*Ribes* spp.), California huckleberry (*Vaccinium ovatum*), salmonberry (*Rubus spectabilis*), red elderberry (*Sambucus racemosa* var. *racemosa*), common snowberry (*Symphoricarpos albus* var. *laevigatus*), burning bush (*Euonymus occidentalis*), leather fern (*Polypodium scouleri*), and Dudley's sword fern (*Polystichum dudleyi*). Associated forbs and grasses included species such as alum root (*Heuchera micrantha*), Douglas iris (*Iris douglasiana*), common trillium (*Trillium chloropetalum*), American vetch (*Vicia americana* ssp. *americana*), California fetid adderstongue (*Scoliopus bigelovii*), Hooker's fairybell (*Prosartes hookeri*), largflower fairybell (*Prosartes smithii*), showy rock montia (*Montia parvifolia*), California mistmaiden (*Romanzoffia californica*), stream violet (*Viola glabella*), Torrey's melicgrass (*Melica torreyana*), clustered lady's slipper (*Cypripedium fasciculatum*),[1] Central coast iris (*Iris longipetala*),[1] California bottlebrush grass (*Elymus californicus*),[2] and great polemonium (*Polemonium carneum*)[3] (data from Consortium of California Herbaria).

---

1 California Rare Plant Rank 4.2.

2 California Rare Plant Rank 4.3.

3 California Rare Plant Rank 2B.2.

Despite their relatively limited distribution, conifer forests in the watershed provided habitat for a wide range of species, including amphibians such as rough-skinned newt (*Taricha granulosa*) and California slender salamander (*Batrachoseps attenuatus*); small mammals such as western gray squirrel (*Sciurus griseus*); carnivores such as mountain lion (*Puma concolor*) and grizzly bear (*Ursus arctos californicus*); and birds such as Steller's Jay (*Cyanocitta stelleri*), Hermit Thrush (*Catharus guttatus*), Chestnut-backed Chickadee (*Poecile rufescens*), Pine Siskin (*Spinus pinus*), Northern Saw-whet Owl (*Aegolius acadicus*), and Marbled Murrelet (*Brachyramphus marmoratus*; Burroughs 1928, data from Global Biodiversity Innovation Facility). The federally threatened and state endangered Marbled Murrelet, a seabird which nests in the canopy of old-growth and mature/late seral conifer trees in close proximity to the ocean, has been documented in Douglas-fir forests on the western side of Pilarcitos Canyon; critical habitat has been designated in this area under the Endangered Species Act (USFWS 1997, San Francisco Planning Department 2008, 50 CFR Part 17, Raphael et al. 2018).

Commercial logging by American settlers was widespread throughout the Santa Cruz Mountains in the mid-19th century, and old-growth trees were eliminated in many areas (see page 64). While early logging operations in the watershed were not nearly as extensive as in other parts of the region, they did have a substantial impact on the redwood forests along Kings Mountain Ridge (Stephens 1856a, Oberlander 1953, Stanger 1967), and may

Hermit thrush (*Catharus guttatus*). (Photo by Becky Matsubara, courtesy CC BY 2.0)

Looking south towards old growth Douglas-fir forest at the southern end of Fifield Ridge. *(Photo courtesy SFPUC)*

have resulted in a short-term decrease in overall conifer extent within the watershed. As the logging industry declined and SVWC bought up watershed lands, however, conifer forests began to expand. The lack of regular fires and other disturbances drove the encroachment of conifers into areas formerly occupied by shrublands and hardwood forests (and to a lesser extent, grasslands). In addition, intentional planting of species not indigenous to the watershed, such as Monterey cypress (*Hesperocyparis macrocarpa*) and Monterey pine (*Pinus radiata*), as well as native species such as Douglas-fir and coast redwood, has further expanded the distribution of conifer forests.

By 1946-48, conifer forests occupied an estimated 9% of the watershed, and by 2015-17 they had increased to 15% (see Fig. 8.1). Much of the increase in conifer forest has occurred in the areas adjacent to the historical "nodes" of redwood forest on Kings Mountain Ridge and Douglas-fir forest on the west side of Pilarcitos Canyon (see Fig. 8.2). Douglas-fir forest, for instance, has expanded northwards along the ridge south of Montara Mountain, and eastward along both the western and eastern slopes of Cahill Ridge, displacing chaparral and hardwood forest communities. Conifer forest has also expanded northward and eastward along Kings Mountain Ridge, similarly displacing shrubland and hardwood forest communities. Planted conifer trees have displaced native habitats in the northwestern part of the watershed, along the eastern slope of Kings Mountain Ridge, and in other areas. For instance, areas mapped as "Monterey Cypress Grove" in the modern vegetation mapping (Schirokauer et al. 2003) currently occupy at least 542 ac within the watershed.

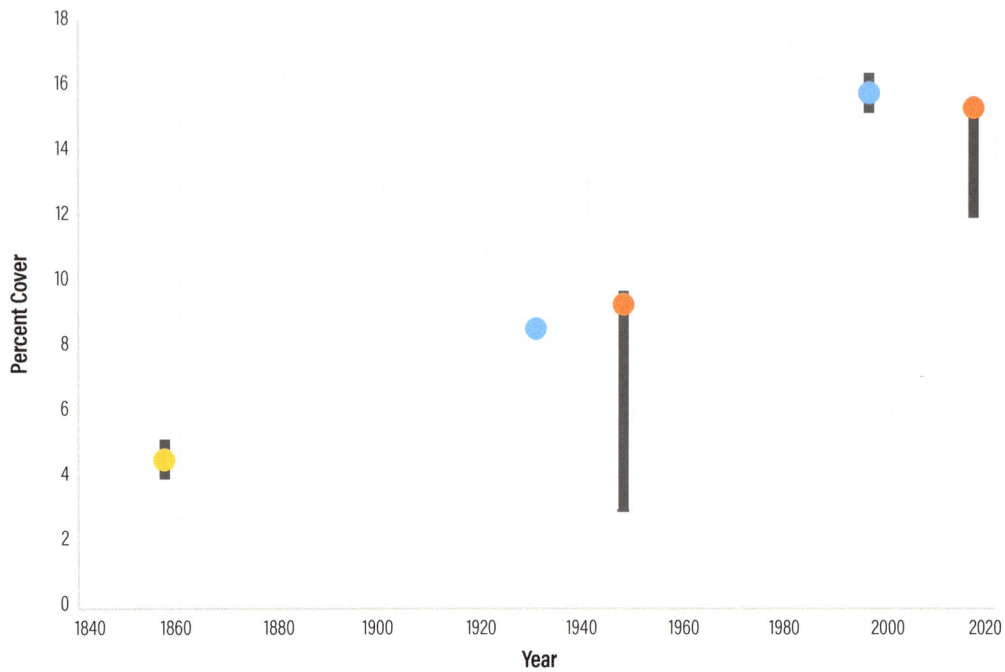

Figure 8.1. This diagram shows the change in estimated conifer forest extent over time throughout the study area, based on the three quantitative analyses described in Chapter 4: the aerial imagery point analysis (ca. 1947 and ca. 2016), vegetation mapping comparison (ca. 1930 and ca. 1995), and GLO survey analysis (ca. 1857). Gray boxes around the ca. 1947 and ca. 2016 points represent the range in the conifer forest frequency classified by four mappers for 200 overlapping points in the aerial imagery point analysis (see pages 27-29). The gray box around the ca. 1995 point represents the uncertainty associated with the forest (hardwood or conifer) class in the modern vegetation mapping (see pages 30-31). The gray box around the ca. 1857 point represents the uncertainty associated with GLO points classified as hardwood/conifer forest (see pages 32-34).

GLO Survey Analysis

Vegetation Mapping Comparison

Aerial Imagery Point Analysis

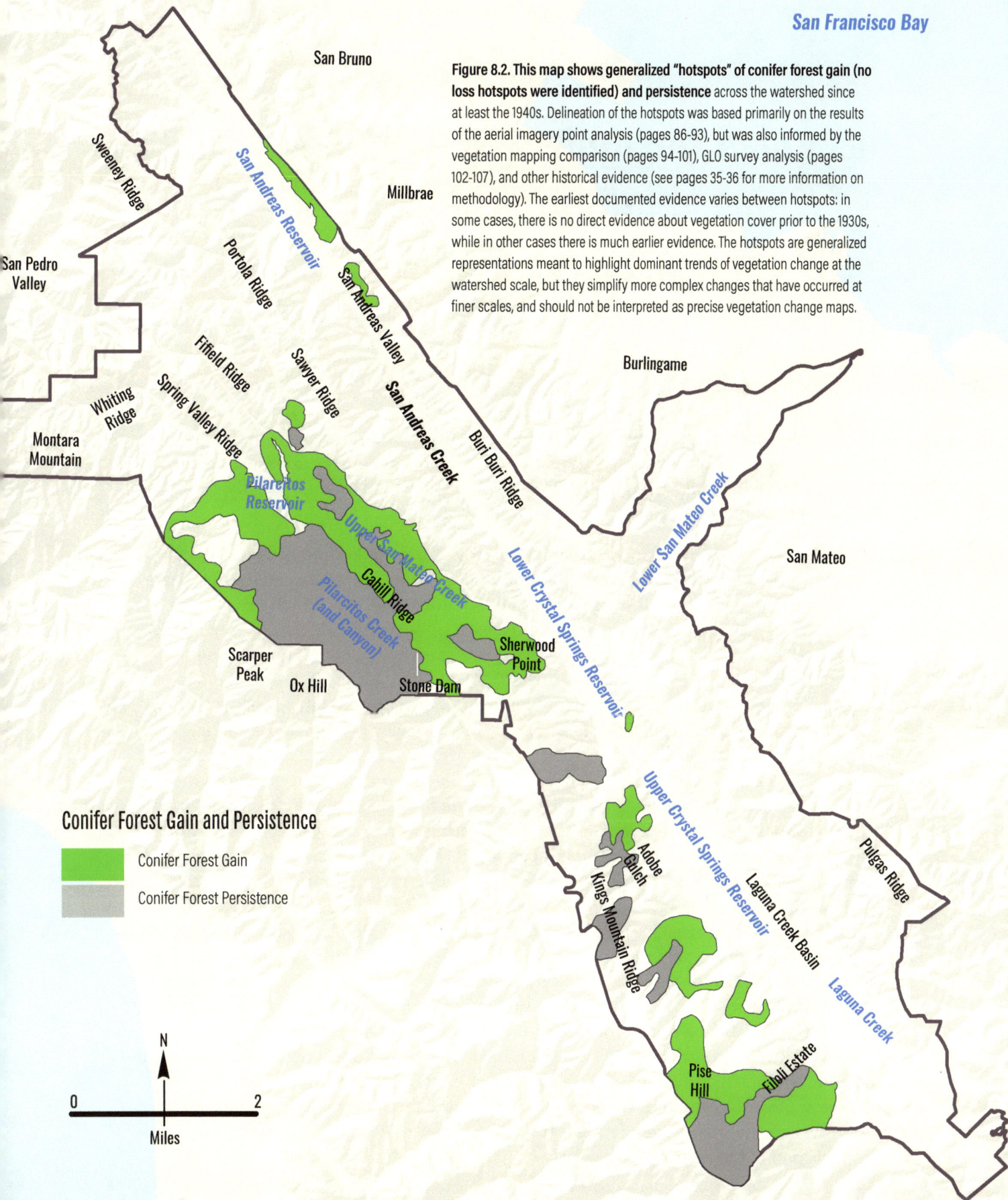

San Francisco Bay

San Bruno

Millbrae

Figure 8.2. This map shows generalized "hotspots" of conifer forest gain (no loss hotspots were identified) and persistence across the watershed since at least the 1940s. Delineation of the hotspots was based primarily on the results of the aerial imagery point analysis (pages 86-93), but was also informed by the vegetation mapping comparison (pages 94-101), GLO survey analysis (pages 102-107), and other historical evidence (see pages 35-36 for more information on methodology). The earliest documented evidence varies between hotspots: in some cases, there is no direct evidence about vegetation cover prior to the 1930s, while in other cases there is much earlier evidence. The hotspots are generalized representations meant to highlight dominant trends of vegetation change at the watershed scale, but they simplify more complex changes that have occurred at finer scales, and should not be interpreted as precise vegetation change maps.

Burlingame

Sweeney Ridge

San Pedro Valley

San Andreas Reservoir

Portola Ridge

San Andreas Valley

San Andreas Creek

Buri Buri Ridge

Lower San Mateo Creek

San Mateo

Fifield Ridge

Sawyer Ridge

Whiting Ridge

Spring Valley Ridge

Pilarcitos Reservoir

Upper San Mateo Creek

Lower Crystal Springs Reservoir

Montara Mountain

Cahill Ridge

Pilarcitos Creek (and Canyon)

Sherwood Point

Scarper Peak

Ox Hill

Stone Dam

Upper Crystal Springs Reservoir

Adobe Gulch

Kings Mountain Ridge

Laguna Creek Basin

Pulgas Ridge

Conifer Forest Gain and Persistence

Conifer Forest Gain

Conifer Forest Persistence

Laguna Creek

Pise Hill

Filoli Estate

N

0   2

Miles

Redwood-dominated forests occupied the upper portions of Kings Mountain Ridge in the southwestern corner of the watershed

Logging in the mid-19th century virtually eliminated the old growth redwood forest

Redwood-dominated forests occupied the southwestern part of the watershed historically, along the upper portions of Kings Mountain Ridge. As noted in Chapter 7, Friar Francisco Palóu observed these forests from the location where his expedition entered the Laguna Creek Basin in November of 1774, describing "high mountains... all grown over with red-wood, white and live-oaks, and other trees" (Palóu et al. 1969).

Commercial logging operations, beginning as early as the 1830s, removed nearly all of the old-growth redwood trees from this part of the Peninsula (see page 64). While there is limited data about the pre-logging extent of redwood forests within the study area, several mid-19th century sources provides clues. GLO surveyor C. C. Tracy, for instance, reported "enter[ing] redwoods" as he crossed eastward over Kings Mountain Ridge in the southwestern corner of the watershed in 1852 (Tracy 1853b). Surveyor Thomas Stephens, while surveying along Kings Mountain Ridge approximately 1.5 miles south of the watershed boundary in 1856, described the area "beginning here and extending for two miles [north], to the top of the mountain" as "densely timbered with redwood and live oak, principally the former" (Stephens 1856a). He noted that "great quantities have been cut and used, leaving the ground a chereau de friese [chevaux de frise] of stumps and tree tops." (A "Chevaux de frise" is a French term for an anti-cavalry barrier, historically made of logs and pointed stakes. In this context, it likely refers to "slash" and other debris from logging operations.) Approximately 2 miles further north, he reported that he was "along edge of woods." After continuing on for another 0.8 miles, Stephens encountered a "U.S. Coast Survey Signal from Peases," referring to the Pise Hill U.S. Coast Survey station marker established in 1854. (James Pease was an early settler in the region; "Pise" is likely a misinterpretation of "Pease"; Trevenon 2008). A 1910 U.S. Coast and Geodetic Survey publication describes the location of the Pise Hill station as "on the top of the range of redwood hills, about a quarter of a mile from the northwest terminus of the redwood forest" (Duvall and Baldwin 1910). While it is unclear from this description as to whether the station was north or south of the "northwest terminus of the redwood forest," Stephen's survey notes suggest that the station was north of the forest edge. As he continued northwards along the ridge, Stephens (1856a) reported that "there are no longer any trees on this part of the mountain."

Conifer Forests

Despite the early impacts of logging, observers throughout the late 19th and 20th centuries continued to report second-growth redwood forests in this part of the watershed (Fig. 8.3). Describing the view from the Crystal Springs area in the 1860s, for instance, Brace (1869) wrote, "On the far hills, toward the south [view from Crystal Springs/San Mateo Creek canyon], we can see the gigantic trunks of the superb 'Red Wood' (*Sequoia sempervirens*)." Witnesses in the 1907-14 court case described "groves of redwood" at the top of Kings Mountain Ridge in the southwest corner of the watershed, and noted that these were "second growth redwoods" (parcel 194; see Appendix B). These redwoods extended north into the "Husing" tract, which also had "quite a grove of redwoods." A 1912 map of the Husing tract labels upper portions of the slope as "heavy timber," middle portions of the slope as "redwoods and oaks," and lower portions of the slope as "oaks" and "small oaks" (Fig. 8.4). Groves of redwoods also occupied portions of canyons further north along the ridge: parcel 50, for example, contained a "deep gulch in which are some few redwood well up the mountain, and firs."

Redwoods on Kings Mountain Ridge were affected by the massive fires of September 1889, which burned thousands of acres throughout the watershed (see page 52). The *San Francisco Chronicle* (1889b), for instance, reported, "Those woods over there on the O'Connor ranch were blazing mighty lively. You see some of these redwoods are pretty tall. Well, the flames roared up fifty feet higher than the top of the tallest of them." The O'Connor ranch extended from the top of Kings Mountain Ridge near the southwestern corner of the watershed down to the Laguna Creek Basin (Easton 1868). Long-time resident Nate Comstock, who "had lived in the mountains nearby for thirty-two years," stated that "in all his time in the mountains [he had] never seen such heavy fires" (*San Francisco Chronicle* 1889b). This fire may have in turn spurred re-sprouting and secondary growth of redwoods, as they are a fire-adapted species that commonly exhibits robust regeneration after fire (Griffith 1992, Lazzeri Aerts and Russell 2014).

While logging resulted in a decrease in the extent of redwood forests on Kings Mountain Ridge during the mid-19th century, since the 1930s conifer forests have expanded substantially in this area. Both the aerial imagery point analysis and the vegetation mapping comparison show many areas where hardwood forest, and to a lesser extent shrubland and grassland, have converted to conifer forest in this part of the watershed. Interestingly, while the 1930s Wieslander mapping shows nearly all of the conifer forest in this area dominated by redwood, with just a small patch dominated by Douglas-fir, the modern vegetation mapping indicates that the vast majority of the conifer forest in this part of the watershed today is dominated by Douglas-fir, with a small amount of remaining redwood. Oberlander (1953) noted that oak-madrone stands along Kings Mountain Ridge were being outcompeted by Douglas-fir in a number of areas.

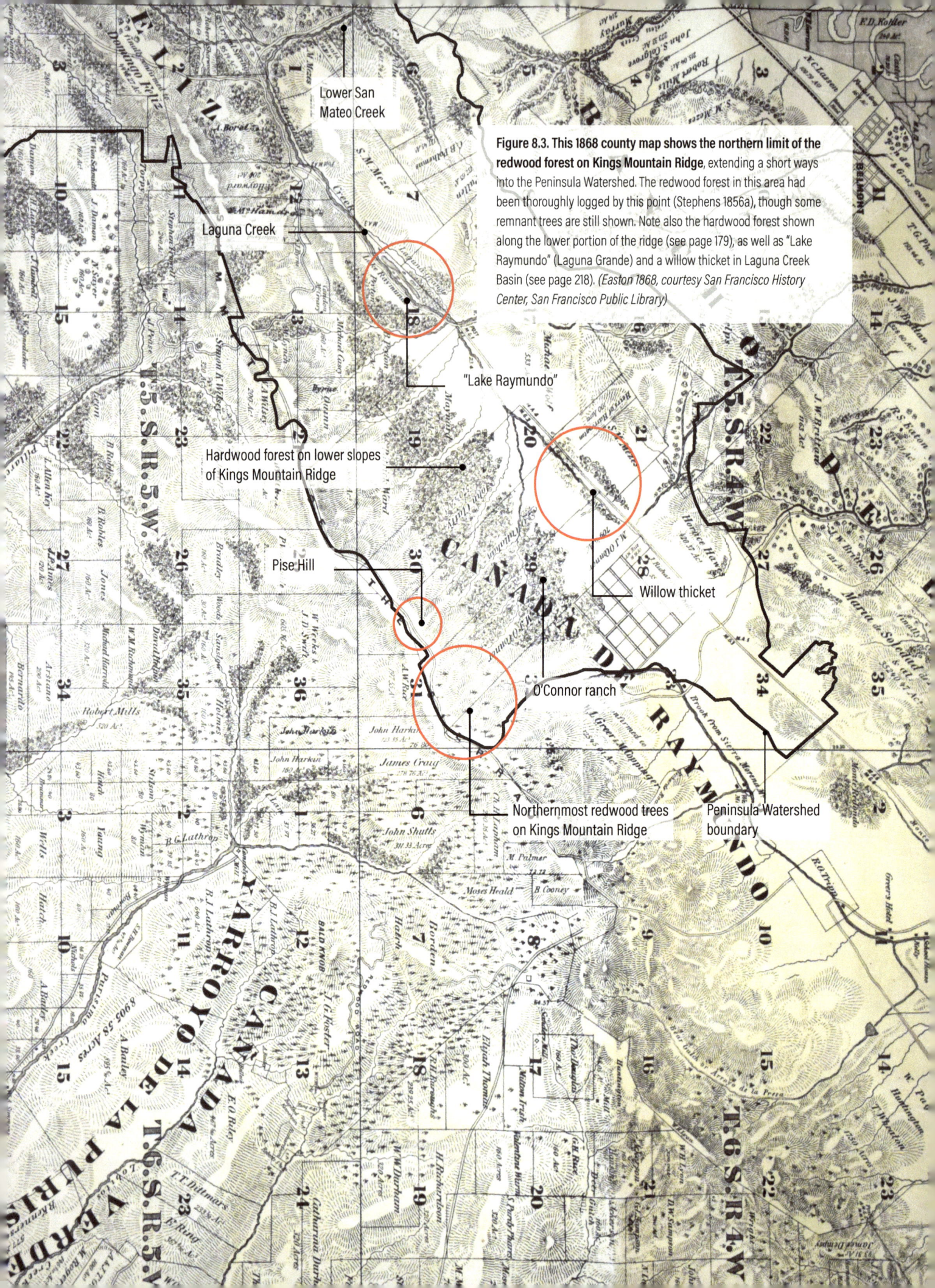

Figure 8.3. **This 1868 county map shows the northern limit of the redwood forest on Kings Mountain Ridge**, extending a short ways into the Peninsula Watershed. The redwood forest in this area had been thoroughly logged by this point (Stephens 1856a), though some remnant trees are still shown. Note also the hardwood forest shown along the lower portion of the ridge (see page 179), as well as "Lake Raymundo" (Laguna Grande) and a willow thicket in Laguna Creek Basin (see page 218). *(Easton 1868, courtesy San Francisco History Center, San Francisco Public Library)*

Lower San Mateo Creek

Laguna Creek

"Lake Raymundo"

Hardwood forest on lower slopes of Kings Mountain Ridge

Pise Hill

Willow thicket

O'Connor ranch

Northernmost redwood trees on Kings Mountain Ridge

Peninsula Watershed boundary

Figure 8.4. "Map of the land of E.A. Husing" along Kings Mountain Ridge, 1912. The western (upslope) portion of the parcel is labeled "heavy timber and "redwoods and oaks." The eastern (downslope) portions of the parcel are labeled "oaks" and "small oaks." *([Box] Maps 001, 1890-1913, [item] 623, courtesy SFPUC)*

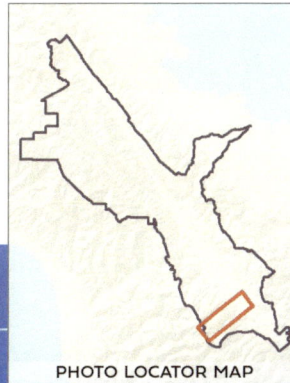

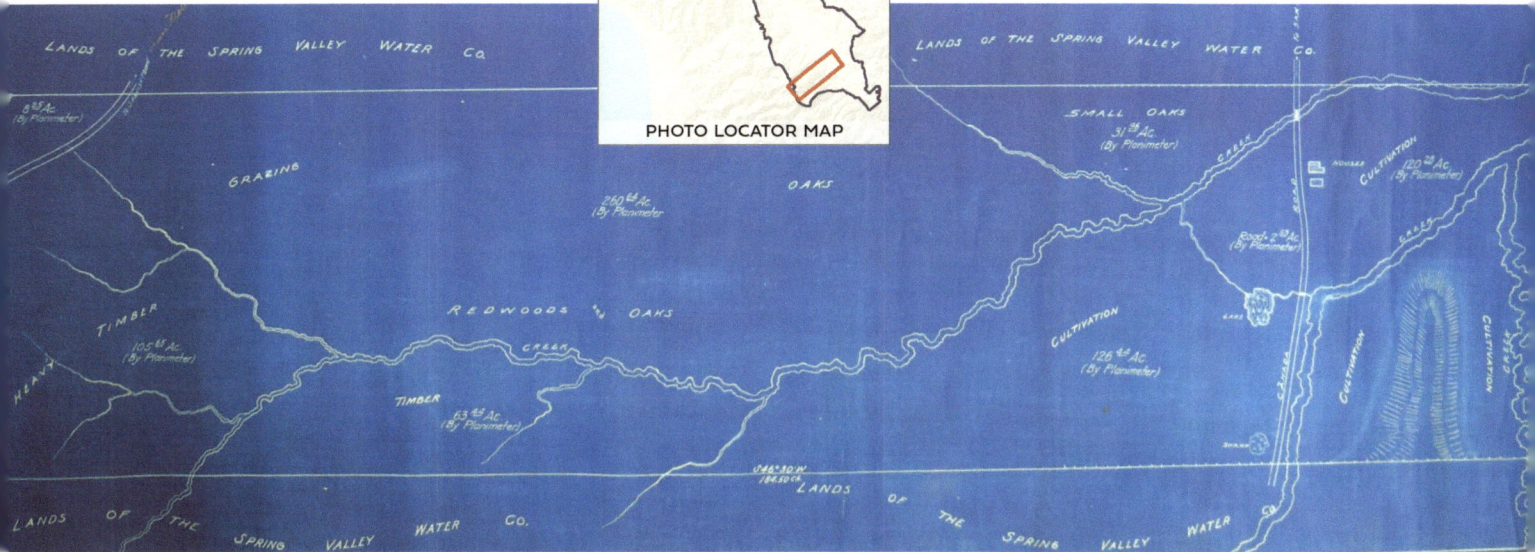

PHOTO LOCATOR MAP

Timber mill in the Santa Cruz Mountains, late 19th century. *(1964-071_OVLange, Photo courtesy San Mateo County Historical Association)*

Douglas-fir-dominated forests occupied hillslopes on the western side of Pilarcitos Canyon

These forests have persisted, and Douglas-fir has expanded substantially into surrounding areas

The greatest concentration of conifer forest in the watershed historically was located around Pilarcitos Canyon. The western side of Pilarcitos Canyon supported a large stand dominated by Douglas-fir along with smaller groves of redwoods. Scattered groves of Douglas-fir, intermixed with shrubland and hardwood forest, were also present to a lesser extent along the crest of Cahill Ridge on the east side of Pilarcitos Canyon.

Situated in a relatively remote portion of the watershed, the conifer forests in this area escaped the notice of the early Spanish explorers passing through San Andreas Valley. Thus, the earliest descriptions of these habitats come from observers in the mid-19th century. One early glimpse comes from Calvin Brown, chief engineer for the Spring Valley Water Company during the construction of Pilarcitos Dam in the 1860s, who described the vegetation and wildlife in Pilarcitos Canyon downstream of the dam:

> Both sides of the gorge or huge ravine... were formed by lofty ridges varied in slopes and outlines by multifarious swelling masses of hills, their sides in some places clothed in tall trees of fir, alder, madrone, and bay, and in others by a lower growth of oaks and bushes with alternations of bare patches... Looking down the ravine from the site of the dam the eye wandered over a long reach of forest both in the bottom and upon the slopes of the ridges... Until its discovery as a promising water source it had remained a secluded and safe retreat for coyotes, pumas, big wildcats, and the formidable grizzly bears... Deer also abounded in this region, and occasionally were still to be seen, as well as smaller game, its whole native fauna in variety and abundance testifying to the wild solitude of the locality. (Brown in Lawrence 1922b)

GLO surveyors in the mid-19th century also noted the presence of extensive conifer forests on the western side of Pilarcitos Canyon, intermixed with oak, madrone, and other hardwood trees. Just to the northeast of Scarper Peak, for instance, surveyor William Lewis (1860) described a "fine grove of large red wood trees"; this location corresponds fairly closely to an existing grove of redwoods shown in the modern vegetation mapping. Four years later, surveyor A. W. Von Schmidt (1864b) described "timber oak, redwood, madrona [sic], and pine" in the area just south of Pilarcitos Dam. Further south, in the area just southeast of Ox Hill, he again noted "timber, pine, oak and madrona [sic]." The references to pine are questionable, as later sources give no indication of pines in this area, and may represent a misidentification of Douglas-fir. This appears particularly likely in light of the surprising lack of references to Douglas-fir in GLO descriptions of this area, despite other mid-19th through early 20th century evidence of extensive stands of fir (e.g., Brown in Lawrence 1922b, Davy 1895, SVWC vs. San Francisco 1916; see below), as well as later documentation of old growth Douglas-fir trees that, based on their size, would likely have been present during this period (e.g., Oberlander 1953).

Scattered fir trees and smaller groves of conifer forest were also present on the eastern side of Pilarcitos Canyon, though shrubland was the dominant vegetation type in this area (see page 149). As noted previously, for instance, GLO surveyor Robert Matthewson (1861) descibed "chaparral... [with] wild plum... lilac... fir... oak" while surveying along Cahill Ridge in 1861, and scattered conifers are visible in mid-19th through early 20th century drawings and photographs of this region (see Fig. 6.6, Fig. 6.7, and Fig. 6.8).

Massive old growth and mature/late seral Douglas-firs and other trees in this part of the watershed created unique climatic conditions, maintaining a relatively cool and moist understory year round. High levels of precipitation in this area (relative to other parts of the watershed; see Fig. 1.2) were substantially augmented by fog drip (see Fig. 1.3). On Cahill Ridge, for instance, Oberlander (1956) found that summer fog drip could contribute over 50 in of precipitation in just over a month, and reported that "the condensation appeared to provide the conditions for the orchids *Epipactis gigantea* and *Eburophyton austinae* [*Cephalanthera austiniae*, phantom orchid], since these plants were found exclusively in these moist ridge tops." Botanist J. Burtt Davy further described the understory conditions and plant composition of the old growth Douglas-fir forests while traveling south between Montara Mountain and Scarper Peak in 1895:

> Following the summit of this ridge [south]... we... reached the *Pseudotsuga* [Douglas-fir]... trees on which is growing in such luxuriance *Polypodium Scouleri* in great masses to a height of 25 ft. up the trunks... *Antitrichia curtipendula* var. *gigantea* grows on these trees in great masses + to great heights up the trees, forming broad paddings or cushions on the branches. The soil is rich with decayed vegetable matter + the whole area where there are trees is kept humid by the ocean fogs, the highest row of trees having the richest growth of mosses... The undergrowth is dense + consists largely of *Rubus parviflorus*, *Sambucus callicarpa* [*Sambucus racemosa* var. *racemosa*, Pacific red elderberry] + *Corylus* [*Corylus cornuta*, beaked hazelnut], with some *Euonymus* [*Euonymus occidentalis*, burning bush]. (Davy 1895)

Early 20th century sources, such as parcel descriptions from the 1907-14 SVWC court case, attest to the persistence of conifer forests on the western side of Pilarcitos Canyon. The hillslope southeast of Pilarcitos Dam supported "a heavy growth of fir, oak, madrone, etc., with alder and bay in the gulches" (parcel 5-2; Fig. 8.5; see Appendix B). The parcel just east of Scarper Peak was "heavily wooded with fir" on the eastern portion, and "covered with brush and scattering firs" on the western portion (parcel 124). The redwood grove noted by GLO surveyor Lewis (1860) was also located in this area: describing the view from Scarper Peak in 1902, for instance, William A. Brewer wrote, "Down the eastern slopes... we behold great patches of virgin forest, the stately redwoods marching up each steep canyon" (Brewer 1903). Tree density was particularly high in ravines and stream canyons: the gulch running through parcel 3 (draining the northeast side of Ox Hill) was "heavily wooded with fir," while other parts of the parcel were "brushy, with occasional fir and hardwood timber" (see Fig. 8.5d). Similarly, parcel 6 to the south was "heavily wooded with fir, and exceedingly precipitous," except for a "central spur" that was "brush-covered, with small clearings" (Fig. 8.6). The timber thinned out further to the south: the hillslope to the west of Stone Dam supported a "growth of fir about half-way to the crest of the Albrecht Ridge" (parcel 5-2), while the upper portions of the hillslope were "covered in dense chaparral and some fir timber" (parcel 60; Fig. 8.7).

Figure 8.5. Douglas-fir-dominated conifer forests on the west side of Pilarcitos Canyon, just southeast of Pilarcitos Dam. A) "View down the valley" south of Pilarcitos Dam, ca. 1865. B) "East slope of Montara Mountains between the Pilarcitos and Stone Dams," 1907-14. C) Scarper Peak area just south of Pilarcitos Dam, with Pilarcitos Cottage in foreground, 1907-14. D) "Panorama looking southwest and south" showing "Douglas Fir and coast live oak - California laurel - madrone woodland," 1932. *(A: Carleton Watkins (American, 1829 - 1916), [View down the Valley], about 1867, Albumen silver print, 84.XC.870.576, The J. Paul Getty Museum, Los Angeles; B: Photo F3725_1180-28, courtesy California State Archives; C: Photo F3725_1180-26, courtesy California State Archives; D: Photo number 286966, 286967, courtesy Wieslander Vegetation Type Mapping Collection, Marian Koshland Bioscience and Natural Resources Library, UC Berkeley, www.lib.berkeley.edu/BIOS/vtm/)*

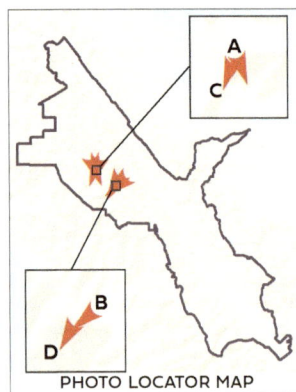

PHOTO LOCATOR MAP

A. ca. 1865

B. 1907-14

C. 1907-14

D. 1932

Wieslander VTM mapping from ca. 1930 documents a large swath of Douglas-fir-dominated conifer forest along the eastern side of Pilarcitos Canyon, extending from just south of Pilarcitos Dam past the point where Pilarcitos Creek exits the study area. Photographs taken by Wieslander VTM surveyors at various locations along Cahill Ridge and Fifield Ridge show these forests, in many places intermixed with coast live oak, California bay, Pacific madrone, coyote brush, and California coffeeberry (Fig. 8.8; see also Fig. 8.5d and Fig. 6.8 on page 157). Oberlander (1953) describes the Douglas-fir trees on "Cahill and Montara ridges in the vicinity of Pilarcitos Canyon" as "virgin trees... 3-5 feet in diameter."

Conifer forests in this part of the watershed have persisted throughout the 20th century, and indeed have expanded substantially into surrounding areas around Scarper Peak west of Pilarcitos Reservoir and along both slopes of Cahill Ridge east of Pilarcitos Canyon (see Fig. 8.2). This encroachment is visible in rephotographs as well as comparison of historical and modern aerial photos and comparison of Wieslander VTM and modern vegetation mapping. To the west and northwest of Pilarcitos Dam, Douglas-fir-dominated conifer forest has encroached into areas historically occupied by coyote brush-dominated shrubland (Fig. 8.9). Douglas-fir forest has also encroached into shrublands on the southern portion of Spring Valley Ridge (Fig. 8.10), on the hills to the east and northeast of Pilarcitos Dam (Fig. 8.11; see also Fig. 6.6 on page 156), and along the western slope of Cahill Ridge on the east side of Pilarcitos Canyon. On the eastern slope of Cahill Ridge in Upper San Mateo Creek Canyon, conifer forest has expanded into areas historically dominated by hardwood forest.

Figure 8.6. (left) View south around Scarper Peak, 1907-14, showing conifer forest on parcels 6 and 60 (1907-14 SVWC court case). *(Photo F3725_1180-34, courtesy California State Archives)*

PHOTO LOCATOR MAP

Figure 8.7. A) Douglas-fir in vicinity of Stone Dam, ca. 1870. B) Upper hillslope to the west of Stone Dam, 1907-14. *(A: Photo D-361, courtesy SFPUC; B: Photo F3725 1180-32, courtesy California State Archives)*

A. ca. 1870

B. 1907-14

PHOTO LOCATOR MAP

Figure 8.8. A) "Looking south from point on [Fifield] ridge, showing east arm of Pilarcitos Lake. Shows Douglas fir, woodland of coast live oak, madrone and California laurel and coastal sagebrush type of *Baccharis pilularis* and *Rhamnus californica [Frangula californica]*," 1932. B) "Cahill Ridge... Panorama looking NW to S. Shows Douglas fir type and coastal sagebrush of *Baccharis pilularis* and *Rhamnus californica*," 1932. (A: Photo number 286968; B: Photo numbers 286963, 286964, 286965; both images courtesy Wieslander Vegetation Type Mapping Collection, Marian Koshland Bioscience and Natural Resources Library, UC Berkeley, www.lib.berkeley.edu/ BIOS/vtm/)

REPHOTOGRAPHY

ca. 1910

2018

PHOTO LOCATOR MAP

Figure 8.9. Rephotograph pair (ca. 1910 to 2018) showing encroachment of conifer forest into coyote brush-dominated shrubland on the west side of Pilarcitos Dam. *(Top: Photo D-1164, courtesy SFPUC; Bottom: Photo by SFEI-ASC, August 2018)*

REPHOTOGRAPHY

1907-14

2018

Figure. 8.10. Rephotograph pair (1907-14 to 2018) showing encroachment of conifer forest into shrubland on the southern portions of Spring Valley and Fifield ridges. *(Top: Photo F3725_1180-30, courtesy California State Archives; Bottom: Photo by SFEI-ASC, August 2018)*

Figure 8.11. Rephotograph pair (1936 to 2018) showing encroachment of conifer forest into shrublands on east side of Pilarcitos Dam. *(Top: Photo D-3093, courtesy SFPUC; Bottom: Photo by SFEI-ASC, August 2018)*

PHOTO LOCATOR MAP

Intentional planting of Monterey cypress, Monterey pine, Douglas-fir, and other species has expanded the distribution of conifer and hardwood forests

Plantings of conifer trees have replaced grasslands, shrublands, and hardwood forest adjacent to these areas

In addition to the expansion of conifer forest into areas formerly occupied by grasslands, shrublands, and hardwood forest through successional processes, deliberate introduction and planting of both native and non-indigenous tree species has led to the expansion of both conifer and hardwood forests in many parts of the watershed. In particular, Monterey cypress and Monterey pine have been planted (and have subsequently expanded) in several locations across the eastern slope of Kings Mountain Ridge and the western slope of Buri Buri Ridge. Planted stands of Douglas-fir and fir (*Abies* spp.) exist in a number of locations as well (see page 66).

Multiple stands of Monterey cypress and Monterey pine were planted along the eastern slope of Kings Mountain Ridge during the late 19th or early 20th century. A number of these stands are described by witnesses in the 1907-14 SVWC court case; in many instances these witnesses refer to "pines and cedars," though given the current prevalence of Monterey cypress in these areas it appears that the references to "cedars" were misidentifications. For instance, one witness noted that parcel 122, just to the north of present-day Highway 92 (see Appendix B), "has been farmed [and] has new growth of cypress and pine trees"; another witness stated that "the upper portion [of the parcel] has been forested with cedars" (Fig. 8.12). Just to the south, the upper portion of parcel 68.3 was likewise "planted with scattering cedar." To the south of Highway 92, witnesses stated that "a number of cedar trees have been planted" (parcel 132.1), that "several open spaces have been forested with cedar and pine," and that there was a "big gulch nicely timbered [with] cypress, laurel, oak, [and] eucalyptus trees in [the] north-easterly corner" (parcel 132.2; Fig. 8.13). Further south, witnesses stated that "the portion [of parcel 134] nearest

Figure 8.12 (above). Early landscape photos showing planted stands likely dominated by Monterey cypress and Monterey pine on Kings Mountain Ridge (black boxes). A) 1907-14 photo showing planted conifer forest on the upper portion of parcel 122 (see Appendix B). B) The same planted stand of conifers on parcel 122 is also visible in this 1907-14 panorama on the center right. *(Photos F3725_1180-17 and F3725_1180-16, courtesy California State Archives)*

Figure 8.13 (below). A rectangular stand of planted Montery cypress is visible in this 1907-14 photo looking south along the eastern slope of Kings Mountain Ridge. The original caption reads, "the cyprus [sic], shown at the right, is in the 160 acre parcel of 132" (see Appendix B). *(Photo F3725_1180-7, courtesy California State Archives)*

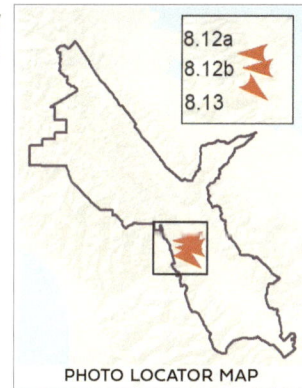

PHOTO LOCATOR MAP

the crest of the ridge has been planted to cedars," and that on the upper portion of [parcel 132.3] a number of pines and cedars have been planted."

The Wieslander VTM mapping shows a number of groves of conifers throughout the eastern slope of Kings Mountain Ridge, including several groves of Monterey cypress (misidentified as *Cupressus abramsiana*, a rare, endangered conifer endemic to the Santa Cruz Mountains). Based on their location relative to other vegetation types present during this period, as well as mid-19th century evidence, it is likely that these areas were originally dominated by shrubland, and in some places by hardwood forest (see chapters 6 and 7).

Monterey cypress, Monterey pine, and other species were also planted in a number of areas formerly dominated by grassland on the western slope of Buri Buri Ridge. Witnesses in the 1907-14 SVWC court case, for instance, describe a number of conifer plantings on the eastern side of San Andreas Reservoir and Lower Crystal Springs Reservoir. The keeper's cottage on the northeastern end of San Andreas reservoir was surrounded by "planted trees" that consisting of "pine and eucalyptus" (Parcel 15). To the south, portions of parcel 218 were described as "forested with cedar trees" (again, likely a misidentification of Monterey cypress), while the eastern portion of parcel 20 had similarly "been forested" with "cypress trees." Further to the south, along the eastern shore of Lower Crystal Springs Reservoir, witnesses reported that "the crest of [Buri Buri] ridge forms a windbreak particularly where the original owners planted groves of trees to increase the wind-break" (parcel 90). At Crystal Springs Country Club, at the perimeter of the watershed northeast of Lower Crystal Springs Reservoir, a 1921 planting plan documented a large quantity of trees to be planted, including hundreds of non-native hardwood trees and 1525 *Cupressus macrocarpa* (*Hesperocyparis macrocarpa*, Monterey cypress) trees. Multiple coast redwood were also to be planted at the golf club (Roeding 1921). F. W. Roeding, superintendent of the SVWC agricultural department, reported planting "quite an area to Monterey pines, cypress, etc... near the northerly end of Crystal Springs Lake" around 1910 (Roeding 1931). Another stand of Monterey cypress were planted on the southeastern side of San Andreas Reservoir in the 1950s by a group of Boy Scouts led by Dr. Bill Friedman (J. Avant, pers. comm.).

In addition to Monterey cypress and Monterey pine, Douglas-fir, fir (*Abies* spp.), and coast redwood have also been deliberately planted in some parts of the watershed. Dingman (2014), for instance, notes the presence of a small Douglas-fir planting north of Scarper Peak (west of Pilarcitos

Dam), reported to be "approximately 30-40 years of age." Numerous other plantings of Douglas-fir and or fir exist in the northwestern part of the watershed around Montara Mountain and along Spring Valley Ridge, Whiting Ridge, and northern Fifield Ridge (S. Simono, pers. comm.; see Fig. 3.13 on page 66). Oberlander (1953) also reports that "several Coast Redwood were planted in the early 1920's in a draw on the northwest slope of Sawyer Ridge." Field assessments of tree spacing, age structure, and other aspects of tree plantings could be employed to further document planted stands of Douglas-fir, coast redwood, and other species (see Chapter 10).

Monterrey cypress grove. *(Photo by KQED Quest, courtesy CC BY 2.0)*

The floor of the hollow is almost all of it lakes, swamps, and stream-beds, all grown over with tule patches and trees.

—FRANCISCO PALÓU DESCRIBING SAN ANDREAS VALLEY IN 1774 (PALÓU ET AL. 1969)

"Falls near Stone Dam." *(BANC PIC 1982.086 ALB, courtesy The Bancroft Library, UC Berkeley)*

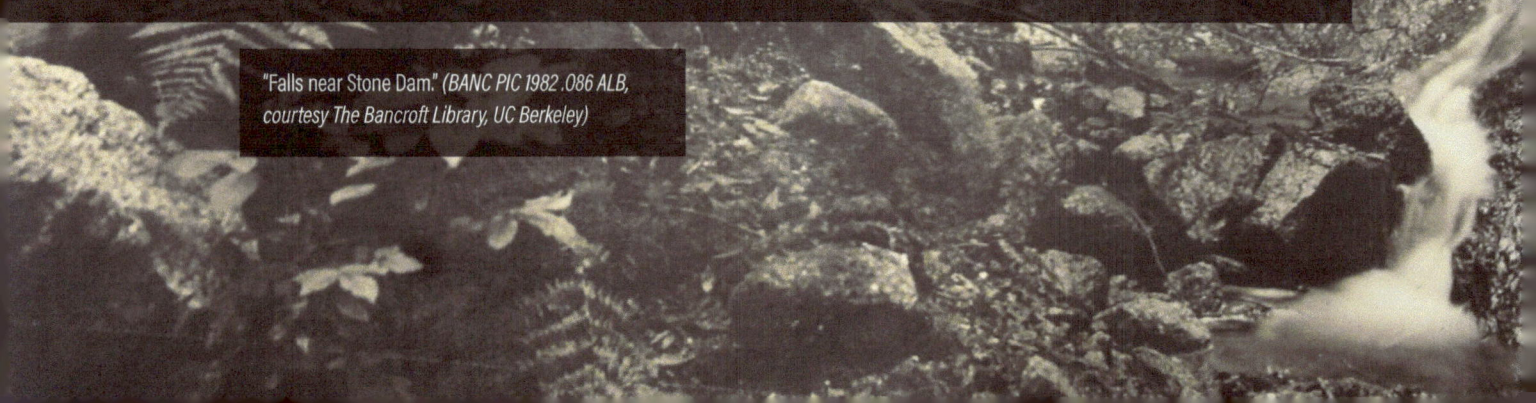

CHAPTER 9

Wetland and Riparian Habitats

Overview

Among the most drastic changes that have occurred within the Peninsula Watershed over the past two centuries are alterations in hydrology and loss of historical wetlands. Hydromodification, including construction of dams and water conveyance structures, occurred unusually early in the watershed (relative to other locations in the region), and had profound impacts on streamflow patterns, sediment dynamics, and the distribution of wetlands and riparian habitats (see pages 67-69 for more information on hydromodification). This chapter provides a brief description of historical hydrologic conditions and wetland/riparian habitat types in the watershed. Because the primary goal of this report was to examine shifts in terrestrial vegetation communities, the level of analysis in this chapter is more limited than in the preceding chapters.

A rich mosaic of wetland and riparian habitat types existed historically in the watershed, including sag ponds, tule and bulrush (*Schoenoplectus* spp.)-dominated freshwater emergent wetlands, tidal marsh and mudflats (at the mouth of San Mateo Creek), willow (*Salix* spp.) thickets, and diverse riparian forest and riparian scrub habitat types. Non-tidal wetlands were concentrated primarily in San Andreas Valley and the Laguna Creek Basin: historical accounts describe a heterogeneous mosaic of wetlands and riparian vegetation prior to substantial land transformation. For example, traveling through the valley in 1774 with the Palóu-Rivera Expedition, Francisco Palóu observed, "The floor of the hollow is almost all of it lakes, swamps, and stream-beds, all grown over with tule patches and trees" (Palóu et al. 1969). San Mateo Creek, San Andreas Creek, Pilarcitos Creek, and other streams within the watershed were characterized by pronounced seasonal variability in streamflow, and supported diverse riparian habitats ranging from dense willow scrub to forests dominated by mature oak and California bay trees (Fig. 9.1).

Wetlands and riparian forests provided habitat for a wide range of wildlife, including reptiles and amphibians such as California newt (*Taricha torosa*), rough-skinned newt (*T. granulosa*), aquatic gartersnake (*Thamnophis atratus*), western pond turtle (*Actinemys marmorata*),[1] foothill yellow-legged frog (*Rana boylii*),[1] California red-legged frog (*Rana draytonii*),[2] and San Francisco garter snake (*Thamnophis sirtalis tetrataenia*);[3] fish such as coho salmon (*Oncorhynchus kisutch*), steelhead (*Oncorhynchus mykiss*), California roach (*Hesperoleucus*

---

1 A CDFW Species of Special Concern.

2 Federally listed as Threatened.

3 Federally and State listed as Endangered.

symmetricus), and riffle sculpin (*Cottus gulosus*; see page 228); and aquatic, wetland, and riparian birds such as Common Loon (*Gavia immer*), Common Merganser (*Mergus merganser*), Virginia Rail (*Rallus limicola*), Wilson's Warbler (*Cardellina pusilla*), Swainson's Thrush (*Catharus ustulatus*), Belted Kingfisher (*Megaceryle alcyon*), and numerous other species (Burroughs 1928, data from Global Biodiversity Information Facility). Early observers commented on the immense numbers of birds that flocked to wetlands in the San Andreas Valley. Father Crespí, for instance, traveling through San Andreas Valley in November of 1769, reported that the "lake" in the valley supported "countless ducks, cranes, geese, and other fowl" (Crespí and Brown 2001).

Common Merganser (*Mergus merganser*). (Photo by Alan D Wilson, courtesy CC BY 2.5)

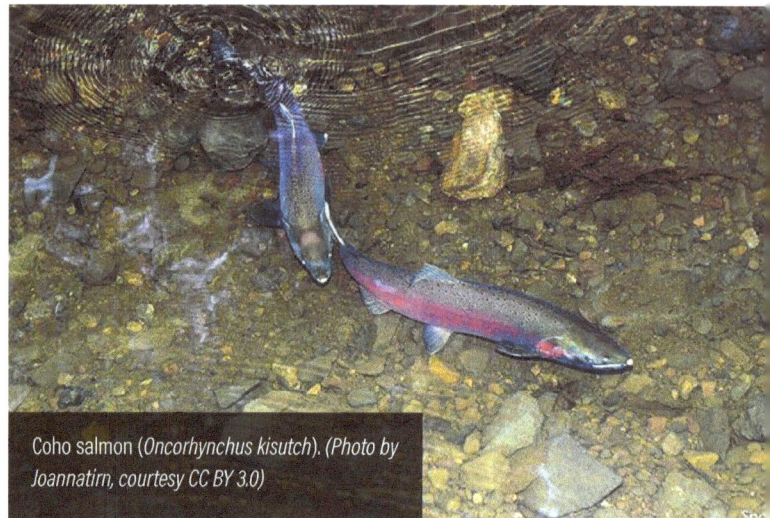

Coho salmon (*Oncorhynchus kisutch*). (Photo by Joannatirn, courtesy CC BY 3.0)

California newt (*Taricha torosa*). (Photo by Steve Jurvetson, courtesy CC BY 2.0)

Historical Hydrology

Seasonal streamflow within the watershed varied substantially

Streamflow within the watershed varied seasonally, though early evidence suggests that the larger creeks in the watershed generally maintained at least a minimal baseflow during the dry season. On San Mateo Creek, for instance, an 1874 report found that "in Summer the water of San Mateo creek sinks low in its gravel bed, but in the Winter it is such a turbulent and fierce little stream that it makes nothing of sweeping down half a dozen bridges during the season" (*San Francisco Chronicle* 1874). Accounts from the 1880s and 90s suggest that there was little to no dry season surface flow (Mendell 1885, Dandridge 1889, *San Francisco Call* 1913), but these observations were made after major dams were constructed at San Andreas (1868) and Crystal Springs (1877-1890), which would have dramatically altered flow regimes downstream.

Streamflow in Pilarcitos Creek also varied significantly by season. In the mid-1800s it "flowed with a rapid and large volume [in the winter months]... In the long dry season, however, it dwindled down to a feeble run" (Brown in Lawrence 1922b). Tributaries to Pilarcitos Creek (and Pilarcitos Reservoir) were intermittent: an 1891 newspaper article reports that "later in the season... the small streams which feed the lake run dry" (*San Francisco Call* 1891). Laguna Creek may have maintained more dry season flow than other streams in the watershed: an account of the creek during the 1880s describes it as "a busy creek that sang and gurgled with a wealth of pure water even in the late summer" (Burke 1926).

Figure 9.1. 1868 county map showing confluence of San Andreas, San Mateo, and Laguna creeks to the west of present-day Lower Crystal Springs Dam. *(Easton 1868, courtesy San Francisco History Center, San Francisco Public Library)*

Wetland Habitats

Sag ponds and freshwater wetlands were dominant and persistent features in San Andreas Valley and the Laguna Creek Basin

Within San Andreas Valley and the Laguna Creek Basin, a series of open water/wetland complexes occurred along the path of the San Andreas Fault, in the areas now occupied by the San Andreas and Crystal Springs reservoirs. The lowest portions of the valley supported persistent sag ponds, which are bodies of freshwater that form in depressions within a fault zone. These sag ponds were often bordered by extensive wetland and riparian areas, including tule-dominated freshwater emergent marsh, willow thickets, and willow riparian scrub.

Laguna Grande, a large sag pond located along Laguna Creek at the site of current Upper Crystal Springs Reservoir, measured approximately 300 m wide by 1700 m long (Wackenreuder 1855, Scowden 1875) and occupied over 100 ac. Described by early observers as a "very long lake" (Font 1776) or a "grassy lagoon" (Tracy 1852), Laguna Grande is a prominent feature in depictions of the valley in Mexican diseños and other early maps (Fig. 9.2). It is variably characterized as a single large lake (e.g., Wackenreuder 1855) or as a mosaic of tule marsh and open water (e.g., Stevens 1856, Scowden 1875, Evans 1873), likely reflecting seasonal variability in the extent of inundation. The lake was large and deep enough to maintain perennial open water: testimony from an 1883 SVWC court case describes it as a "large deep hole" with water at least 4-5 ft deep, and in some places 10 or more feet deep, with "always a lot of water in it in summer months" (Kilsby 1883). Fed by a combination of springs and perennial flows from Laguna Creek (Burke 1926), Laguna Grande supported "clear and cold" water year-round (Kilsby 1883), and likely provided important rearing habitat for native salmonids (Leidy et al. 2005a,b).

Further north, sag ponds and freshwater emergent wetlands were also prominent along San Andreas Creek historically, and are consistently recorded in late 18th through mid-19th century sources (Figure 9.3). Traveling through the area in November 1774, for instance, Captain Francisco Rivera described a "small-sized lake" (Rivera et al. 1969); two years later, traveling with the Anza Expedition in March of 1776, Pedro Font described a "stream or long narrow lake" along the valley (Font and Brown 2011). These descriptions are consistent with the depiction in several early maps (e.g., Schussler 1867, Easton 1868). Similar to Laguna Grande, these ponds were sometimes depicted as a mosaic of wetlands and open water habitat types. An early U.S. Coast Survey map (USCS 1869) shows several sag ponds with marsh vegetation symbology, while Hoffman (1867) shows "tules chaparral and lagunas" stretching across the valley floor between San Andreas and Crystal Springs.

With the construction of the Crystal Springs and San Andreas dams in the late 19th century (see page 67), much of the vast mosaic of wetlands in San Andreas Valley and the Laguna Creek Basin—sag ponds, emergent marsh, willow thickets, and riparian forest—was lost.

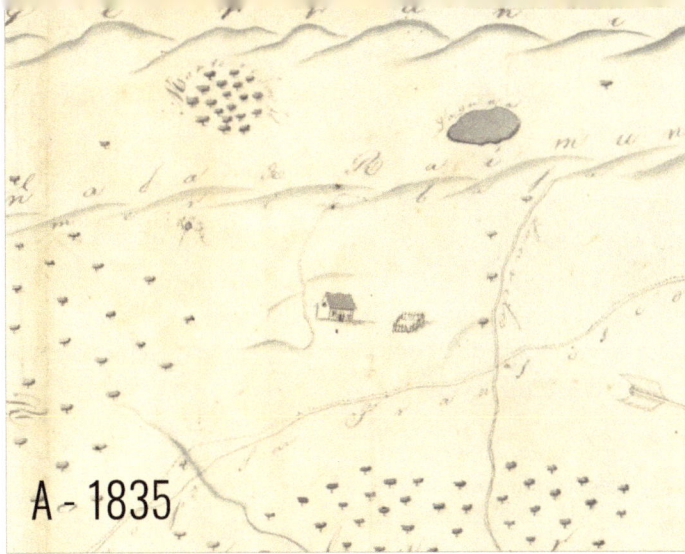

Figure 9.2. **Early maps of Laguna Grande,** a large sag pond that was located along Laguna Creek at the site of current Upper Crystal Springs Reservoir. While the depictions of Laguna Grande vary considerably, all of the maps show a persistent body of water in this location. A) A ca. 1835 diseño of Pulgas Rancho. B) An 1856 survey map of Pulgas Rancho. C) An 1855 county map. D) An 1875 map showing "proposed Crystal Springs Reservoir." *(A: Land Case Map A-121, courtesy The Bancroft Library, UC Berkeley; B: Stevens 1856, courtesy Bureau of Land Management; C: Wackenreuder 1855, courtesy San Francisco Department of Public Works; D: Photo G4363.S28N44 1875, courtesy The Bancroft Library, UC Berkeley)*

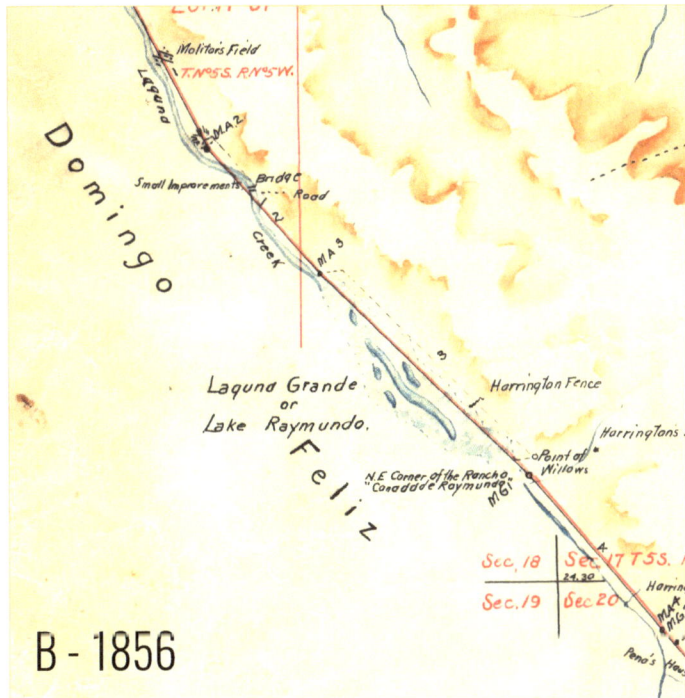

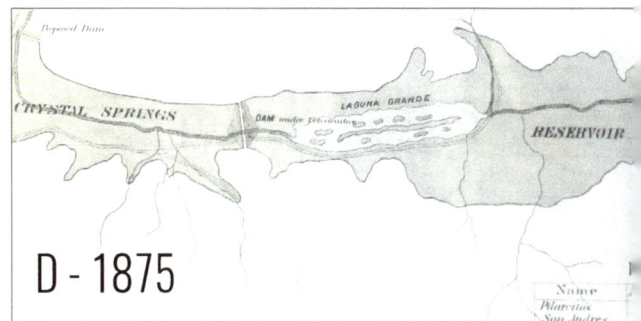

PHOTO LOCATOR MAP

Though most of the historical wetland area was eliminated, "novel" wetland habitats have formed along the margins of the reservoirs in some areas. In addition, remnants of some historical wetland habitats have persisted: for example, a large willow thicket still exists at the head of Upper Crystal Springs Lake where Laguna Creek flows into it, in much the same location as the willow thicket that existed there historically.

A - 1855

Figure 9.3. A chain of sag ponds, surrounded by emergent freshwater wetlands, was located along San Andreas Creek, as depicted in these early maps. A) An 1855 county map. B) An 1864 map of Buri Buri Rancho, showing "tules, chaparral and lagunas" along San Andreas Creek. C) An 1869 U.S. Coast Survey map of the San Francisco Peninsula. D) An 1868 county map. (A: Wackenreuder 1855, courtesy San Francisco Department of Public Works; B: U.S. Surveyor General's Office 1864, courtesy Bureau of Land Management; C: USCS 1869, courtesy David Rumsey Map Collection; D: Photo 912.794 Sa. 588, 50496, courtesy San Francisco History Center, San Francisco Public Library)

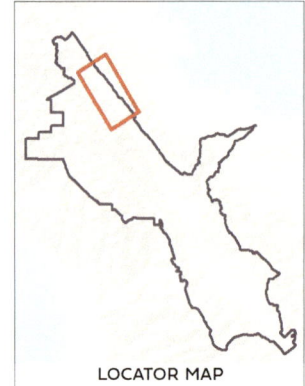

LOCATOR MAP

B - 1864

C - 1869

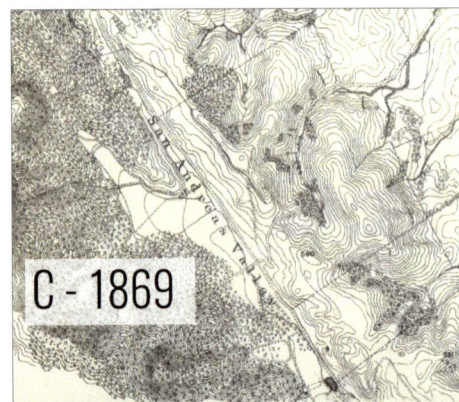

D - 1868

Springs were a common hydrological feature

Historically, numerous springs and seeps existed throughout the watershed, and in many cases they supported small wetland features. For instance, on Sneath's ranch near present-day Sneath Lane to the northwest of the watershed, springs were described as "so numerous... that in every field there is a trough into which pours constantly, throughout the year, a stream of pure cold water from a pipe connecting with a contiguous spring" (Alley 1883). Springs were also observed near the Jepson Laurel (the oldest and largest laurel in California), midway along the Sawyer Camp portion of the Crystal Springs Trail (Potter 1923), and in Upper San Mateo Creek Canyon between Fifield and Sawyer ridges (Reilly 1902). A 1902 map of the Fifield Dairy depicts numerous springs along the slopes of Fifield and Sawyer ridge (see Fig. 5.13 on page 132). Springs along faults are known to shift (both in terms of flow rate and location) in response to seismic activity (D. Freyberg pers. comm.).

Many water features on the landscape were described as spring-fed. For instance, describing San Andreas Valley in 1874, a reporter noted that "there are innumerable springs on the hill-sides, their waters flowing down the ravines and forming tiny lakes" (Vanderlip 1980). Botanist J. Burtt Davy observed in 1895 that "the [Crystal Springs and San Andreas] lakes are fed by the innumerable streamlets which flow into them," and reported finding Harlequin lotus (*Hosackia gracilis*; California Rare Plank Rank 4.2) in "a springy place" near San Andreas Reservoir (Davy 1895). Many of these springs likely supported small wetland complexes characterized by species such as fountain thistle (*Cirsium fontinale*), seep monkeyflower (*Erythranthe guttata*), and split awn sedge (*Carex tumulicola*). The type locality of fountain thistle, for instance, was collected in 1886 "at Crystal Springs... growing among the various springs and streamlets at the north side of the reservoir from which San Francisco is supplied with water" (Greene 1887); later collectors recorded the species "around spring[s] on serpentine," "along brooklet[s] on serpentine," and "in [a] moist gully, fed from spring seepage" around the Crystal Springs reservoirs (data from Consortium of California Herbaria). Split awn sedge was described as occupying "damp or springy" places (data from Consortium of California Herbaria).

Seep monkeyflower (*Erythranthe guttata*), Crystal Springs. *(Photo by SFPUC)*

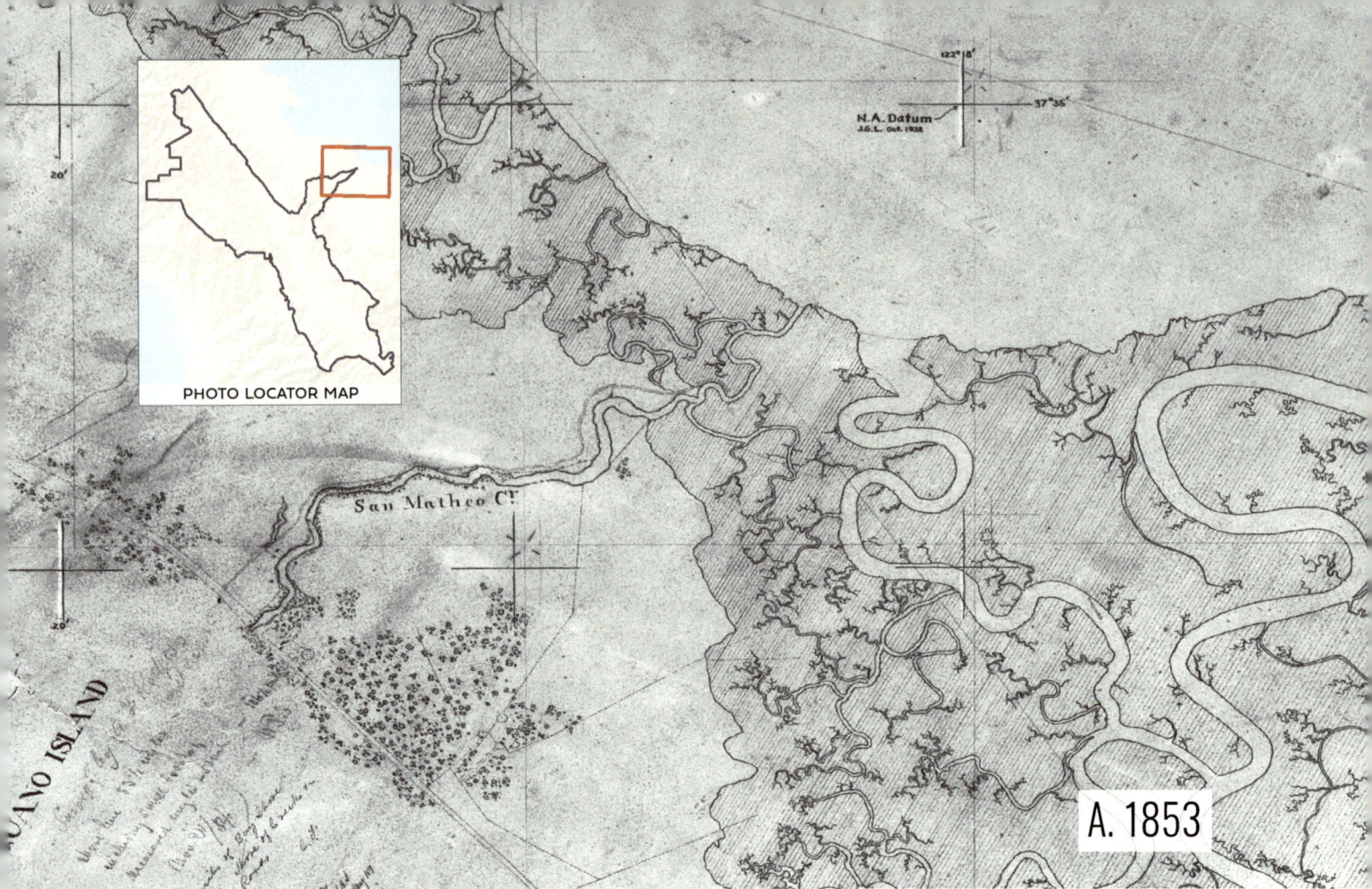

Figure 9.4. The mouth of San Mateo Creek was surrounded by tidal marsh, pictured here (A, above) in an 1853 U.S. Coast Survey topographic sheet. By the late 19th century, the marsh had been diked and reclaimed (B, below). *(A: Rodgers 1853; B: Rodgers and Westdahl 1898; courtesy NOAA)*

Tidal marsh and mudflats existed at the outlet of San Mateo Creek

A strip of tidal marsh historically occupied the area around the mouth of San Mateo Creek where it entered the Bay on the easternmost side of the study area (Fig. 9.4A). The width (i.e., inland extent) of the marsh at the mouth of San Mateo Creek was relatively narrow (~500 m wide), particularly in comparison with the width of the marsh just to the south around present-day Foster City (~5,500 m; Rodgers 1853).

The tidal marsh plant community was dominated by pickleweed (*Salicornia pacifica*). Surveying near the mouth of San Mateo Creek in 1856, for instance, GLO surveyor Thomas Stephens wrote, "The precise demarcation between the Tide Marsh and the firm land is well defined by the growth of a peculiar species of aquatic plant, a salt grass which covers the flat along the edges of the Bay, and on its inner margin grows slightly above the elevation of ordinary high tide" (Stephens 1856b). Similarly, in 1861 surveyor Aaron Van Dorn described the marsh near mouth of San Mateo Creek as "covered by a thick green sward of salt grass (Glasswort or Samphire)" (Van Dorn 1861).[4] Tidal influence was estimated to persist up to a mile upstream from its mouth (Mendell 1885). The combined flows from San Mateo, Laguna, and San Andreas creeks may have supplied a significant quantity of sediment to establish and maintain the marsh.

Efforts to reclaim the tidal marsh around Lower San Mateo Creek commenced in the mid- to late 19th century. Mendell (1884-5) reported that an earthen dam had been constructed across the mouth of the creek, shutting it off to tidal influence and causing fluvial flows to "spread over and deposit their silt upon the adjoining marsh"; these modifications were "successful in raising and reclaiming the marsh." The 1897-8 U. S. Coast Survey descriptive report accompanying T-sheet #2310 (Fig. 9.4B) further summarizes the impacts of marsh reclamation efforts in this area, including the subsequent subsidence and land use trajectory:

> The area westward of Hayward's Landing [just to the east of San Mateo Creek] is enclosed by the only effective dyke within the limits of this sheet. The old sloughs within this dyke still exist and contain water, but it stands at a lower level than that of the adjacent sloughs and is gradually being freshened by winter rains. The marsh land within is solidifying and sinking below the level of the surrounding marsh in a natural state. It is first utilized for grazing cattle and soon becomes fit for cultivation. I have been informed that such reclaimed land is very productive. (Westdahl and Rodgers 1898)

---

4 While the contemporary usage of "saltgrass" generally refers to *Distichlis spicata*, the terms "glasswort" and "samphire" in Van Dorn's description indicate that he was likely referring to pickleweed, *Salicornia pacifica*.

Riparian Habitats

A variety of riparian habitat types existed in the watershed, including willow thickets, willow-dominated riparian forest, willow riparian scrub, oak and California bay-dominated riparian forest, and mixed riparian forests comprised of California bay, maple (*Acer* spp.), alder (*Alnus* spp.), and a variety of other species. These habitat types varied within and across different stream reaches, and are discussed in more detail below.

Streams in San Andreas Valley and the Laguna Creek Basin supported a range of willow- and oak- dominated riparian habitats

In addition to the sag ponds and emergent marshes described above, San Andreas Creek supported a dense riparian corridor dominated by willow forest, willow scrub, and live oak (Fig. 9.5). As noted in Chapter 7, while traveling along San Andreas Creek in 1776, Pedro Font described "a great amount of woods in the hollow... [with] live oaks, madrons [sic], redwoods, cottonwoods, and other kinds as well." He also observed "a great amount of brush shoots on the bank of the stream" (Font and Brown 2011).[5] GLO surveyors and other observers in the mid- to late 19th century also documented extensive stands of willow and live oak (Addison 1857, Lewis 1860, Matthewson 1861, Von Schmidt 1864b); the riparian vegetation along San Andreas Creek was described as a "very dense growth of underbrush and willow" (*Daily Alta California* 1877) and a "mass of brush" (Maddox 1890).

Laguna Creek to the south also supported extensive stands of willow and oak riparian forest, along with several distinct willow thickets (Fig. 9.6). Three significant willow thickets were described by early travelers, including one at the foot of Laguna Grande, one at the head of Laguna Grande ("thickly overgrown with willows") and one where Laguna Creek meets current day Upper Crystal Springs Reservoir [Steven 1856]; the area was called an "impassable swamp" by one GLO surveyor (Tracy 1852).

---

5 Some of the trees described by Font may not have been riparian (i.e., directly associated with San Andreas Creek). A range of early evidence suggests that, prior to reservoir construction, San Andreas Valley supported extensive stands of (non-riparian) oak woodlands and other hardwood forest types (see page 174).

Figure 9.5. Riparian corridor along San Andreas Creek. A) Riparian scrub is visible along San Andreas Creek in this 1883 photo, looking to the southeast down San Andreas Valley. B) A 1913 view of the same general area, showing riparian scrub. *(A: BANC PIC 1954.016 -- fALB, courtesy The Bancroft Library, UC Berkeley; B: Lawson 1914, courtesy USGS)*

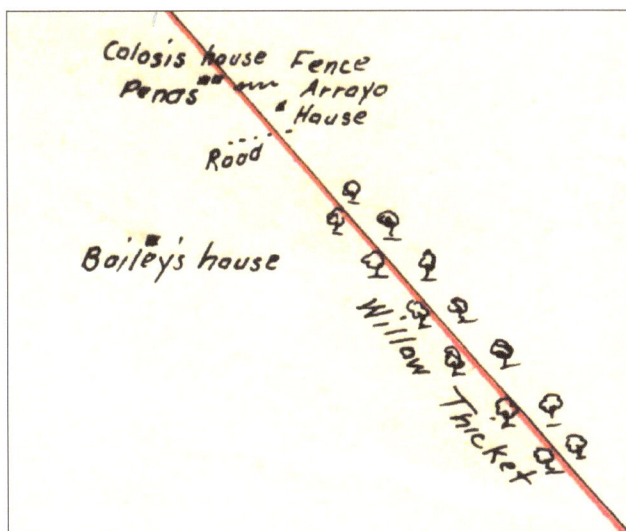

Figure 9.6. **A large willow thicket** is shown near the upstream end of present-day Upper Crystal Springs Reservoir in this 1856 plat map of Rancho Cañada de Raymundo. *(Photo O 239, 168, B-8, courtesy Bureau of Land Management)*

Pilarcitos Creek supported a diverse mixed riparian forest

Within Pilarcitos Canyon downstream of the present-day reservoir, Pilarcitos Creek supported a diverse riparian forest comprised of oak, madrone, Douglas-fir, big leaf maple (*Acer macrophyllum*), California buckeye (*Aesculus californica*), willow, California bay, alder, and other species (Davy 1895, Von Schmidt 1864a, *Daily Alta California* 1887; Fig. 9.7). An 1887 newspaper article, for instance, described the road through Pilarcitos Canyon downstream of the dam as "lined on either side with any variety of trees and smaller brush from the gigantic fir to the smaller willow and wild laurel" (*Daily Alta California* 1887). Botanist J. Burtt Davy echoed this account in 1895, describing the bottom of Pilarcitos Canyon south of the dam as dominated by "an abundance of *Salix* scrub, with *Ribes* [spp.]," and further downstream by "*Umbellularia* [California bay], *Acer* [maple], *Alnus* [alder]" (Davy 1895).

Figure 9.7. Early 20th century photos showing mixed riparian forest along Pilarcitos Creek. (top) "Falls near Stone Dam." (bottom) "Pilarcitos Creek." (*Both photos 1982 .086 ALB, courtesy The Bancroft Library, UC Berkeley*)

An oak- and bay-dominated riparian forest lined Lower San Mateo Creek

Descending below the modern-day Lower Crystal Springs Dam, Lower San Mateo Creek supported a different sort of riparian vegetation: a riparian forest dominated by large oak and bay trees (Fig. 9.8), some reportedly of "extraordinary size" (Taylor 1862). The riparian canopy also included sycamore (*Platanus racemosa*), California buckeye, holly-leaf cherry, cottonwood (*Populus* spp.), Oregon ash (*Fraxinus latifolia*), and maple (Font and Brown 2011, California Farmer 1859). While traveling with the Anza Expedition in March of 1776, for instance, Pedro Font reported that "the arroyo of San Mattheo [sic] has many laurels and ash trees on its banks" (Font and Brown 2011). Author Benjamin Parke Avery (1878) described Lower San Mateo Creek as "a little trout-stream embowered with chestnut oaks, with densely-leaved and aromatic bay trees, with tall, straight alders rooted in the very water, and with many flowering shrubs, its lower banks curtained by hanging vines or edged with mosses and tufted grass."

Figure 9.8. Photos by Carleton Watkins, 1875, showing the riparian forest along Lower San Mateo Creek. Part of a collection titled "Sun Sketches." *(A: BANC PIC 1974.006:15--ffALB; B: BANC PIC 1974.006:17--ffALB; C: BANC PIC 1974.006:14--ffALB; all courtesy The Bancroft Library, UC Berkeley)*

Fish Populations

Early surveys indicate that a number of native fish species, including coho salmon (*Oncorhynchus kisutch*), rainbow trout/steelhead (*O. mykiss*), California roach (*Hesperoleucus symmetricus*), and riffle sculpin (*Cottus gulosus*), all used the upper reaches of San Mateo Creek and other tributaries prior to reservoir construction. Steelhead, for instance, were "collected in the head waters of San Matteo [sic] creek" in 1855 (Girard 1858). Specimens of coho salmon, collected from Upper San Mateo Creek by Alexander Agassiz in 1860, are the earliest records of this species from any Bay Area watershed (Agassiz 1860, Leidy et al. 2005a). Riffle sculpin was collected from San Mateo Creek in 1854, but has not been observed since (Girard 1858, Leidy et al. 2005a); threespine stickleback (*Gasterosteus aculeatus*) was also potentially identified by early collectors (Leidy et al. 2005a). Other native fish species such as Sacramento sucker (*Catostomus occidentalis*) and tule perch (*Hysterocarpus traskii*) were also likely common in watershed streams historically (EDAW Inc. 2002). The watershed's productive wetland and aquatic habitats supported huge numbers of fish: one report in 1861 claimed that "in about three hours' fishing in the west branch of San Mateo creek, near Crystal Springs, [fishermen] caught 130 mountain trout" (*Sacramento Daily Union* 1861). The vast mosaic of wetlands on the floor of San Andreas Valley and the Laguna Creek Basin likely provided extremely high quality rearing habitat for salmonids.

Dam construction in the late 19th century altered the hydrology of the watershed and blocked access to much of the spawning and rearing habitat in the upper watershed for anadromous fish (Leidy et al. 2005a). Coho salmon have been extirpated from the watershed, though anadromous populations of steelhead persist below Lower Crystal Springs Dam and Stone Dam and are monitored by SFPUC. Healthy populations of adfluvial trout remain in all west Bay reservoirs, and use the reservoirs as a surrogate for the ocean while spawning in both the tributaries and on the banks of reservoirs (A. Brinkerhoff pers. comm.). Reservoirs contain both native species as well as invasive species such as largemouth bass (*Micropterus salmoides*), bluegill (*Lepomis macrochirus*), bullhead (*Ameiurus* spp.), catfish (Family Ictaluridae), carp (Family Cyprinidae), and goldfish (*Carassius auratus*). Creeks within the watershed are dominated by native species. Both Upper and Lower San Mateo Creek, for instance, support rainbow trout/steelhead, Sacramento sucker, and sculpin; Lower San Mateo Creek also supports three-spined stickleback. Pilarcitos Creek (below Pilarcitos Dam) supports healthy populations of rainbow trout/steelhead, sculpin, and three-spined stickleback (A. Brinkerhoff pers. comm.)

(Top) Rainbow trout (*Oncorhynchus mykiss; photo courtesy of NOAA, NMFS Southwest Fisheries Center*); (bottom left) Riffle sculpin (*Cottus gulosus; photo by Rober Tabor, courtesy USFWS Pacific Region*); (bottom right) Sacramento sucker (*Catostomus occidentalis; photo by Velo Steve, courtesy of CC BY 2.0*).

CHAPTER 10

Conclusions

The Peninsula Watershed has experienced widespread changes over the past two centuries resulting from a complex set of environmental and land use drivers. The distribution of vegetation communities in the watershed today reflects both persistence and change over time. Examination of these patterns and trajectories reveals important lessons about the future resilience and vulnerability of the watershed's ecosystems and the diverse plant and animal species they support. The trend towards increasing dominance of woody vegetation types in many parts of the watershed; the substantial loss of grassland, wetland, and chaparral habitats; and the spread of invasive species and plant pathogens represent fundamental shifts that should be central considerations when evaluating management and restoration priorities.

Despite far-reaching landscape changes over the past two centuries, many of the Peninsula Watershed's vegetation communities remain relatively intact, and continue to provide important ecological functions. The watershed contains 49 sensitive plant communities, and supports 25 special status plant species and numerous potentially locally rare plant species (Nomad Ecology 2020). Seventeen special status animal species have been recorded in and around the watershed (CNDDB 2012). The significant natural and cultural resources of the Peninsula Watershed have been widely recognized, and to a large extent preserved by SFPUC's stewardship. Nevertheless, the Peninsula Watershed's ecosystems

Lower Crystal Springs Reservoir. (Photo by SFEI-ASC, August 2018)

will be subject to a number of threats over the coming decades—climate change, non-native species invasions, plant pathogens, the risk of catastrophic wildfire, increasing urbanization along the periphery, seismic activity—that will demand difficult management decisions and proactive restoration measures.

One of the primary applications of historical ecology is in identifying appropriate restoration targets and strategies. Information about the historical habitat mosaics in the Peninsula Watershed—their distribution, structure, and composition—as well as the plant and animal species that they supported and the physical and ecological processes that sustained them, provides a functional understanding about the potential of the landscape to support native habitats and provide desired ecological functions and ecosystem services. In addition, an understanding of the ecological processes that supported habitat diversity and complexity in the past, and of the drivers of vegetation changes over the past two centuries, can help in anticipating future landscape change and identifying the factors that will maximize ecological resilience in the response to climate change and other factors. This study provides a foundation for developing a vision for the future watershed that integrates landscape trajectories, projected effects of climate change, and management options to sustain desired resources and functions over time.

This chapter provides an initial set of management considerations and recommendations based on the results of the historical ecology and landscape change research, as well as ideas for future research directions to build on the findings presented in this report. Additional management implications and future research directions will no doubt be identified as this research is integrated with other analyses and with on-the-ground knowledge of land managers in the Peninsula Watershed.

Management Considerations and Recommendations

The protection and stewardship of the Peninsula Watershed's diverse and valuable ecosystems over the coming decades will face unprecedented challenges. While many uncertainties exist, several fundamental principles of sustainable ecosystem management provide guidance in navigating these challenges (Dale et al. 2000; Chapin et al. 2009a,b; Beller et al. 2019):

- Assess ecosystem vulnerabilities and identify critical thresholds, and use this information to prioritize management actions;

- Minimize ecosystem exposure to stressors such as habitat fragmentation, invasive species, or insect pests and pathogens;

- Build adaptive management frameworks that foster innovation and facilitate learning and advancement of practical scientific knowledge;

- Protect and restore qualities that confer landscape resilience, such as large contiguous open spaces and ecosystem diversity, complexity, connectivity, and redundancy;

- Anticipate and plan for future ecosystem changes, such as vegetation shifts, altered water availability, and increased wildfire risk;

- Consider a site's natural potential as well as the regional social and environmental context when making management decisions and setting conservation priorities.

These general principles, combined with information about historical ecosystem functioning and patterns of landscape persistence and change, provide an important framework for management and restoration decision-making (Hobbs et al. 2014, Higgs et al. 2014). In many locations throughout the Peninsula Watershed, current ecosystem function still approximates historical function in many respects, and protection of these ecosystems should be prioritized. In some locations, landscape change has drastically altered ecosystem function, and major constraints make restoration of the historical ecosystem infeasible. Many other locations fall somewhere between these two extremes, and restoration efforts in these areas may succeed in recovering some of the lost ecosystem function. Managers should thus consider a site's historical trajectory, as well as potential future changes, when prioritizing conservation and restoration efforts and making decisions about potential interventions. Within the context of the sustainable ecosystem management principles discussed above, the findings from this historical ecology and landscape change investigation suggest a number of specific management priorities, offered below.

Serpentine grasslands.
(Photo courtesy SFPUC)

Place particular emphasis on conservation and restoration of ecosystems that support high levels of native biodiversity, are regionally rare, or have declined within the watershed over time

Analysis of landscape change over the past two centuries shows that the vegetation types that have declined the most within the watershed include grasslands, certain shrubland communities, old growth and mature/late seral forests, and wetlands. Preservation and restoration of these habitat types should be prioritized.

Prioritize preservation and restoration of native grasslands. Grassland extent declined by an estimated 56% over the past 70-80 years, far more than any other vegetation type in the watershed, due to a combination of urban development, intentional tree planting, and encroachment by *Baccharis* and other woody vegetation. As with grasslands throughout California, the spread of non-native annual grasses has altered the composition of the Peninsula Watershed's grasslands, though a surprisingly large proportion are still dominated by native perennial grasses (S. Simono pers. comm.). Grasslands within the watershed support numerous rare and sensitive plant species (Nomad Ecology 2020) and a number of special status wildlife species (see page 111). Serpentine grasslands within the watershed support an especially high number of rare and endemic species (Nomad Ecology 2020), and have been more resistant to encroachment of woody vegetation than non-serpentine grasslands.

Serpentine grasslands.
(Photo courtesy SFPUC)

The disproportionate loss of grasslands within the watershed, and the high ecological value of the remaining grasslands, warrants particular emphasis on the preservation and restoration of this habitat type. These efforts can build upon the demonstrated success of recent projects. For example, SFPUC recently completed an effort to restore and enhance nearly 40 ac of native grassland in the Adobe Gulch area (an area historically dominated by grassland that had recently converted to *Baccharis* shrubland) through its Bioregional Habitat Restoration (BHR) Program (Winzler & Kelly 2010, AECOM 2020).

Protect and restore diverse, rare, and/or sensitive shrubland communities throughout the watershed. As noted in Chapter 6, shrublands within the watershed include a diverse array of coastal scrub and chaparral communities that differ widely in terms of their ecology and historical trajectory. While the overall extent of shrublands has remained relatively stable over time, certain chaparral and coastal scrub communities appear to have declined substantially relative to their historical distribution (see page 162). Sensitive chaparral communities dominated by species such as giant chinquapin (*Chrysolepis chrysophylla* var. *minor*), brittle leaf manzanita (*Arctostaphylos crustacea* ssp. *crustacea*), and Kings Mountain manzanita (*A. regismontana*) face a number of threats, including forest encroachment, senescence due to lack of periodic fire, and *Phytophthora* pathogens.

In addition, many areas support old, highly diverse stands of coastal scrub that differ substantially from the early-successional stands of *Baccharis*-dominated scrub that have recently invaded native grasslands. These mature stands of coastal scrub likely provide high quality wildlife habitat and support uncommon plant assemblages, and should be protected.

Protect old growth and mature/late seral forests throughout the watershed.
Though encroachment of hardwood and conifer forests has displaced other habitat types in many parts of the watershed, mature forests were a significant component of the landscape historically, and continue to provide valuable wildlife habitat. Oak woodlands support some of the highest levels of biodiversity of any ecosystem in the state (Tietje et al. 2005, Swiecki and Bernhardt 2006), and have experienced significant declines regionally and statewide (Mensing 2006, Whipple et al. 2011). Old growth and mature/late seral Douglas-fir forests on the western side of Pilarcitos Canyon have a relatively limited distribution within the watershed (as opposed to early successional stages of Douglas-fir forest), and provide important plant and wildlife habitat for many species.

One of the major threats facing oak woodlands in the watershed is Sudden Oak Death (SOD), which has resulted in substantial tree mortality over the past several decades (see page 75). In addition to the direct effects of tree mortality, increased fuel loads resulting from SOD-induced tree mortality may exacerbate the risk of high severity fires, potentially altering successional pathways and resulting in a range of other ecological effects (Forrestel et al. 2015). SFPUC has established and implemented decontamination protocols to help control the spread of *Phytophthora ramorum* (the pathogen that causes SOD) and other plant pathogens, pests, and invasive species (SFPUC n.d.), and is examining the possibility of targeted removal of California bay laurel (a primary vector for the spread of *P. ramorum*) in oak-dominated areas as an SOD control measure (Swiecki 2020, M. Ingolia pers. comm.). SFPUC land managers should continue to explore and evaluate options for monitoring, prevention, and control of SOD infections and other plant pathogens.

Look for opportunities to restore some of the lost ecological functions historically provided by wetland and riparian habitats in San Andreas Valley and the Laguna Creek Basin prior to reservoir construction. These habitats represented some of the most significant perennial freshwater wetlands on the Peninsula: sag ponds, emergent marshes, willow thickets, and other habitats in these areas supported a huge abundance and diversity of wildlife, including waterfowl, amphibians, fish, and other taxa. Though options for restoring these habitats are limited by reservoir operations, there may be opportunities to create or restore wetlands along the margins of the reservoirs. SFPUC has already restored more than 10 ac of mitigation wetlands adjacent to the Peninsula Watershed reservoirs through its Bioregional Habitat Reserve (BHR) Program (SFPUC 2020, E. Natesan pers. comm.).

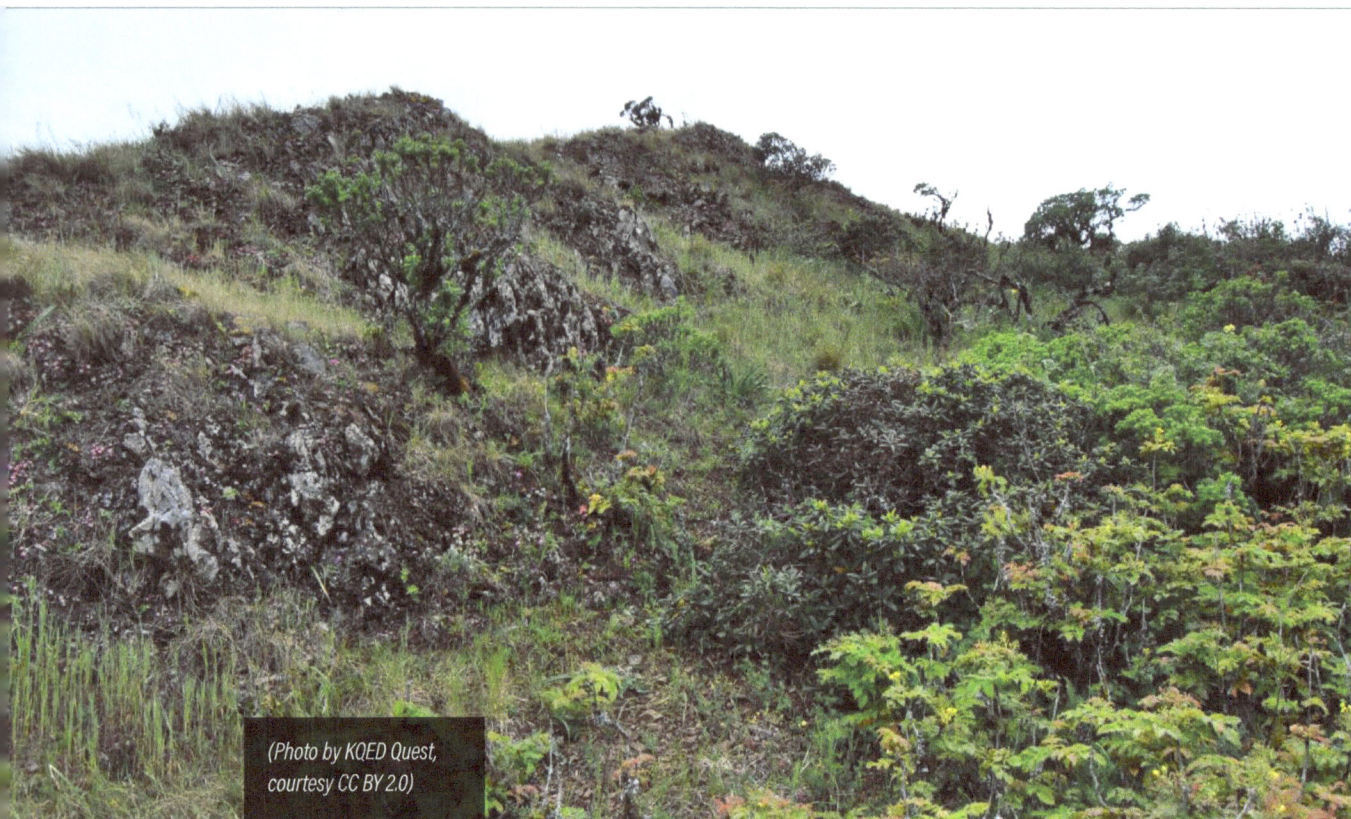

(Photo by KQED Quest, courtesy CC BY 2.0)

Identify opportunities to reduce habitat fragmentation

Large, contiguous patches of habitat, and connections between adjacent patches, are fundamental elements of landscape resilience (Fischer and Lindenmayer 2007, Beller et al. 2015). Large habitat patches support large and diverse plant and animal populations with complex and overlapping ecological functions, and help to sustain landscape-scale physical and biological processes (Connor and McCoy 1979, Peterson et al. 1998). Connections between patches promote gene flow and wildlife movement, support migration or recolonization following disturbance, and facilitate range shifts in response to climate change (Tewksbury et al. 2002, Standish et al. 2014).

The Peninsula Watershed supports large, contiguous habitat patches, and is regionally recognized as an essential core habitat in the Conservation Lands Network (Bay Area Open Space Council 2019). Several highways or other types of infrastructure within or along the margins of the watershed, however, may present significant barriers for wildlife movement or other elements of landscape connectivity. The Nature Conservancy's Omniscape Explorer tool, for instance, identifies the areas around Highway 92 and Highway 35 (Skyline Blvd) as potentially having a significant impact on "regional habitat connectivity" (TNC n.d.). Highway 280 and the watershed's reservoirs are also identified as major connectivity barriers. Further analysis should be conducted to examine the impacts of habitat fragmentation in these areas and to explore potential mitigation options.

Implement an invasive species control program

Invasive species have substantially altered the composition of a number of plant communities in the Peninsula Watershed over the past 250 years (see page 74), with significant impacts on native biodiversity and ecological function. For instance, though native perennial grasslands have persisted to a surprising extent within the watershed, invasive plants have become established in many locations, and represent a major threat to native grassland communities (Nomad Ecology 2009).

In some cases, where a population of an invasive plant is just beginning to establish in a new location, it may be feasible and cost-effective to attempt complete eradication (Rejmánek and Pitcairn 2002, Simberloff 2003). Successful control in these cases often depends on "early detection and rapid response" (Reaser et al. 2020). In cases where an invasive species is already widely established, however, effective management will likely require ongoing maintenance treatment, multiple complementary control techniques (which may include both direct and indirect measures), long-term monitoring, and adaptive management depending on observed success (DiTomaso et al. 2007). An invasive species control program and management plan should be developed to guide and prioritize invasive species management efforts in the watershed.

Broadleaf Stonecrop *(Sedum spathulifolium)*. *(Photo by David A. Hoffmann, courtesy CC BY 2.0)*

Consider removing planted trees from areas that formerly supported native grasslands or shrublands

Stands of planted trees (and associated naturalized populations), including both native species such as Douglas-fir and non-indigenous species such as Monterey cypress (*Hesperocyparis macrocarpa*), Monterey pine (*Pinus radiata*), and eucalyptus (*Eucalyptus* spp.), exist in a number of locations within the watershed (see page 66).[1] Intentional tree planting since the mid-19th century has contributed to the decline of native grassland and shrubland ecosystems (see pages 178 and 210), and may have produced a number of other deleterious ecological effects. Invasive cultivars of Monterey pine, for instance, have been shown to rapidly colonize northern coastal scrub communities, resulting in a reduction in native shrub cover and species richness (Steers et al. 2013). Eucalyptus stands are often associated with increased fire hazard, reduced water availability, and reduced understory plant cover and diversity (Fork et al. 2015, Wolf and DiTomaso 2016). Peninsula Watershed managers should consider the feasibility of removing or thinning stands of planted trees and associated naturalized populations (or, at a minimum, containing further spread) in areas where they have displaced native grassland or shrubland habitats, such as Buri Buri Ridge and Kings Mountain Ridge.

---

1 Though Monterey cypress and Monterey pine are both native to certain areas of California, naturalized populations outside of the natural ranges are considered invasive (Cal-IPC 2006).

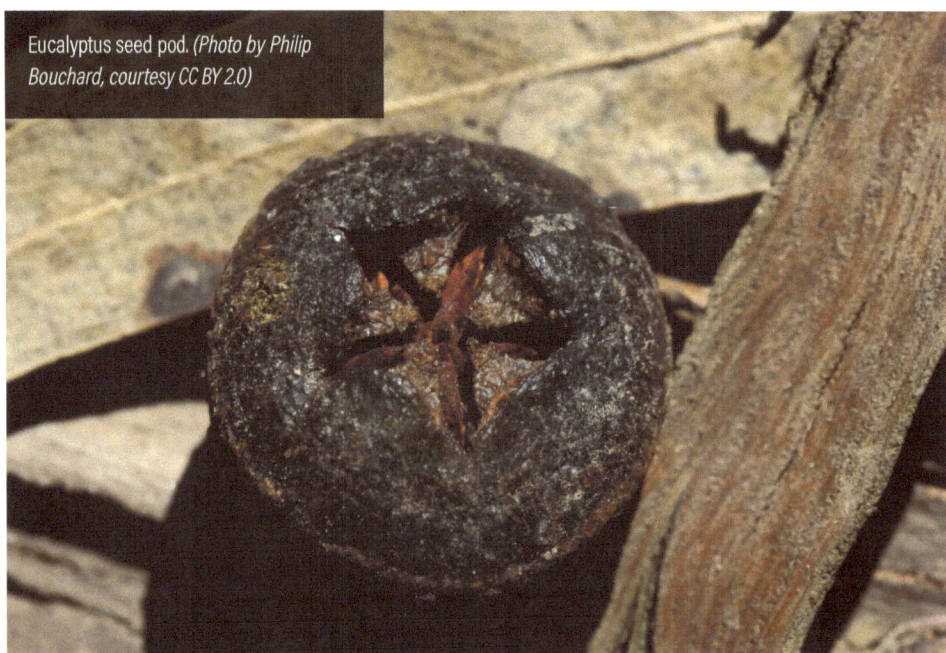

Eucalyptus seed pod. *(Photo by Philip Bouchard, courtesy CC BY 2.0)*

Identify opportunities to reintroduce periodic disturbances to disturbance-adapted ecosystems or to emulate the ecological effects of those disturbances

Ecosystems in many parts of the Peninsula Watershed, like those in many parts of the world, evolved in a landscape characterized by periodic disturbances that created a shifting mosaic of vegetation communities in dynamic equilibrium (Turner 2010, Newman 2019). Fires, herbivory, floods, and other ecological and physical processes played a key role in creating landscape heterogeneity at multiple scales and maintaining early successional vegetation communities such as grasslands. As disturbance regimes have changed over the past two centuries, ecosystems have responded through vegetation shifts and structural changes that have contributed to losses of some vegetation communities (and expansion of others) and declines of associated ecological functions.

Research in the field of disturbance ecology suggests that, to the degree possible, land managers should seek to emulate natural disturbance patterns and the "historical range of variability" experienced by native ecosystems—while acknowledging the inherent "nonstationarity"[2] of those systems—in order to sustain disturbance-adapted ecosystems and the plants and animals they support (Landres et al. 1999, Betancourt 2012, Safford et al. 2012a, Newman 2019). Indeed, as the climate changes, the goal of maintaining ecological function and integrity by restoring key biological and physical processes, rather than specific suites of species, is likely to gain importance (Safford et al. 2012b). On the other hand, myriad constraints, including proximity to urban areas, impacts associated with creating fire lines, prevalence of non-native species, air quality regulations, and others make full restoration of historical fire regimes or other disturbance patterns impractical, and raise serious questions about the ecological benefits of such interventions. For instance, while a number of chaparral species are fire-dependent (see page 164), the use of prescribed burning to maintain populations of those species in an urban setting like the Bay Area is extremely challenging, and can in fact be counterproductive if applied inappropriately (Parker 1987, Parker 1990, Keeley 2002). Given these challenges and complexities, prescribed fire in chaparral is not examined here.

In California grasslands, studies investigating the efficacy and ecological benefits of prescribed fire and grazing to promote native biodiversity and control woody encroachment have found mixed results. Both grazing

---

2 "Stationarity" refers to "the concept that ecosystems fluctuate within an unchanging envelope of variability" (Safford et al. 2012a). The concept of "nonstationarity" recognizes that the range of climatic variability and other environmental conditions have changed in the past and will continue to change in the future.

and prescribed fire can be effective tools for controlling coyote brush encroachment into grasslands if timed appropriately (McBride and Heady 1968, Tyler et al. 2007, Hopkinson et al. 2020), and indeed, over the long term, prescribed fire and grazing (along with targeted mechanical thinning and herbicide treatments) may be among the few tools available to maintain the watershed's grasslands and prevent or slow succession to woody vegetation. However, the effects of fire and grazing on native and non-native species abundance and diversity are variable, and depend heavily on site-specific factors like existing species composition, substrate, microclimate, and management history, as well as management variables such the timing, frequency, and intensity of the disturbance (Stahlheber and D'Antonio 2013, Bartolome et al. 2014, SFEI 2020). In addition, the effects of prescribed fire and grazing frequently differ between taxonomic groups, with positive responses among some plant species or guilds and negative or neutral responses among others, and in some cases substantial increases in non-native species cover (Hatch et al. 1999, D'Antonio et al. 2000, Hayes and Holl 2003, Stahlheber and D'Antonio 2013, Carlsen et al. 2017, A. Forrestel pers. comm.).

Given the uncertainties associated with the effects of these management interventions in California's coastal grasslands, and the high existing habitat quality of many of the Peninsula Watershed's grasslands, efforts to introduce prescribed fire or grazing in these ecosystems should proceed with caution. Reintroduction of these types of disturbances should be conducted through controlled experiments as part of an adaptive management framework, applied in conjunction with a robust monitoring program and an invasive species management program and supplemented with seeding of native grasses or forbs. Experimental prescribed burns or grazing (emulating native herbivory) may be most appropriate in grasslands that have recently experienced or are currently experiencing woody encroachment or have recently been restored, such as along portions of the western slope of Pulgas Ridge, the western slope of Sawyer Ridge, or in the Adobe Gulch area (see Fig. 5.1 on page 110). In contrast, the introduction of disturbances should probably be avoided in grasslands with a high proportion of existing native species that are not in immediate danger of coyote brush invasion, where fire or grazing may facilitate the spread of non-native species.

Consider the feasibility of re-introducing species that have been (or may have been) extirpated from the watershed

Human activities over the past two and a half centuries, including hunting, predator control, fire suppression, grazing, reservoir construction, and a range of land and water uses, have resulted in the extirpation (or presumed extirpation) of a number of native plant and animal species from the watershed. With the removal or lessening of some (though not all) of these anthropogenic stressors over the past century, as well as increasing attention on protection and restoration of the watershed's natural resources, ecosystems within the watershed may once again be able to support some of these extirpated species. Habitat condition assessments or feasibility analyses should be conducted to determine whether current conditions are now suitable to support stable populations of extirpated species; in cases where conditions appear suitable, re-introduction of extirpated species could be considered.

A number of plant species documented in the vicinity of the watershed historically have not been observed in recent surveys, and some of these species may be extirpated from the watershed. For instance, of the 45 rare or special status plant species targeted in the 2018 rare plant survey within the watershed (Nomad Ecology 2020), 16 species were not observed and may be extirpated. Successful reintroduction of some of these species may be possible; for instance, the Creekside Center for Earth Observation is currently collaborating with SFPUC to reintroduce San Mateo thornmint (*Acanthomintha duttonii*) to serpentine grasslands on the eastern edge of the watershed using seed from the last remaining population in Edgewood Park (Niederer and Schwind 2020).

Reintroduction of vertebrate species is complicated by many factors, and existing land use constraints may preclude successful reintroduction. However, an assessment of the feasibility of native wildlife reintroduction would be a worthwhile endeavor, particularly in cases where the extirpation occurred as a result of hunting pressures. Several bird species, for instance, such as mountain quail (*Oreortyx pictus*) and roadrunner (*Geococcyx californianus*), appear to have been extirpated from the watershed during the late 19th or early 20th centuries (see page 78). Beaver (*Castor canadensis*) are presumed to have been present within the watershed historically, and likely played an important role as ecosystem engineers in wetland and riparian habitats (Lanman et al. 2013). Tule elk (*Cervus canadensis nannodes*), a large herbivore endemic to California that was historically abundant in grasslands throughout the central coast ranges and was nearly hunted to extinction, has been successfully reintroduced in a number of areas, including Point Reyes National Seashore. Experimental reintroductions have found that tule elk, like non-native grazers, can be effective in suppressing shrub encroachment (Johnson and Cushman 2007). Continued efforts should also be made to increase the amount and quality of habitat available to native fish, including anadromous fish such as steelhead (*Oncorhynchus mykiss*).

Future Research Directions

There are a number of potential future research directions, aside from the experimental approaches for management recommended in the previous section, that could be undertaken to build upon the findings from the historical ecology research, or to examine unanswered questions raised during the course of this study, including:

Conduct oral interviews with tribal leaders. While the information presented here documents the historical ecology of the Peninsula Watershed from a Western scientific perspective based largely on archival data sources, it would be enlightening to learn more about the perspectives of tribal leaders with regards to Native land management, Indigenous science and Traditional Ecological Knowledge, and natural resource management and restoration. Of particular interest would be the perspective of the Indigenous community on Native use of fire and its role in contemporary ecosystem management.

Conduct phytolith sampling and analysis to provide further information about the historical extent of grasslands, particularly in the northern part of the watershed. Mid-19th century information indicates that the northwestern portion of the watershed was dominated by shrublands, with grasslands prevalent along ridgelines (see pages 110 and 145). However, it is unknown whether these grasslands were of "natural" origin or the product of clearing for grazing and other land uses, or whether grasslands were more extensive in this area prior to the mid-19th century. Targeted analysis of phytoliths—silica-based structures that form in the tissues of some plant species and persist for long periods in the soil—could be used to help resolve this question.

A limitation of this approach is that phytolith analysis cannot be used to determine the presence or absence of forb-dominated grasslands, because forbs do not produce phytoliths. Thus, high phytolith content would be positive evidence of grassland presence, but low phytolith content would not be evidence that forb-dominated grasslands were absent. However, previous studies have found that coastal prairies likely had a greater ratio of grasses to forbs than inland grasslands (Evett and Bartolome 2013), and thus phytolith analysis in the northwestern portion of the watershed might be expected to provide evidence of grasslands if they were in fact prevalent. Indeed, researchers in nearby Quiroste Valley and other coastal sites have used phytolith analysis to provide compelling evidence for the presence of grasslands maintained by anthropogenic fire prior to European settlement (Evett and Cuthrell 2013, Evett and Cuthrell 2017; see page 51).

Obtain tree cores to determine the age structure of Douglas-fir trees around Pilarcitos Canyon. While early records indicate that a large stand of mature Douglas-fir forest occupied the hillslope on the west side of Pilarcitos Canyon, over the past two centuries Douglas-fir has encroached into many of the surrounding shrubland and hardwood forest habitats (see Fig. 8.2 on page 195 and discussion on pages 200-209). However, the precise extent of the old growth or mature/late seral stand that existed historically, and the timing of the Douglas-fir expansion, is unknown. Dendrochronology (the use of annual growth rings to determine tree age) could be used to more accurately differentiate the historical "node" of Douglas-fir forest from more recently established trees, and to estimate the rate of expansion.

Quantify water use by different vegetation types in the watershed. Along with impacts on wildlife support, wildfire risk, and other factors, one of the major potential implications of the vegetation shifts documented in this study is changes in water use and availability. Examination of water use and fog interception by the different vegetation types in the watershed could be used to evaluate the effects of documented vegetation shifts—in particular the substantial expansion of Douglas-fir—on streamflow and water availability (both for ecosystems and the water supply reservoirs within the watershed).

Analyze satellite imagery of drought-based tree mortality. In the recent severe drought in California, some regions experienced significant tree mortality, while others were only minimally impacted. The variable responses were likely due in part to differences in the amount of seasonally variable water that is stored in soils and weathered bedrock, as well as differences in stand density and other factors (Hahm et al. 2019, McLaughlin et al. 2020). It is important to understand the effects of drought on tree mortality, as climate change is expected to increase the intensity and frequency of drought in California.

Satellite imagery could be used to evaluate tree mortality in the Peninsula Watershed following the recent California drought (as well as to evaluate tree mortality associated with Sudden Oak Death). Because the effect of drought on the health and structure of ecosystems, as well as their flammability, depends not only on the composition of the vegetation community but also on access to subsurface water, this satellite analysis could be accompanied by installation of a monitoring network of observation wells to try to better understand the distribution and movement of ground water. Mortality patterns could be compared among vegetation communities within the watershed, to identify where in the watershed ecosystems are most sensitive or resilient to drought. Mortality patterns could also be compared between the Peninsula Watershed and other regions in the state, to examine whether the Peninsula Watershed is more or less sensitive to drought than other similar ecosystems.

Model vegetation succession and disturbance processes to better understand the links between observed vegetation shifts and hypothesized drivers. While the research presented in this report provides strong anecdotal evidence for understanding the drivers of landscape change, quantitative models would help to determine the relative importance of different drivers and to identify physical factors (e.g., slope, aspect, soil type, groundwater depth) that predict where particular vegetation types have been more or less susceptible to certain types of change. The dataset produced as part of the aerial imagery point analysis—consisting of 5,000 randomly distributed points classified according to vegetation cover in the historical (1946-48) and modern (2015-17) aerial imagery (see page 27)—would be highly valuable as an input to future modeling efforts.

Integrate findings about past vegetation changes with projected future vegetation shifts under climate change. Current climate models predict that average annual maximum temperature in San Mateo County will increase by 4.4-7.2 °F (2.4-4 °C) by 2070-2099 over 1961-1990 levels. Wildfire risk will also intensify, with annual area burned projected to increase by 36-492% by 2070-2099 over 1961-1990 levels depending on emissions scenario, climate model, and population growth scenario (Cal-Adapt 2020). These and other climate-related changes will have profound, albeit complex and sometimes opposing, effects on vegetation distribution and ecosystem function, and may further intensify current trajectories of vegetation change.

For example, research suggests that encroachment of coyote brush into coastal grasslands may accelerate under a warmer climate (Zavaleta 2001, Cornwell et al. 2012, Kidder 2015). In the area around Cahill Ridge and Sawyer Ridge within the Peninsula Watershed, Douglas-fir extent is projected to increase by 315 to 588 ha by 2080 under future climate change (Dingman 2014). Projections of vegetation change on the San Francisco Peninsula developed by Cornwell et al. (2012) indicate that grassland extent will likely decrease substantially under climate change (far more than any other vegetation type, and proportionally more than in other regions of the Bay Area); conversely, shrublands dominated by chamise (which is drought-tolerant and resprouts following fire) is forecast to increase substantially. Historical ecology provides the foundation for interpreting and understanding projections of vegetation shifts under future climate change, and for understanding how these projected future changes relate to past patterns of vegetation change and persistence.

Create a "resilience framework" to evaluate the future resilience (or vulnerability) of different zones and/or vegetation types within the watershed. This resilience framework could be used to synthesize information about the vulnerability of different zones/ vegetation types to different types of stressors or disturbances, such as catastrophic fire, warming, drought, succession/encroachment, disease/insects, and seismic risk. Arranging this information along a set of disturbance "axes" could be helpful in comparing the vulnerability of different ecosystems in the watershed to particular stressors and in identifying ecosystems that may be particularly sensitive to climate change or other disturbances (and thus should potentially be the subject of targeted conservation or management efforts).

Develop an integrated landscape management plan or resilience strategy.
The future research directions discussed above center around major
themes of landscape change (e.g., climate change, wildfire, directional
vegetation change), all of which influence the health and structure of
ecological communities and their ability to provide ecosystem services
(e.g., carbon sequestration, biodiversity support, water availability, sense
of place). In the face of multiple drivers of change, multiple management
constraints, and considerable uncertainty about future trajectories, the
watershed would benefit from an integrated landscape management plan
or resilience strategy that synthesizes information about past landscape
change, current management activities (e.g., fuel reduction projects, invasive
plant management, construction/mitigation projects), and future climate
projections to identify multi-benefit management and restoration strategies.
The development of this integrative vision could build alignment among
stakeholders for a coordinated set of management strategies and actions that
can maintain the health of the watershed in the coming decades.

Hardwood riparian forest along Pilarcitos Creek, with big leaf maple (*Acer macrophyllum*), California bay (*Umbellularia californica*), red alder (*Alnus rubra*), red elderberry (*Sambucus racemosa*), coast live oak (*Quercus agrifolia*), beaked hazelnut (*Corylus cornuta*), and California buckeye (*Aesculus californica*). *(Photo courtesy SFPUC)*

CHAPTER 11
References

ABI (Association for Biodiversity Information). 2003. *Classification of the Vegetation of Point Reyes National Seashore Golden Gate National Recreation Area, Samuel P. Taylor, Mount Tamalpais, and Tomales State Parks, Marin, San Francisco, and San Mateo Counties, California.* Association for Biodiversity Information, California Native Plant Society, California Department of Fish and Game, Wildlife and Habitat Analysis Branch, California Heritage Program. Sacramento: National Park Service. *Courtesy of Heath Bartosh.*

Addison JE. 1857. *Field Notes of the Final Survey of the Rancy San Mateo.* San Francisco: U.S. Survey-General's Office. *Courtesy of Bureau of Land Management.*

AECOM. 2020. *Adobe Gulch Grassland Bioregional Habitat Restoration Site Annual Monitoring Report: 2019.* Prepared for San Francisco Public Utilities Commission. Oakland, CA.

Agassiz A. 1860. [Record of *Oncorhynchus kisutch*, collected "San Mateo Creek."] Ichthyology record #68471. Museum of Comparative Zoology, Harvard University.

Alexander PW, CP Hamm. 1916. *History of San Mateo County from the Earliest of Times with a Description of Its Resources and Advantages; and the Biographies of Its Representative Men.* Burlingame, CA: Burlingame Publishing Co.

Alley BF. 1883. *History of San Mateo County, California, Including Its Geography, Topography, Geology, Climatography, and Description.* San Francisco: B. F. Alley.

Anderson MK. 2005. *Tending the Wild: Native American Knowledge and the Management of California's Natural Resources.* 1st ed. Berkeley: University of California Press.

Avery BP. 1878. *Californian Pictures in Prose and Verse.* New York: Hurd and Houghton.

Axelrod DI. 1978. The Origin of Coastal Sage Vegetation, Alta and Baja California. *American Journal of Botany* 65(10):1117-1131.

Babal M. 1990. *The Top of the Peninsula: A History of Sweeney Ridge and the San Francisco Watershed Lands, San Mateo County, California.* Historic Resource Study. National Park Service.

Bartolome JW, BH Allen-Diaz, S Barry, LD Ford, M Hammond, P Hopkinson, F Ratcliff, S Spiegal, MD White. 2014. Grazing for Biodiversity in Californian Mediterranean Grasslands. *Rangelands* 36(5):36–43.

Bartolome JW, P Hopkinson, M Hammond, N Schowalter. 2012. *Point Pinole Regional Shoreline Restoration of Coastal Prairie Using Prescribed Burning, 2010 and 2011: Second and Third Year Report to the East Bay Regional Park District.* Department of Environmental Science, Policy, and Management, University of California, Berkeley.

Baxter JW, VT Parker. 1999. Canopy Gaps, Zonation and Topography Structure: A Northern Coastal Scrub Community on California Coastal Bluffs. *Madroño* 46(2):69-79.

Bay Area Open Space Council. 2019. *The Conservation Lands Network 2.0 Report.* Berkeley, CA.

Beisner BE, DT Haydon, K Cuddington. 2003. Alternative Stable States in Ecology. *Frontiers in Ecology and the Environment* 1(7):376–82.

Beller EE, EN Spotswood, AH Robinson, MG Anderson, ES Higgs, RJ Hobbs, KN Suding, ES Zavaleta, JL Grenier, RM Grossinger. 2019. Building Ecological Resilience in Highly Modified Landscapes. *BioScience* 69(1):80-92.

Beller EE, RM Grossinger, JL Grenier. 2015. *Landscape Resilience Framework: Operationalizing Ecological Resilience at the Landscape Scale*. 752. Richmond, CA: San Francisco Estuary Institute-Aquatic Science Center.

Betancourt J. 2012. Reflections on the Relevance of History in a Nonstationary World. In *Historical Environmental Variation in Conservation and Natural Resource Management*, eds. John A. Wiens, Gregory D. Hayward, Hugh D. Safford, Catherine M. Giffen. Chichester, UK: John Wiley & Sons, Ltd.

Bilotta GS, RE Brazier, PM Haygarth. 2007. The Impacts of Grazing Animals on the Quality of Soils, Vegetation, and Surface Waters in Intensively Managed Grasslands. *Advances in Agronomy* 94:237–80.

Bond W, J Keeley. 2005. Fire as a Global 'Herbivore': The Ecology and Evolution of Flammable Ecosystems. *Trends in Ecology & Evolution* 20(7):387–94.

Bowman JN. 1947. *The Area of the Mission Lands. Courtesy of The Bancroft Library, UC Berkeley.*

Brabb EE, RW Graymer, DL Jones. 1998. *Geology on the Onshore Part of San Mateo County, California: A Digital Database*. 98–137. U.S. Geological Survey.

Brace LB. 1869. *The New West: Or, California in 1867-1868*. Vol. 8. New York: G. P. Putnam & Son.

Brewer WA. 1903. Notes and Correspondence: Scarper Peak, in San Mateo County. In *The Sierra Club Bulletin* 4(3):243–45. San Francisco: The Sierra Club.

Bromfield D. 1893. Survey of a Road from the Summit Springs House to the Main Spanishtown Road. *Courtesy of San Mateo County Public Works and Recorder.*

Bromfield D. 1894. Official Map of San Mateo County, California. San Francisco: Schmidt Label & Lith. Co. *Courtesy of the Library of Congress.*

Brown C. 1853. *Domingo Feliz, claimant for Feliz Rancho, 1 square league, in San Mateo County. BANC MSS Land Case Files 148 ND. p. 10. U.S. District Court, Northern District. Courtesy of The Bancroft Library, UC Berkeley.*

Brown AK. 2005. *Reconstructing early historical landscapes in the Northern Santa Clara Valley*. Russell K. Skowronek. Santa Clara, CA: Santa Clara University.

Brown LB, B Allen-Diaz. 2009. Forest Stand Dynamics and Sudden Oak Death: Mortality in Mixed-Evergreen Forests Dominated by Coast Live Oak. *Forest Ecology and Management* 257(4):1271–80.

Brylski P. 2008. California Wildlife Habitat Relationships System: Dusky-Footed Woodrat, Neotoma Fuscipes. Sacramento: California Department of Fish and Wildlife, California Interagency Wildlife Task Group.

Buordo EA. 1956. A Review of the General Land Office Survey and of its Use in Quantitative Studies of Former Forests. *Ecology* 37:754–768.

Burcham LT. 1961. Cattle and Range Forage in California: 1770-1880. *Agricultural History Society* 35(3):140–49.

Burke WF. 1926. On the Way to Carey's. *San Francisco Water* 5(4):12–16. San Francisco: Spring Valley Water Company. *Courtesy of San Francisco Public Library.*

Burroughs O. 1928. Bird Life at San Andres [sic]. *San Francisco Water* 7(2):11. San Francisco: Spring Valley Water Company. *Courtesy of San Francisco Public Library.*

Cal-Adapt. 2020. Climate Tools. Available: https://cal-adapt.org/tools/. Accessed: April 19, 2020.

Cal-IPC (California Invasive Plant Council). 2006. California Invasive Plant Inventory. Cal-IPC Publication 2006-02. Berkeley, CA. Available: www.cal-ipc.org.

California Farmer and Journal of Useful Sciences. 1859. "Woodside." July 22. *Courtesy of the California Digital Newspaper Collection.*

California Oak Mortality Task Force. 2019. Sudden Oak Death, Diagnosis and Management: Best Management Practices. California Oak Mortality Task Force.

Callaway RM, CM D'Antonio. 1991. Shrub Facilitation of Coast Live Oak Establishment in Central California. *Madroño* 38(3):158–69.

Callaway RM, FW Davis. 1993. Vegetation Dynamics, Fire, and the Physical Environment in Coastal Central California. *Ecology* 74(5):1567–78.

Callaway RM, FW Davis. 1998. Recruitment of *Quercus Agrifolia* in Central California: The Importance of Shrub-dominated Patches. *Journal of Vegetation Science* 9(5):647–56.

Carlsen TM, EK Espeland, LE Paterson, DH MacQueen. 2017. Optimal Prescribed Burn Frequency to Manage Foundation California Perennial Grass Species and Enhance Native Flora. *Biodiversity and Conservation* 26(11):2627–56.

Carranco L, JT Labbe. 2003. *Logging the Redwoods.* Caldwell, Idaho: Caxton Printers.

Caziarc DS. 2012. *The Invasion of California Grasslands: Past, Present, and Future Implications.* California Polytechnic State University.

CCH2 Portal. 2020. Available: http//:www.cch2.org/portal/index.php. Accessed: December 05, 2020.

Chapin FS, GP Kofinas, C Folke, SR Carpenter, P Olsson, N Abel, R Biggs, RL Naylor, E Pinkerton, DMS Smith, W Steffen, B Walker, OR Young. 2009a. Resilience-Based Stewardship: Strategies for Navigating Sustainable Pathways in a Changing World. In *Principles of Ecosystem Stewardship*, eds. F. Stuart Chapin et al. New York: Springer.

Chapin FS, SR Carpenter, GP Koflnas, C Folke, N Abel, WC Clark, P Olsson, DMS Smith, B Walker, OR Young, F Berkes, R Biggs, JM Grove, RL Naylor, E Pinkerton, W Steffen, FJ Swanson. 2009a. Ecosystem Stewardship: Sustainability Strategies for a Rapidly Changing Planet. *Trends in Ecology and Evolution* 25(4):241-249.

CHTA (California Highway Transportation Agency). 1968. *A Progress Report on Interstate 280 'Junipero Serra' Freeway.* San Mateo: California Highway Transportation Agency, San Francisco Division of Highways.

Ciardi G. 2019. *Fire Hazard Severity Reduction and Integrated Vegetation Management on the Crystal Springs Watershed.* Fire Safe San Mateo County.

Clar RC. 1959. *California Government and Forestry: From Spanish Days until the Creation of the Department of Natural Resources in 1927.* Sacramento: Division of Forestry, Department of Natural Resources, State of California.

Clements FE. 1916. *Plant Succession: An Analysis of the Development of Vegetation.* Washington: Carnegie Institution of Washington. *Courtesy of the Cornell University Library of the New York State College of Agriculture.*

Clements FE. 1936. Nature and Structure of the Climax. *The Journal of Ecology* 24(1):252–84.

Cloud JJ. 1877. *Official Map of the County of San Mateo, California, Showing New Boundary Line and Deliniating the Lines of Cities, Towns, Private Claims, Ranchos, Water Works, and Rail Roads.* San Francisco: Britton, Rey & Co. Lith. *Courtesy of Stanford University David Rumsey Map Center.*

CNDDB (California Natural Diversity Database). 2012. California Department of Fish and Game, Biogeographic Data Branch.

Collins BD, DR Montgomery. 2001. Importance of Archival and Process Studies to Characterizing Pre-Settlement Riverine Geomorphic Processes and Habitat in the Puget Lowland. *Water Science and Application* 4:227–243.

Connor EF, ED McCoy. 1979. The Statistics and Biology of the Species-Area Relationship. *The American Naturalist* 113(6):791-833.

Cornwell WK, SA Stuart, A Ramirez, CR Dolanc, JH Thorne, DD Ackerly. 2012. *Climate Change Impacts on California Vegetation: Physiology, Life History, and Ecosystem Change.* EC-500-2012-023. Berkeley: California Energy Commission.

Costansó M, FM Stanger, AK Brown. 1969. The Journals of Portolá, Costansó, and Crespí, October 1 to November 15. In *Who Discovered the Golden Gate? The Explorers' Own Accounts,* eds. Frank M. Stanger, Alan K. Brown. San Mateo, CA: San Mateo County Historical Association.

Cowart A, R Byrne. 2013. A Paleolimnological Record of Late Holocene Vegetation Change from the Central California Coast. *California Archaeology* 5(2):337–52.

Crespí J, AK Brown. 2001. *A Description of Distant Roads: Original Journals of the First Expedition into California, 1769-1770.* San Diego: San Diego State University Press.

Cuthrell RQ. 2013a. Archaeobotanical Evidence for Indigenous Burning Practices and Foodways at CA-SMA-113. *California Archaeology* 5(2):265–90.

Cuthrell RQ. 2013b. An Eco-Archaeological Study of Late Holocene Indigenous Foodways and Landscape Management Practices at Quiroste Valley Cultural Preserve, San Mateo County. Dissertation. Berkeley: University of California, Berkeley.

D'Antonio CM, C Malmstrom, SA Reynolds, J Gerlach. 2007. Ecology of Invasive Non-native Species in California Grassland. In *California Grasslands: Ecology and Management,* eds. Mark Stromberg, Jeffrey Corbin, Carla D'Antonio. Berkeley, CA: University of California Press.

D'Antonio C, S Bainbridge, C Kennedy, J Bartolome, S Reynolds. 2000. *Ecology and Restoration of California Grasslands with Special Emphasis on the Influence of Fire and Grazing on Native Grassland Species.* University of California, Berkeley.

Daily Alta California. 1860. "Letter from Murphy's." June 10. *Courtesy of the California Digital Newspaper Collection.*

Daily Alta California. 1877. "Fire in San Mateo County." September 17. *Courtesy of the California Digital Newspaper Collection.*

Daily Alta California. 1887. "Mammoth Reservoirs: Sources of the Spring Valley Company's Water Supply." June 30. *Courtesy of the California Digital Newspaper Collection.*

Daily Alta California. 1889. "Spring Valley Property Damaged in San Andreas Valley: The Fire Still Spreading." September 18. *Courtesy of the California Digital Newspaper Collection.*

Dale VH, S Brown, RA Haeuber, NT Hobbs, N Huntly, RJ Naiman, WE Riebsame, MG Turner and TJ Valone. 2000. Ecological Principles and Guidelines for Managing the Use of Land. *Ecological Applications* 10(3):639-670.

Dandridge PP. 1889. *Resurvey of Rancho Las Pulgas.* Book CA-G9. San Francisco: U.S. Survey-General's Office. *Courtesy of Bureau of Land Management, Sacramento CA.*

Davidson G. 1869. Report of Prof. George Davidson, of the U.S. Coast Survey. In *San Francisco Water Company Reports by Engineers and Others on a Permanent Supply of Pure Fresh Water to the City of San Francisco.* San Francisco: San Francisco Water Company. *Courtesy of the California State Library.*

Davis GJ. 1929. "G. J. Davis to G. A. Elliott." December 2. Box MB 027. *Courtesy of the San Francisco Public Utilities Commission Archives.*

Davis GJ. 1951. "Letter to H. C. Melone, Superintendent, Agricultural Division, San Francisco Water Department." January 8. Permanent Collection, SFPUC Archives. *Courtesy of the San Francisco Public Utilities Commission Archives.*

Davis FW, DD Baldocchi, CM Tyler. 2016. Oak Woodlands. In *Ecosystems of California*, eds. Harold Mooney, Erika Zavaleta. Oakland: University of California Press.

Davy JB. 1895. "California, U.S.A. - 1895 J. Burtt Davy Field Notes." Jepson Herbarium, University of California, Berkeley.

Dingman JR. 2014. *Modeling the Impacts of Climate Change on the Douglas-Fir Forest within the San Francisco Peninsula Watershed Using GIS and LiDAR.* Dissertation. University of California, Berkeley.

DiTomaso JM, SF Enloe, MJ Pitcairn. 2007. Exotic Plant Management in California Annual Grasslands. In *California Grasslands: Ecology and Management*, eds. Mark Stromberg, Jeffrey Corbin, Carla D'Antonio. Berkeley, CA: University of California Press.

Dukes JS, MR Shaw. 2007. Responses to Changing Atmosphere and Climate. In *California Grasslands: Ecology and Management*, eds. Mark Stromberg, Jeffrey Corbin, Carla D'Antonio. Berkeley, Calif: University of California Press.

Dunne JA, VT Parker. 1999. Species-Mediated Soil Moisture Availability and Patchy Establishment of Pseudotsuga Menziesii in Chaparral. *Oecologia* 119(1):36–45.

Duvall CR, AL Baldwin. 1910. *Triangulation in California, Part II, Volume 2.* U.S. Coast and Geodetic Survey. U.S. Government Printing Office.

Easton AS. 1868. *Official Map of the County of San Mateo, California, Including City and County of San Francisco with All New Additions of Cities Towns and Villas Delineating the Lines of Ranchos, Private Claims, Water Works, Rail Roads.* San Francisco: Britton & Rey's Lithographic Steam Press. Item no. 912.794 Sa. 588, 50496. *Courtesy of San Francisco Public Library.*

eBird. 2017. *eBird: An Online Database of Bird Distribution and Abundance [web application]*. Cornell Lab of Ornithology, Ithaca, New York. Available: http://www.ebird.org. Accessed: July 1, 2019.

EDAW, Inc. 2002. *Final Peninsula Watershed Management Plan*. San Francisco Public Utilities Commission.

Edmonds J. 2014. *The Teagues of San Mateo County*. Lake Wildwood Association.

Edwards SW. 2007. Rancholabrean Mammals of California and Their Relevance for Understanding Modern Plant Ecology. In *California Grasslands: Ecology and Management*, eds. Mark Stromberg, Jeffrey Corbin, Carla D'Antonio. Berkeley, Calif: University of California Press.

Egan D, EA Howell, eds. 2005. *The Historical Ecology Handbook: A Restorationist's Guide to Reference Ecosystems*. The Science and Practice of Ecological Restoration. Washington, DC: Island Press.

Emery NC, CM D'Antonio, CJ Still. 2018. Fog and Live Fuel Moisture in Coastal California Shrublands. *Ecosphere* 9(4):1-19.

Engelhardt Z. 1924. *San Francisco or Mission Dolores*. Chicago: Franciscan Herald Press.

Evans AS. 1873. *A La California: Sketches of Life in the Golden State*. San Francisco: A. S. Bancroft & Company. *Courtesy of the Internet Archive.*

Evett RR, JW Bartolome. 2013. Phytolith Evidence for the Extent and Nature of Prehistoric Californian Grasslands. *The Holocene* 23(11):1644–49.

Evett RR, RQ Cuthrell. 2013. Phytolith Evidence for a Grass-Dominated Prairie Landscape at Quiroste Valley on the Central Coast of California. *California Archaeology* 5(2):319–35.

Evett RR, RQ Cuthrell. 2017. Testing Phytolith Analysis Approaches to Estimate the Prehistoric Anthropogenic Burning Regime on the Central California Coast. *Quaternary International* 434(April):78–90.

Fischer J, DB Lindenmayer. 2007. Landscape Modification and Habitat Fragmentation: A Synthesis. *Global Ecology and Biogeography* 16(3):265-280.

Font P, AK Brown. 2011. *With Anza to California, 1775-1776: The Journal of Pedro Font, O.F.M.* v. 1. Norman, OK: Arthur H. Clark Co.

Ford LD, GF Hayes. 2007. Northern Coastal Scrub and Coastal Prairie. In *Terrestrial Vegetation of California*, eds. Michael Barbour, Todd Keeler-Wolf, Allan A. Schoenherr. Berkeley: University of California Press.

Fork S, A Woolfolk, A Akhavan, E Van Dyke, S Murphy, B Candiloro, T Newberry, S Schreibman, J Salisbury, K Wasson. 2015. Biodiversity Effects and Rates of Spread of Nonnative Eucalypt Woodlands in Central California. *Ecological Applications* 25(8):2306-2319.

Forrestel AB, BS Ramage, T Moody, MA Moritz, SL Stephens. 2015. Disease, Fuels and Potential Fire Behavior: Impacts of Sudden Oak Death in Two Coastal California Forest Types. *Forest Ecology and Management* 348:23-30.

Fredricks D. 2008. The Arguellos and Rancho de Las Pulgas. *The Daily Journal*, August 4, 2008. Available: https://web.archive.org/web/20110524045655/http://www.smdailyjournal.com/article_preview.php?id%3D95950.

Garbelotto PI.2017. Final Progress Report on Work Conducted Between 2009 and 2014 for the SFPUC. Prepared for San Francisco Public Utilities Commission.

George MR, MF Alonso. 2008. Oak Woodland Vegetation Dynamics: A State and Transition Approach. In *Proceedings of the Sixth California Oak Symposium: Today's Challenges, Tomorrow's Opportunities,* eds. Adina Merenlender, Douglas McCreary, Kathryn L. Purcell. Gen. Tech. Rep. PSW-GTR-217. Albany, CA: U.S. Department of Agriculture, Forest Service, Pacific Southwest Research Station.

Gifford-Gonzalez D, CM Boone, RE Reid. 2013. The Fauna from Quiroste: Insights into Indigenous Foodways, Culture, and Land Modification. *California Archaeology* 5(2):291-317.

Girard C. 1858. Fishes. In *Explorations and Surveys for a Railroad Route from the Mississippi River to the Pacific Ocean, v. 10.* War Department. Washington: Beverley Tucker, Printer.

Giusti, GA, DD McCreary, RB Standiford. 2005. *A Planner's Guide for Oak Woodlands.* Vol. 3491. UCANR Publications.

Greene, EL. 1887. Studies in the Botany of California and Parts Adjacent. *Bulletin of the California Academy of Sciences* 2(6):41-448.

Greenlee JM, JH Langenheim. 1990. Historic Fire Regimes and Their Relation to Vegetation Patterns in the Monterey Bay Area of California. *American Midland Naturalist* 124(2):239–53.

Griffin JR. 1978. Maritime Chaparral and Endemic Shrubs of the Monterey Bay Region, California. *Madroño* 25(2):65-81.

Griffith RS. 1992. *Sequoia sempervirens.* Fire Effects Information System. U.S. Department of Agriculture, Forest Service, Rocky Mountain Research Station, Fire Sciences Laboratory.

Hahm WJ, DN Dralle, DM Rempe, AB Bryk, SE Thompson, TE Dawson, WE Dietrich. 2019. Low Subsurface Water Storage Capacity Relative to Annual Rainfall Decouples Mediterranean Plant Productivity and Water Use From Rainfall Variability. *Geophysical Research Letters* 46(12):6544-6553.

Hall NT, RH Wright, KB Clahan. 1999. Paleoseismic Studies of the San Francisco Peninsula Segment of the San Andreas Fault Zone near Woodside, California. *Journal of Geophysical Research: Solid Earth* 104(B10):23215–36.

Hanson WD. 2005. *San Francisco Water and Power, A History of the Municipal Water Department and Hetch Hetchy System.* San Francisco: San Francisco Water and Power, City and County of San Francisco.

Harrison SP, JH Viers. 2007. Serpentine Grasslands. In *California Grasslands: Ecology and Management,* eds. Mark Stromberg, Jeffrey Corbin, Carla D'Antonio. Berkeley, Calif: University of California Press.

Hatch DA, JW Bartolome, JS Fehmi, DS Hillyard. 1999. Effects of Burning and Grazing on a Coastal California Grassland. *Restoration Ecology* 7(4):376–81.

Hayes GF, KD Holl. 2003. Cattle Grazing Impacts on Annual Forbs and Vegetation Composition of Mesic Grasslands in California. *Conservation Biology* 17(6):1694–1702.

Hayward GD, TT Veblen, LH Suring, B Davis. 2012. Challenges in the Application of Historical Range of Variation to Conservation and Land Management. In *Historical Environmental Variation in Conservation and Natural Resource Management,* eds. John A. Wiens, Gregory D. Hayward, Hugh D. Safford, Catherine M. Giffen. Chichester, UK: John Wiley & Sons, Ltd.

Hermann RK, DP Lavender. 1990. Pseudotsuga Menziesii (Mirb.) Franco. In *Silvics of North America, Vol. 1*, eds. Russell M. Burns, Barbara H. Honkala. Washington, DC: U.S. Department of Agriculture, Forest Service.

Higgs E, DA Falk, A Guerrini, M Hall, J Harris, RJ Hobbs, ST Jackson, JM Rhemtulla, W Throop. 2014. The Changing Role of History in Restoration Ecology. *Frontiers in Ecology and the Environment* 12(9):499-506.

Hoag BL. 1973. The Deathshead Tree. *La Peninsula*, February 1973. San Mateo County Historical Association Museum. *Courtesy of the San Mateo County History Museum.*

Hobbs RJ, E Higgs, CM Hall, P Bridgewater, FS Chapin III, EC Ellis, JJ Ewel, LM Hallett, J Harris, KB Hulvey, ST Jackson, PL Kennedy, C Kueffer, L Lach, TC Lantz, AE Lugo, J Mascaro, SD Murphy, CR Nelson, MP Perring, DM Richardson, TR Seastedt, RJ Standish, BM Starzomski, KN Suding, PM Tognetti, L Yakob, L Yung. 2014. Managing the Whole Landscape: Historical, Hybrid, and Novel Ecosystems. *Frontiers in Ecology and the Environment* 12(10):557-564.

Hobbs RJ, HA Mooney. 1986. Community Changes Following Shrub Invasion of Grassland. *Oecologia* 70(4):508–13.

Hoffman CF. 1867. *Map of the Region Adjacent to the Bay of San Francisco.* State Geological Survey of California. *Courtesy of the David Rumsey Map Collection.*

Holland RF. 1986. *Preliminary Descriptions of the Terrestrial Natural Communities of California.* Sacramento: California Department of Fish and Game, Natural Heritage Division.

Holland VL, DJ Keil. 1995. *California Vegetation.* Dubuque, Iowa: Kendall/Hunt Pub. Co.

Hoover MB, DE Kyle, eds. 2002. *Historic Spots in California.* 5th ed. Stanford, CA: Stanford University Press.

Hopkinson P, M Hammond, JW Bartolome, L Macaulay. 2020. Using Consecutive Prescribed Fires to Reduce Shrub Encroachment in Grassland by Increasing Shrub Mortality. *Restoration Ecology* 28(4):850-858.

Horton TR, TD Bruns, VT Parker. 1999. Ectomycorrhizal fungi associated with Arctostaphylos contribute to Pseudotsuga menziesii establishment. *Canadian Journal of Botany*, 77:93-102.

Hsu W, A Remar, E Williams, Adam McClure, S Kannan, R Steers, C Schmidt, JW Skiles. 2012. *The Changing California Coast: Relationships between Climatic Variables and Coastal Vegetation Succession.* American Society for Photogrammetry and Remote Sensing Annual Conference 2012.

Hynding A. 1982. *From Frontier to Suburb: The History of San Mateo and the Peninsula.* First. Belmont, Calif.: Star Publishing.

Jackson RD, JW Bartolome. 2007. Grazing Ecology of California Grasslands. In *California Grasslands: Ecology and Management*, eds. Mark Stromberg, Jeffrey Corbin, Carla D'Antonio. Berkeley, Calif: University of California Press.

Jackson ST, RJ Hobbs. 2009. Ecological Restoration in the Light of Ecological History. *Science* 325(5940):567–69.

Johnson BE, JH Cushman. 2007. Influence of a Large Herbivore Reintroduction on Plant Invasions and Community Composition in a California Grassland. *Conservation Biology* 21(2):515–26.

Johnson BE, RR Everett, KG Lightfoot, CJ Stiplen. 2010. *Exploring the Traditional Use of Fire in the Coastal Mountains of Central California.* Final Report to the Joint Fire Science Program JFSP Project No. 10-1-09-3. Paicines, CA: National Park Service.

Johnson RF. 1955. *The Valley of the Lakes.* San Mateo, CA: College of San Mateo. *Courtesy of the San Mateo County History Museum.*

Jones WC. 1853. *Argument in the Case of Arguello and Others, Claiming Rancho de las Pulgas. Made Before the Board, July 18, 1853 by William Carey Jones, of Counsel for Claimants. George Kerr & Co.* United States Commission on Land Claims in California. *Courtesy of The Bancroft Library, UC Berkeley.*

Kauffman E. 2003. Climate and Topography. In *Atlas of the Biodiversity of California.* Sacramento: Calif. Dept. of Fish and Game.

Keeley JE. 1991. Seed Germination and Life History Syndromes in the California Chaparral. *The Botanical Review* 57(2):81–116.

Keeley JE. 2002. Native American Impacts on Fire Regimes of the California Coastal Ranges. *Journal of Biogeography* 29(3):303–20.

Keeley JE. 2005. Fire History of the San Francisco East Bay Region and Implications for Landscape Patterns. *International Journal of Wildland Fire* 14(3):285–96.

Keeley JE, FW Davis. 2007. Chaparral. In *Terrestrial Vegetation of California,* eds. Michael Barbour, Todd Keeler-Wolf, Allan A. Schoenherr. Berkeley: University of California Press. Berkeley: University of California Press.

Kelly M, B Allen-Diaz, N Kobzina. 2005. Digitization of a Historic Dataset: the Wieslander California Vegetation Type Mapping Project." *Madroño* 52(3):191-201.

Kennedy PG, WP Sousa. 2006. Forest Encroachment into a Californian Grassland: Examining the Simultaneous Effects of Facilitation and Competition on Tree Seedling Recruitment. *Oecologia* 148(3):464–74.

Kidder A. 2015. *On Baccharis pilularis DC. (coyote brush, Asteraceae) Water Relations during Succession into Coastal Grasslands in a Changing Climate.* Dissertation. University of California, Berkeley.

Killam D, A Bui, S LaDochy, P Ramirez, J Willis, W Patzert. 2014. California Getting Wetter to the North, Drier to the South: Natural Variability or Climate Change? *Climate* 2(3):168–80.

Kilsby JS. 1883. *Spring Valley Water Works (Plaintiff and Appellant) vs. San Mateo Water Works et al. (Defendants and Respondents).* Supreme Court of California.

Kneese GA. 1927. *Official Map of San Mateo County, California.* County Surveyor. *Courtesy of San Mateo County Planning Department.*

Krippendorff K. 2004. *Content Analysis: An Introduction to its Methodology.* Thousand Oaks: Sage Publications.

Landres PB, P Morgan, FJ Swanson. 1999. Overview of the Use of Natural Variability Concepts in Managing Ecological Systems. *Ecological Applications* 9(4):1179–88.

Langenheim JH, JM Greenlee, B Benson, P Ritter. 1983. *Vegetation, Fire History, and Fire Potential of Big Basin Redwoods State Park, California.* California Department of Parks and Recreation.

Lanman CW, K Lundquist, H Perryman, JE Asarian, B Dolman, RB Lanman, MM Pollock. 2013. The Historical Range of Beaver (*Castor canadensis*) in Coastal California: an Updated Review of the Evidence. *California Fish and Game* 99(4):193-221.

Lawrence WB. 1917. "WBL to S. P. Eastman." February 10. Box MB 061. *Courtesy of the San Francisco Public Utilities Commission Archives.*

Lawrence WB. 1922a. Crystal Springs, Past and Present. *San Francisco Water* 1(1):11-14. San Francisco: Spring Valley Water Company. *Courtesy of San Francisco Public Library.*

Lawrence WB. 1922b. Pilarcitos. *San Francisco Water* 1(3):8-10. San Francisco: Spring Valley Water Company. *Courtesy of San Francisco Public Library.*

Lawrence WB. 1923. Predatory Animal Control. *San Francisco Water* 2(3):14-15. San Francisco: Spring Valley Water Company. *Courtesy of San Francisco Public Library.*

Lawrence WB. 1925. Fire Prevention on Watersheds. *San Francisco Water* 4(2):14-16. San Francisco: Spring Valley Water Company. *Courtesy of San Francisco Public Library.*

Lawson AC. 1895. *Sketch of the Geology of the San Francisco Peninsula. Extract from the fifteenth annual report of the Survey, 1893-94.* Washington, DC: U.S. Geological Survey.

Lawson AC. 1914. *Geologic Atlas of the United States, San Francisco Folio: Tamalpais, San Francisco, Concord, San Mateo, and Haywards [sic] Quadrangle, California.* U.S. Geologic Survey. San Francisco Folio No. 193.

Lazzeri-Aerts R, W Russell. 2014. Survival and Recovery Following Wildfire in the Southern Range of the Coast Redwood Forest. *Fire Ecology* 10(1):43–55.

Leidy RA, G Becker, BN Harvey. 2005a. Historical Status of Coho Salmon in Streams of the Urbanized San Francisco Estuary, California. *California Fish and Game* 91(4):219–54.

Leidy RA, GS Becker, BN Harvey. 2005b. *Historical Distribution and Current Status of Steelhead/Rainbow Trout (Oncorhynchus Mykiss) in Streams of the San Francisco Estuary, California.* Oakland, CA: Center for Ecosystem Management and Restoration.

Leventhal A, L Field, H Alvarez, R Cambra. 1994. The Ohlone Back from Extinction. In *The Ohlone Past and Present: Native Americans of the San Francisco Bay Region*, ed. Lowell John Bean. Ballena Press Anthropological Papers, no. 42. Menlo Park, CA: Ballena Press.

Lewis WJ. 1859a. *A Transcript of the Field Notes of the Survey of the Exterior Boundaries of T. 5 S. R. 5 W. Mt. Diablo Meridian, in the State of California.* U.S. Department of the Interior, Bureau of Land Management Rectangular Survey, California. Book R192. *Courtesy of Bureau of Land Management, Sacramento, California.*

Lewis WJ. 1859b. *Field Notes of the Final Survey of the Rancho San Pedro, Francisco Sanchez.* San Francisco: U.S. Survey-General's Office. Book G12: *Courtesy of Bureau of Land Management, Sacramento, California.*

Lewis, WJ. 1860. *A Transcript of the Field Notes of the Survey of the First Standard South, Mount Diablo Meridian, State of California.* U.S. Department of the Interior, Bureau of Land Management Rectangular Survey, California. Book R192. *Courtesy of Bureau of Land Management, Sacramento, California.*

Lightfoot KG, O Parrish. 2009. *California Indians and Their Environment: An Introduction.* California Natural History Guides. Berkeley: University of California Press.

Lightfoot KG, RQ Cuthrell, CJ Striplen, MG Hylkema. 2013a. Rethinking the Study of Landscape Management Practices among Hunter-Gatherers in North America. *American Antiquity* 78(2):285–301.

Lightfoot KG, RQ Cuthrell, CM Boone, R Byrne, AS Chavez, L Collins, A Cowart, RR Evett, PVA Fine, D Gifford-Gonzalez, MG Hylkema, V Lopez, TM Misiewicz, REB Reid. 2013b. Anthropogenic Burning on the Central California Coast in Late Holocene and Early Historical Times: Findings, Implications, and Future Directions. *California Archaeology* 5(2):371–90.

Lightfoot KG. 2006. *Indians, Missionaries, and Merchants: The Legacy of Colonial Encounters on the California Frontiers.* Berkeley; London: University of California Press.

Maddox BM. 1890. *Spring Valley Water Work (Plaintiff and Respondent)s vs. Mary A. Drinkhouse (Defendant and Appellant).* Supreme Court of California.

Martinez M. 1852. *Cañada de Raimundo [sic], María Luisa Greer and Manuel Coppinger, claimants. BANC MSS Land Case Files 75 ND. p. 14. U.S. District Court, Northern District. Courtesy of The Bancroft Library, UC Berkeley.*

Matthewson RC. 1858. *Field Notes of the Unofficial Survey of the Rancho Buri Buri, Jose de La Cruz Sanchez, et al., Confirmee.* San Francisco: U.S. Survey-General's Office. Book G7. *Courtesy of Bureau of Land Management, Sacramento, California.*

Matthewson RC. 1861. *Field Notes of the Survey of the Rancho Feliz, D. Feliz, Confirmee.* San Francisco: U.S. Survey-General's Office, Bureau of Land Management. Book G7. *Courtesy of Bureau of Land Management, Sacramento, California.*

McBride JR, HF Heady. 1968. *Invasion of Grassland by Baccharis Pilularis DC.* Berkeley, CA: University of California, Berkeley.

McBride JR. 1974. Plant Succession in the Berkeley Hills, California. *Madroño* 22(7):317–29.

McLaughlin BC, R Blakey, AP Weitz, X Feng, BJ Brown, DD Ackerly, TE Dawson, SE Thompson. 2020. Weather Underground: Subsurface Hydrologic Processes Mediate Tree Vulnerability to Extreme Climatic Drought. *Global Change Biology* 26(5):3091-3107.

McLean DD. 1930. *Game Bulletin No. 2: The Quail of California.* Sacramento: State of California Department of Natural Resources, Division of Fish and Game.

Mendell GH. 1885. *Report of Lieutenant-Colonel George H. Mendell, Corps of Engineers, Bvt. Col., U.S.A., Officer in Charge for the Fiscal Year Ending June 30, 1885, with Other Documents Relating to the Works.* San Francisco: U.S. Army Corps of Engineers. *Courtesy of the HathiTrust Digital Library.*

Mensing SA. 1998. 560 Years of Vegetation Change in the Region of Santa Barbara, California. *Madroño* 45(1):1–11.

Mensing SA. 2006. The History of Oak Woodlands in California, Part II: The Native American and Historic Period. *The California Geographer* 46:1-31.

Milliken R, LH Shoup, BR Ortiz. 2009. *Ohlone/Costanoan Indians of the San Francisco Peninsula and Their Neighbors, Yesterday and Today.* San Francisco: National Park Service, Golden Gate National Recreation Area.

Milliken R. 2009. *Ethnohistory and Ethnogeography of the Coast Miwok and Their Neighbors, 1783-1840.* San Francisco: National Park Service, Golden Gate National Recreation Area, Cultural Resources and Museum Management Division.

Milliken R, J Johnson, D Earle, N Smith, P Mikkelsen, P Brandy, J King. 2010. *Introduction to the Contact-Period Native California Community Distribution Model.* [Digital map of ethnographic regions.] Far Western Anthropological Research Group, Inc.

Minnich RA. 2008. *California's Fading Wildflowers: Lost Legacy and Biological Invasions.* Berkeley: University of California Press.

Mooney HA, EL Dunn. 1970. Convergent Evolution of Mediterranean-Climate Evergreen Sclerophyll Shrubs. *Evolution* 24(2):292-303.

Moore and DePue. 1878. *The Illustrated History of San Mateo County.* 1st ed. Woodside, CA: Gilbert Richards Publications.

MRLCC (Multi-Resolution Land Characteristics Consortium). 2001. *National Land Cover Database.* Multi-Resolution Land Characteristics Consortium (MRLCC).

Neuman. 1874a. *Gustave Touchard to Spring Valley Water Works. Survey Parcel 37, Map 4.* Drawer "Neuman's Field Notes." *Courtesy of the San Mateo County Public Works and Recorder.*

Neuman. 1874b. *Isaac Freidlander to Spring Valley Water Works. Survey Parcel 38, Map 3.* Drawer "Neuman's Field Notes." *Courtesy of the San Mateo County Public Works and Recorder.*

Neuman. 1874c. *James Byrnes to Spring Valley Water Works. Survey Parcel 36, Map 3.* Drawer "Neuman's Field Notes." *Courtesy of the San Mateo County Public Works and Recorder.*

Neuman. 1887. *Robert Sherwood and Eliza N. Sherwood to Spring Valley Water Works.* Survey Parcel 92, Map 3. Drawer "Neuman's Field Notes." *Courtesy of the San Mateo County Public Works and Recorder.*

Newman EA. 2019. Disturbance Ecology in the Anthropocene. *Frontiers in Ecology and Evolution* 7(May):147.

Newman EA, JB Potts, MW Tingley, C Vaughn, SL Stephens. 2018. Chaparral Bird Community Responses to Prescribed Fire and Shrub Removal in Three Management Seasons. *Journal of Applied Ecology* 55(4):1615–25.

Niederer C, C Schwind. 2020. *San Mateo Thornmint (*Acanthomintha duttonii*) Restoration Project.* Creekside Center for Earth Observation. Prepared for California Department of Fish and Wildlife.

Nomad Ecology. 2009. [Inventory of Non-Indigenous Plant Species within the Peninsula Watershed, GIS layer.].

Nomad Ecology. 2020. *Botanical Resources Survey Report Peninsula Watershed San Mateo County, California.*

Oberlander GT. 1953. *The Taxonomy and Ecology of the Flora of the San Francisco Watershed Reserve.* Dissertation. Stanford University.

Oberlander GT. 1956. Summer Fog Precipitation on the San Francisco Peninsula. *Ecology* 37(4):851–52.

Odion DC, FW Davis. 2000. Fire, Soil Heating, and the Formation of Vegetation Patterns in Chaparral. *Ecological Monographs* 70(1):149–69.

Palóu F, FM Stanger, AK Brown. 1969. The Journals of Rivera and Palóu, November 26 to December 5. In *Who Discovered the Golden Gate? The Explorers' Own Accouts*, eds. Frank M. Stanger, Alan K. Brown. San Mateo, CA: San Mateo County Historical Association.

Palóu F, HE Bolton. 1966. *Historical Memoirs of New California, Vol. 2*. New York: Russel & Russel.

Panich L, J Holson, A Vanderslice. 2009. *Historic Context and Archaeological/Architectural Survey Report for the Habitat Reserve Program, Alameda, San Mateo, Santa Clara, and Tuolumne Counties, California*. San Francisco: Environmental Science Associates.

Parker VT. 1987. Effect of Wet-Season Management Burns on Chaparral Regeneration: Implications for Rare Species. In *Conservation and Management of Rare and Endangered Plants*, eds. Thomas S. Elias, Jim Nelson. Calif. Native Plant Society, Sacramento, CA

Parker VT. 1990. Problems Encountered While Mimicking Nature in Vegetation Management: an Example from a Fire-Prone Vegetation. In *Ecosystem Management: Rare species and significant habitats. Proc. 15th Ann. Natural Areas Conf*, eds. Richard S. Mitchell, Charles J. Sheviak, Donald J. Leopold. *New York State Museum Bulletin* 471:231-234.

Parker VT. 2020. Chaparral of California. In *Encyclopedia of the World's Biomes*, eds. Michael Goldstein, Dominick DellaSala. Elsevier.

Pausas JG, JE Keeley. 2014. Evolutionary Ecology of Resprouting and Seeding in Fire-Prone Ecosystems. *New Phytologist* 204(1):55–65.

Pavlik BM, PC Muick, SG Johnson, M Popper. 2002. *Oaks of California*. Oakland, CA: Cachuma Press, Inc.

Peninou EP. 2000. *A History of the San Francisco Viticultural District Comprising the Counties of Alameda, Monterey, San Benito, San Francisco, San Mateo, Santa Clara, and Santa Cruz with Grape Acreage Statistics and Directories of Grape Growers*. The Wine Librarians Association.

Petaluma Courier. 1889. "Some Facts." September 18. *Courtesy of Newspapers.com*.

Peterson G, CR Allen, CS Holling. 1998. Ecological Resilience, Biodiversity, and Scale. *Ecosystems* 1(1):6-18.

Phytosphere Research. 2013. *SFPUC Sudden Oak Death Management Project Update*. Prepared for San Francisco Public Utilities Commission.

Pictometry, Inc. 2015. *Aerial Orthophoto Mosaic of the SFPUC Peninsula Watershed*. Provided to SFPUC Natural Resources Division.

Plant RE, MP Vayssieres, SE Greco, MR George, TE Adams. 1999. A Qualitative Spatial Model of Hardwood Rangeland State-and-Transition Dynamics. *Journal of Range Management* 52(1):51.

Poinsot W, RP Wong. 2008. *Operational Strategy for the Fire Management Plan: Golden Gate National Recreation Area*. National Park Service.

Postel MP. 1994. *San Mateo: A Centennial History*. San Francisco: Scottwall Asspciates. *Courtesy of the San Mateo Public Library*.

Postel MP. 2010. *Historic Resource Study for Golden Gate National Recreation Area in San*

Mateo County: Sweeney Ridge, Rancho Corral de Tierra (and the Montara Lighthouse Station), Mori Point, Phleger Estate, Milagra Ridge. San Francisco: National Park Service, Golden Gate National Recreation Area.

Potter AC. 1923. The Jepson Laurel Forty Years Ago. *San Francisco Water* 2(3):5-6. San Francisco: Spring Valley Water Company. *Courtesy of San Francisco Public Library.*

Radeloff VC, DJ Mladenoff, HS He, et al. 1999. Forest Landscape Change in the Northwestern Wisconsin Pine Barrens from Pre-European Settlement to the Present. *Canadian Journal of Forest Research* 29:1649–1659.

Raphael MG, GA Falxa, AE Burger. 2018. Chapter 5: Marbled Murrelet. In *Synthesis of Science to Inform Land Management Within the Northwest Forest Plan Area, Volume 1,* eds. Thomas A. Spies, Peter A. Stine, Rebecca Gravenmier, Jonathan W. Long, and Matthew J. Reilly. U.S. Forest Service General Technical Report PNW-GTR-966.

Reaser JK, SW Burgiel, J Kirkey, KA Brantley, SD Veatch, J Burgos-Rodríguez. 2020. The Early Detection of and Rapid Response (EDRR) to Invasive Species: a Conceptual Framework and Federal Capacities Assessment. *Biological Invasions* 22:1-19.

Reilly MF. 1902. *D-82, Map of the Property of W. F. Fifield, Esq. Situated in Township 4 South, Range 5 West, Mt. Diablo Mern.* San Mateo County. *Courtesy of San Francisco Public Utilities Commission.*

Rejmánek M, MJ Pitcairn. 2002. When is Eradication of Pest Plants a Realistic Goal? In *Turning the Tide: the Eradication of Invasive Species,* eds. C.R. Veitch, M.N. Clout. Gland, Switzerland: IUCN SSC Invasive Species Specialist Group.

Richards G. 1973. *Crossroads: People and Events of the Redwoods of San Mateo County.* Woodside, CA: Gilbert Richards Publications. *Courtesy of the California State Railroad Museum Library.*

Rivera F, FM Stanger, AK Brown. 1969. The Journals of Rivera and Palou, November 26 to December 5." In *Who Discovered the Golden Gate? The Explorers' Own Accounts,* eds. Frank M. Stanger, Alan K. Brown. San Mateo, CA: San Mateo County Historical Association.

Rodgers AF. 1853. *T-433: San Francisco Bay between Point San Matheo and Guano Island.* U.S. Coast and Geodetic Survey. *Courtesy of National Oceanic and Atmospheric Administration.*

Rodgers AF, F Westdahl. 1898. *T-2310: Resurvey of San Francisco Bay, California: San Mateo to San Carlos.* U.S. Coast and Geodetic Survey. *Courtesy of National Oceanic and Atmospheric Administration.*

Roeding FW. 1921. *Planting Plan for Crystal Springs Country Club, 1921. Courtesy of the San Francisco Public Utilities Commission Archives.*

Roeding FW. 1931. "F.W. Roeding to Dr. E. Meinecke [letter]." February 26. *Courtesy of the San Francisco Public Utilities Commission Archives.*

Rundel PW. 2007. "Sage Scrub." In *Terrestrial Vegetation of California,* eds. Michael Barbour, Todd Keeler-Wolf, Allan A. Schoenherr. Berkeley: University of California Press. Berkeley: University of California Press.

Russell WH, JR McBride. 2003. Landscape Scale Vegetation-Type Conversion and Fire Hazard in the San Francisco Bay Area Open Spaces. *Landscape and Urban Planning* 64(4):201–8.

Russo TA, AT Fisher, DM Winslow. 2013. Regional and Local Increases in Storm Intensity in the San Francisco Bay Area, USA, between 1890 and 2010. *Journal of Geophysical Research: Atmospheres* 118(8):3392–3401.

Sacramento Daily Union. 1861. "Trout in San Mateo Creek." February 27. *Courtesy of California Digital Newspaper Collection.*

Safford HD. 1995. Woody Vegetation and Succession in the Garin Woods, Hayward Hills, Alameda County, California. *Madroño* 42(4):470–89.

Safford, HD, EC Underwood, NA Molinari. 2018. Managing Chaparral Resources on Public Lands. In *Valuing Chaparral: Ecological, Socio-Economic, and Management Perspectives*, eds. Emma Underwood, Hugh Safford, Nicole Molinari, Jon Keeley. Cham: Springer International Publishing.

Safford HD, GD Hayward, NE Heller, JA Wiens. 2012a. Historical Ecology, Climate Change, and Resource Management: Can the Past Still Inform the Future? In *Historical Environmental Variation in Conservation and Natural Resource Management*, eds. John A. Wiens, Gregory D. Hayward, Hugh D. Safford, Catherine M. Giffen. Chichester, UK: John Wiley & Sons, Ltd.

Safford, HD, JA Wiens, GD Hayward. 2012b. The Growing Importance of the Past in Managing Ecosystems of the Future. In *Historical Environmental Variation in Conservation and Natural Resource Management*, eds. John A. Wiens, Gregory D. Hayward, Hugh D. Safford, Catherine M. Giffen. Chichester, UK: John Wiley & Sons, Ltd.

Safford HD, JED Miller. 2020. An Updated Database of Serpentine Endemism in the California Flora. *Madroño* 67(2):85-104.

San Francisco Board of Supervisors. 1875. Water Supplies. In *San Francisco Municipal Reports for the Fiscal Year 1874-5, Ending June 30, 1875, 613-723*. San Francisco: Spaulding & Barto, Printers.

San Francisco Call. 1891. "Fish and Game: A Rush of Anglers to Lake Pilarcitos." May 6. *Courtesy of California Digital Newspaper Collection.*

San Francisco Call. 1895. "Will Call on the Grand Jury: Mayor Sutro Says the Spring Valley Company Should Be Prosecuted." March 15. *Courtesy of California Digital Newspaper Collection.*

San Francisco Call. 1897. "The Angler: Why the Trout of San Andreas Lake Are Small This Year," May 29. *Courtesy of California Digital Newspaper Collection.*

San Francisco Call. 1912a. "Sea Lions Scare Society Matrons." July 7. *Courtesy of California Digital Newspaper Collection.*

San Francisco Call. 1912b. "Streams Are Stocked with Trout." July 7. *Courtesy of California Digital Newspaper Collection.*

San Francisco Call. 1913. "Gale Blows in Wake of Drenching Rains." December 30. *Courtesy of California Digital Newspaper Collection.*

San Francisco Chronicle. 1874. "A Great Project: Plans of the Spring Valley Water Works." September 5. *Courtesy of ProQuest Historical Newspapers.*

San Francisco Chronicle. 1882. "Fishing." June 17. *Courtesy of ProQuest Historical Newspapers.*

San Francisco Chronicle. 1889a. "A Fiery Scourge: San Mateo County All Ablaze." September 17. *Courtesy of ProQuest Historical Newspapers.*

San Francisco Chronicle. 1889b. "Swept by Fire: Blazing Forests in San Mateo, Thousands of Acres Laid Waste" September 29. *Courtesy of ProQuest Historical Newspapers.*

San Francisco Chronicle. 1897. "Spring Valley the Loser: San Mateo County Case Decided against the Water Company." April 15. *Courtesy of ProQuest Historical Newspapers.*

San Francisco Daily Herald. 1853. *Decision of the U.S. Land Commission in the Las Pulgas Case, October 16, 1853, sec. Supplement. BANC MSS 92/817 c. Courtesy of The Bancroft Library, UC Berkeley.*

San Francisco Planning Department. 2008. *Final Program Environmental Impact Report Volume 3 of 8 for the San Francisco Public Utilities Commission's Water System Improvement Program.* San Francisco Planning Department File No. 2005.0159E, State Clearinghouse No. 2005092026.

San Mateo County Historical Association. 1967. "La Peninsula." Vol. 14, No. 3. San Mateo, CA: San Mateo County Historical Association.

San Mateo Gazette. 1859. "Our County." April 16. *Courtesy of the San Mateo Public Library.*

Sanborn Map Co., Inc. 2017. [Digital orthoimagery of San Mateo, CA.] Colorado Springs, CO.

Sandel B, WK Cornwell, DD Ackerly. 2012. Mechanisms of Vegetation Change in Coastal California, with an Emphasis on the San Francisco Bay Area. In *Climate Change Impacts on California Vegetation: Physiology, Life History, and Ecosystem Change,* eds. William K. Cornwell, Stephanie A. Stuart, Aaron Ramirez, Christopher R. Dolanc, James H. Thorne, David D. Ackerly. Publication number: CEC-500-2012-023. California Energy Commission.

Santa Cruz Weekly Sentinel. 1871. "Around the Circle." June 17. *Courtesy of Newspapers.com.*

Sawyer J, T Keeler-Wolf, J Evens. 2009. *A Manual of California Vegetation, 2nd edition.* California Native Plant Society. Sacramento, CA.

Schirokauer D, T Keeler-Wolf, J Meinke, P van der Leeden. 2003. *Plant Community Classification and Mapping Project Final Report: Point Reyes National Seashore, Golden Gate National Recreation Area, San Francisco Water Department Watershed Lands, Mount Tamalpais, Tomales Bay, and Samuel P. Taylor State Parks.* San Francisco: National Park Service.

Schriver M, RL Sherriff, JM Varner, L Quinn-Davidson, Y Valachovic. 2018. Age and Stand Structure of Oak Woodlands along a Gradient of Conifer Encroachment in Northwestern California. *Ecosphere* 9(10):e02446.

Schussler H. 1867. *Spring Valley Water Works. Courtesy of San Mateo County Historical Association.*

Schussler H. 1875. Report of H. Schussler, Chief Engineer, Spring Valley Water Works. In *San Francisco Pamphlets: Reports of the President, Chief Engineer, and Secretary of the Spring Valley Water Works,* 2:2–24. Spring Valley Water Works.

Scott JM. 2008. "Sudden Oak Death Takes Toll on Crystal Springs Trees." *East Bay Times,* July 14, 2008.

Scowden TR. 1875. *City Water Supply Map of Spring Valley Reservoirs.* San Francisco: Geo H Baker Engraver & Lithographer. *Courtesy of The Bancroft Library, UC Berkeley.*

SFPUC (San Francisco Public Utilities Commission). 2020. *Water Enterprise Environmental Stewardship Policy Implementation Report, June 23, 2020.*

Simberloff D. 2003. Eradications: Preventing Invasions at the Outset. *Weed Science* 51(2):247-253.

Soranno PA, KS Cheruvelil, EG Bissell, MT Bremigan, JA Downing, CE Fergus, CT Filstrup, EN Henry, NR Lottig, EH Stanley, CA Stow, P Tan, T Wagner, KE Webster. 2014. Cross-scale Interactions: Quantifying Multiscaled Cause–Effect Relationships in Macrosystems. *Frontiers in Ecology and the Environment* 12(1):65–73.

Stahlheber KA, CM D'Antonio. 2013. Using Livestock to Manage Plant Composition: A Meta-Analysis of Grazing in California Mediterranean Grasslands. *Biological Conservation* 157(January):300–308.

Standish RJ, RJ Hobbs, MM Mayfield, BT Bestelmeyer, KN Suding, LL Battaglia, V Eviner, CV Hawkes, VM Temperton, VA Cramer, JA Harris, JL Funk, PA Thomas. 2014. Resilience in Ecology: Abstraction, Distraction, or Where the Action Is? *Biological Conservation* 177:43-51.

Stanger FM. 1938. A California Rancho under Three Flags: A History of Rancho Buri Buri in San Mateo County. *California Historical Society Quarterly* 17(3):245–59. *Courtesy of The Bancroft Library, UC Berkeley.*

Stanger FM. 1963. *South from San Francisco: San Mateo County, California, Its History and Heritage.* San Mateo, CA: San Mateo County Historical Association.

Stanger FM. 1967. *Sawmills in the Redwoods.* San Mateo, CA: San Mateo County Historical Association. *Courtesy of The Bancroft Library, UC Berkeley.*

Steers RJ, SL Fritzke, JJ Rogers, J Cartan, K Hacker. 2013. Invasive Pine Tree Effects on Northern Coastal Scrub Structure and Composition. *Invasive Plant Science and Management* 6:231-242.

Steinberg PD. 2002. *Quercus agrifolia.* Fire Effects Information System. U.S. Department of Agriculture, Forest Service, Rocky Mountain Research Station, Fire Sciences Laboratory.

Stephens SL, DL Fry. 2005. Fire History in Coast Redwood Stands in the Northeastern Santa Cruz Mountains, California. *Fire Ecology* 1(1):2–19.

Stephens SL, NG Sugihara. 2006. Fire Management and Policy since European Settlement. In *Fire in California's Ecosystems,* eds. Jan W. van Wagtendonk, Neil G. Sugihara, Scott L. Stephens, Andrea E. Thode, Kevin E. Shaffer, Jo Ann Fites-Kaufman, James K. Agee. Oakland, California: University of California Press.

Stephens TS. 1856a. *Field Notes of the Final Survey of the Rancho Cañada de Raymundo, Maria Luisa Greer and Manuela Coppinger, Confirmee.* San Francisco: U.S. Surveyor General's Office. Book G2. *Courtesy of Bureau of Land Management, Sacramento, California.*

Stephens TS. 1856b. *Field Notes of the Final Survey of the Rancho Las Pulgas, M. de la Soledad Ortega de Arguello, confirmee.* San Francisco: U.S. Surveyor General's Office. Book G3. *Courtesy of Bureau of Land Management, Sacramento, California.*

Sternitsky RF. 1937. A Race of *Euphydryas Editha* BDV. (Lepidoptera). *The Canadian Entomologist* 69(9):203-205.

Stevens TS [sic]. 1856. *Plat of the Pulgas Rancho Finally Confirmed to Maria de La Soledad Ortega de Arguello et Al.* U.S. Surveyor General. *Courtesy of Bureau of Land Management.*

Striplen CJ. 2014. *A Dendroecology-Based Fire History of Coast Redwoods (Sequoia Sempervirens) in Central Coastal California.* Dissertation. University of California, Berkeley.

Stromberg MR, P Kephart, V Yadon. 2001. Composition, Invasibility, and Diversity in Coastal California Grasslands. *California Botanical Society* 48(4):18.

Suding K, E Higgs, M Palmer, JB Callicott, CB Anderson, M Baker, JJ Gutrich, KL Hondula, BMH Larson, A Randall, JB Ruhl, KZS Schwartz. 2015. Committing to Ecological Restoration: Efforts around the Globe Need Legal and Policy Clarification. *Science* 348(6235):638–40.

Suding KN, KL Gross, GR Houseman. 2004. Alternative States and Positive Feedbacks in Restoration Ecology. *Trends in Ecology & Evolution* 19(1):46–53.

Suding KN, RJ Hobbs. 2009. Threshold Models in Restoration and Conservation: A Developing Framework. *Trends in Ecology & Evolution* 24(5):271–79.

SVWC vs. San Francisco. 1916. *Spring Valley Water Company, plaintiff, vs. City and County of San Francisco, et al., defendants: abstract of testimony [and oral arguments] taken before Honorable H.M. Wright.* Neal Publishing Co. 8 vols. U.S. Circuit Court, Ninth Judicial Circuit, Northern District of California, and 26 and 96 District Court of U.S., Northern District of California, second division.

SVWW (Spring Valley Water Works). 1875. Report of H. Schussler, Chief Engineer, Spring Valley Water Works. In *San Francisco Pamphlets: Reports of the President, Chief Engineer, and Secretary of the Spring Valley Water Works*, 2:2–24. Spring Valley Water Works.

Sweeley F. 1925. Fire!—Friend or Foe. *San Francisco Water* 4(2):11-14. San Francisco: Spring Valley Water Company. *Courtesy of San Francisco Public Library.*

Swetnam TW, CD Allen, JL Betancourt. 1999. Applied Historical Ecology: Using the Past to Manage for the Future. *Ecological Applications* 9(4):1189–1206.

Swiecki T. 2020. *Management of Phytophthora ramorum/Sudden Oak Death on SFPUC Lands: Identifying Priority Areas for California Bay Removal Treatments.* Phytosphere Research.

Swiecki, TJ, EA Bernhardt. 2006. *A Field Guide to Insects and Diseases of California Oaks.* Gen. Tech Rep. PSW-GTR-197. Albany, CA: Pacific Southwest Research Station, Forest Service, US Department of Agriculture.

Tappe DT. 1942. *Game Bulletin No. 3: The Status of Beavers in California.* Sacramento: State of California Department of Natural Resources, Division of Fish and Game.

Taylor B. 1862. *Prose Writings of Bayard Taylor: Eldorado.* New York: G. P. Putnam.

Teschemaker HF. 1857. "Deposition (No. 2) of H. F. Teschemaker Taken before Commissioner Harry I. Thornton, and Filed April 13th, 1853." In *A New Revelation: The Title to the San Mateo Rancho Explained*, ed. J.C. Maynard. San Francisco. *Courtesy of The Bancroft Library, UC Berkeley.*

Tewksbury JJ, DJ Levey, NM Haddad, S Sargent, JL Orrock, A Weldon, BJ Danielson, J Brinkerhoff, EI Damschen, P Townsend. 2002. Corridors Affect Plants, Animals, and Their Interactions in Fragmented Landscapes. *Proceedings of the National Academy of Sciencies* 99(20):12923-12926.

The Morning Call. 1893. "Rod and Creel: Angling in the Lakes—the Fish Parasite." June 18. *Courtesy of Chronicling America, Library of Congress.*

The Morning Call. 1894. "The Angler: The Truckee River Is Now Free of Sawdust—Bass and Trout Fishing." August 26. *Courtesy of Chronicling America, Library of Congress.*

Thomsen MA, CM D'Antonio, KB Suttle, WP Sousa. 2005. Ecological Resistance, Seed Density and Their Interactions Determine Patterns of Invasion in a California Coastal Grassland: Interaction of Resistance and Seed Density. *Ecology Letters* 9(2):160–70.

Tietje W, K Purcell, S Drill. 2005. Oak Woodlands as Wildlife Habitat. In *A Planner's Guide for Oak Woodlands*, eds. Gregory A. Giusti, Douglas D. McCreary, Richard B. Standiford. University of California Agriculture and Natural Resources Publication 3491.

Timbrook J, JR Johnson, DD Earle. 1982. Vegetation Burning by Chumash. *Journal of California and Great Basin Anthropology* 4(2):163–86.

TNC (The Nature Conservancy). n.d. *Omniscape Explorer.* Available: https://omniscape.codefornature.org/.

Torregrosa A, C Combs, J Peters. 2016. GOES-Derived Fog and Low Cloud Indices for Coastal North and Central California Ecological Analyses." *Earth and Space Science* 3:46– 67.

Tracy CC. 1852. *Field Notes of the Survey of Standard Parallel, Commencing at the Corner of Townships 4 and 5, South of Ranges 1 East and 1 West, and Running East and West from the Standard Meridian in the State of California.* U.S. Department of the Interior, Bureau of Land Management Rectangular Survey, California. Book R231. *Courtesy of Bureau of Land Management, Sacramento, California.*

Tracy CC. 1853a. *Field Notes of the Survey of Townships 1, 2, 3 and 4 South, Range 4, 5 and 6 West, North of the First Correction Line South and Embracing the Peninsula Between the Bay of San Francisco and the Pacific Ocean.* U.S. Department of the Interior, Bureau of Land Management Rectangular Survey, California. Book R254. *Courtesy of Bureau of Land Management, Sacramento, California.*

Tracy CC. 1853b. *Survey of Township South of First Standard South, and West of the Meridian.* U.S. Department of the Interior, Bureau of Land Management Rectangular Survey, California. Book R254. *Courtesy of Bureau of Land Management, Sacramento, California.*

Trimble SW, AC Mendel. 1995. The Cow as a Geomorphic Agent—a Critical Review. *Geomorphology* 13(1–4):233–53.

Turner MG. 2010. Disturbance and Landscape Dynamics in a Changing World. *Ecology* 91(10):2833–49.

Tyler CM, B Kuhn, FW Davis. 2006. Demography and Recruitment Limitations of Three Oak Species in California. *The Quarterly Review of Biology* 81(2):127–52.

Tyler CM, DC Odion, RM Callaway. 2007. Dynamics of Woody Species in the California Grassland. In *California Grasslands: Ecology and Management*, eds. Mark Stromberg, Jeffrey Corbin, Carla D'Antonio. Berkeley, CA: University of California Press.

Unknown. 1912. *Map of the Lands of E. A. Husing, Esq.* San Mateo, CA. *Courtesy of San Francisco Public Utilities Commission Archives.*

USCS (U.S. Coast Survey). 1869. *San Francisco Peninsula.* U.S. Coast Survey. *Courtesy of the David Rumsey Map Collection.*

USDA (United States Department of Agriculture). 2012. *1981-2010 Monthly Average Precipitation and Temperature.* United States Department of Agriculture. Accessed: October 25, 2017.

U.S Surveyor General's Office. 1864. *Plat of the Rancho Buri Buri, Finally Confirmed to Jose de la Cruz Sanchez et al.* U.S. Surveyor General. *Courtesy of Bureau of Land Management, Sacramento, California.*

Valachovic Y, R Standiford, L Quinn-Davidson, M Kelly, MD Potts, C Lee, M Eitzel, R Sherriff, M Varner, M Schriver, K McGown, G Giusti, S Smith, L Arguello, Y Everett. 2016. *Tools for a Changing Landscape: Understanding Disturbance and Vegetation Dynamics in Northern California Oak Woodlands.* University of California, Agriculture and Natural Resources.

Vallano DM, PC Selmants, ES Zavaleta. 2012. Simulated Nitrogen Deposition Enhances the Performance of an Exotic Grass Relative to Native Serpentine Grassland Competitors. *Plant Ecology* 213:1015-1026.

Van Dorn A. 1861. *Transit, Measurement, and Description Notes of Reclamation District No. 23 in San Mateo County. Courtesy of the San Mateo County Public Works and Recorder.*

Vanderlip B. 1980. *The Story of Crystal Springs Canyon. Courtesy of the San Mateo County Historical Association.*

Vasey MC, ME Loik, VT Parker. 2012. Influence of Summer Marine Fog and Low Cloud Stratus on Water Relations of Evergreen Woody Shrubs (Arctostaphylos: Ericaceae) in the Chaparral of Central California. *Oecologia* 170:325–337.

Vasey MC, VT Parker, KD Holl, ME Loik, S Hiatt. 2014. Maritime Climate Influence on Chaparral Composition and Diversity in the Coast Range of Central California. *Ecology and Evolution* 4(18):3662–3674.

Viader J. 1853. *Maria de la Soledad, Ortega de Arguello et als., Claimants for Las Pulgas, 4 Square Leagues, in San Mateo County, Granted December 10th, 1835, to Luis Arguello. BANC MSS Land Case Files 54 ND. p. 54. U.S. District Court, Northern District. Courtesy of the Bancroft Library, UC Berkeley*

Von Schmidt AW. 1864a. *A Transcript of the Field Notes of the Survey of the South and West Boundaries of T. 4 S. R. 5 W., Mount Diablo Meridian, State of California.* U.S. Department of the Interior, Bureau of Land Management Rectangular Survey, California. Book R192. *Courtesy of Bureau of Land Management, Sacramento, California.*

Von Schmidt AW. 1864b. *A Transcript of the Field Notes of the Survey of the Subdividion Lines in Township 4 & 5 South, Range Five West, of the Mount Diablo Meridian, in the State of California.* U.S. Department of the Interior, Bureau of Land Management Rectangular Survey, California. Book R192. *Courtesy of Bureau of Land Management, Sacramento, California.*

Von Schmidt AW. 1864c. *Field Notes of the Survey of the Subdivision lines in Township Five, South, of Range Five West, Mt. Diablo Meridian.* U.S. Department of the Interior, Bureau of Land Management Rectangular Survey, California. Book R192. *Courtesy of Bureau of Land Management, Sacramento, California.*

Wackenreuder V. 1855. *Wheeler's Topographic Map of San Francisco County (W-6-66).* San Francisco: Marriott & Wheeler. *Courtesy of San Francisco Department of Public Works.*

Westdahl F, AF Rodgers. 1898. *Descriptive Report. Topographical Sheet No. 2310.* U.S. Coast and Geodetic Survey.

Whipple AA, RM Grossinger, FW Davis. 2011. Shifting Baselines in a California Oak Savanna: Nineteenth Century Data to Inform Restoration Scenarios. *Restoration Ecology* 19(101):88-101.

Wilder TJ. 1925. *Peninsula Beauty-Spots. San Francisco Water* 4(4):2-4. San Francisco: Spring Valley Water Company. *Courtesy of San Francisco Public Library.*

Williams K, RJ Hobbs, SP Hamburg. 1987. Invasion of an Annual Grassland in Northern California by *Baccharis pilularis* ssp. *consanguinea. Oecologia* 72(3):461–65.

Williams K, RJ Hobbs. 1989. Control of Shrub Establishment by Springtime Soil Water Availability in an Annual Grassland. *Oecologia* 81(1):62–66.

Williamson J, S Harrison. 2002. Biotic and Abiotic Limits to the Spread of Exotic Revegetation Species. *Ecological Applications* 12(1):40–51.

Winzler & Kelly. 2010. *Mitigation and Monitoring Plan: Adobe Gulch Grassland Restoration Site, San Mateo County, California.* Prepared for San Francisco Public Utilities Commission. Eureka, CA.

Wolf KM, J DiTomaso. 2016. Management of Blue Gum Eucalyptus in California Requires Region-specific Consideration. *California Agriculture* 70(1):39-47.

Wrubel E, VT Parker. 2018. Local Patterns of Diversity in California Northern Coastal Scrub. *Ecology and Evolution* 8:7250-7260.

Wyatt RD. 1947. *Historic Names and Places in San Mateo County.* Centennial edition. San Mateo, CA: San Mateo County Historical Association.

Zavaleta ES. 2001. *Influences of Climate and Atmospheric Changes on Plant Diversity and Ecosystem Function in a California Grassland.* Dissertation. Stanford University.

Zavaleta ES. 2006. Shrub Establishment under Experimental Global Changes in a California Grassland. *Plant Ecology* 184(1):53–63.

Zavaleta ES, LS Kettley. 2006. Ecosystem Change along a Woody Invasion Chronosequence in a California Grassland. *Journal of Arid Environments* 66(2):290–306.

Appendices
Appendix A: Vegetation Classification

Crosswalk between simplified vegetation types used throughout the report (see page 22) and Wieslander VTM classification (Manual of California Vegetation 2009 vegetation alliances; Kelly et al. 2005). Note that some of the species assemblages/vegetation alliances documented in the Wieslander VTM mapping represent classification errors (e.g., *Cupressus abramsiana, Arctostaphylos tomentosa, A. canescens*).

Wieslander VTM Classification (MCV 2009 Alliance)	Simplified Vegetation Type
Cupressus abramsiana Woodland Special Stands	Conifer Forest
Pinus radiata Forest Alliance	Conifer Forest
Pseudotsuga menziesii Forest Alliance	Conifer Forest
Pseudotsuga menziesii-Lithocarpus densiflorus Forest Alliance	Conifer Forest
Sequoia sempervirens Forest Alliance	Conifer Forest
Eucalyptus sp. Provisional Alliance	Hardwood Forest
Lithocarpus densiflorus Forest Alliance	Hardwood Forest
Quercus agrifolia Woodland Alliance	Hardwood Forest
Quercus agrifolia-Quercus lobata Provisional Alliance	Hardwood Forest
Quercus wislizeni Woodland Alliance	Hardwood Forest
Avena barbata Semi-Natural Herbaceous Stands	Grassland
Bromus diandrus Semi-Natural Herbaceous Stands	Grassland
Bromus hordeaceus Provisional Semi-Natural Herbaceous Stands	Grassland
Grass sp. Provisional Alliance	Grassland
Vulpia myuros hirsuta Provisional Semi-Natural Herbaceous Alliance	Grassland
Adenostoma fasciculatum Shrubland Alliance	Shrubland
Adenostoma fasciculatum-Arctostaphylos tomentosa Provisional Alliance	Shrubland
Arctostaphylos canescens Provisional Shrubland Alliance	Shrubland
Arctostaphylos tomentosa Provisional Alliance	Shrubland
Artemisia californica Shrubland Alliance	Shrubland
Baccharis pilularis Shrubland Alliance	Shrubland
Ceanothus thyrsiflorus Shrubland Alliance	Shrubland
Heteromeles arbutifolia Shrubland Alliance	Shrubland
Myrica californica Provisional Alliance	Shrubland
Prunus ilicifolia Shrubland Alliance	Shrubland
Pteridium aquilinum var. *pubescens* Provisional Alliance	Shrubland
Rhamnus californica Shrubland Alliance	Shrubland
Aquatic Provisional Habitat	Water
Agriculture Provisional Developed Habitat	Developed/Disturbed
Urban Provisional Developed Habitat	Developed/Disturbed

Crosswalk between simplified vegetation types used throughout the report (see page 22) and NLCD (MRLCC 2001) classifications.

NLCD Land Cover Class	Simplified Vegetation Type
Evergreen Forest	Conifer Forest
Woody Wetlands	Hardwood Forest
Deciduous Forest	Hardwood Forest
Mixed Forest	Hardwood Forest
Herbaceous	Grassland
Emergent Herbaceous Wetlands	Grassland
Hay/Pasture	Grassland
Shrub/Scrub	Shrubland
Developed, Open Space	Developed
Developed, Low Intensity	Developed
Developed, Medium Intensity	Developed
Developed, High Intensity	Developed
Barren Land	Developed
Cultivated Crops	Developed
Open Water	Water

Crosswalk between simplified vegetation types used throughout the report (see page 22) and vegetation mapping units present in the contemporary vegetation mapping for the Peninsula Watershed (Schirokauer et al. 2003). The alliance-level classifications in the contemporary vegetation mapping are questionable or inaccurate in some cases, and thus were not used to analyze changes in vegetation patterns. Most notably, the grassland classes do not include or accurately represent the extensive native perennial grasslands within the Peninsula Watershed (S. Simono pers. comm.).

Shirokauer et al. (2003) Mapping Unit	Simplified Vegetation Type
Coast Redwood Alliance	Conifer Forest
Douglas-fir Alliance	Conifer Forest
Monterey Cypress Grove	Conifer Forest
California Bay Alliance	Hardwood Forest
California Buckeye Alliance	Hardwood Forest
Coast Live Oak Alliance	Hardwood Forest
Eucalyptus spp. Alliance	Hardwood Forest
Tanoak Alliance	Hardwood Forest
California Annual Grassland Weedy Alliance	Grassland
California Annual Grasslands with Native Component	Grassland
Introduced Perennial Grassland	Grassland
Blue/blossom Alliance	Shrubland
California Sagebrush Alliance	Shrubland
Chamise Alliance	Shrubland
Coffeeberry Alliance	Shrubland
Coyote Brush Alliance	Shrubland
Giant Chinquapin Alliance	Shrubland
Holly/leaf Cherry Alliance	Shrubland
Poison Oak Alliance	Shrubland
Arroyo Willow Alliance	Riparian/Wetland
Bulrush – Cattail – Spikerush Marsh Mapping Unit	Riparian/Wetland
Red Alder Alliance	Riparian/Wetland
Rush Alliance	Riparian/Wetland
Willow Mapping Unit	Riparian/Wetland
Beaches or Mudflats	Water
Water	Water
Active Pasture or Agriculture	Developed/Disturbed
Built-up Urban disturbance	Developed/Disturbed
Disturbed	Developed/Disturbed

Coastal scrub and chaparral groupings based on vegetation alliances in the Wieslander VTM (top) and modern vegetation mapping (bottom). The original species assemblages recorded in the Wieslander VTM mapping were assigned shrubland alliances, based on the Manual of California Vegetation classification system, as part of the digitization of the VTM data by researchers at UC Berkeley and UC Davis (Kelly et al. 2005). Species identifications and alliance level classifications are in some cases highly questionable (see page 162), and thus were not used as the basis for analysis. Alliances in the modern vegetation mapping are based on Schirokauer et al. 2003. Alliances in both the Wieslander VTM and modern vegetation mapping are categorized as coastal scrub or chaparral where possible, though in some cases it was not possible to make this determination.

Wieslander VTM Classification (MCV 2009 Alliance)	Coastal scrub or chaparral (or either)
Adenostoma fasciculatum Shrubland Alliance	Chaparral
Adenostoma fasciculatum-Arctostaphylos tomentosa Provisional Alliance	Chaparral
Arctostaphylos canescens Provisional Shrubland Alliance	Chaparral
Arctostaphylos tomentosa Provisional Alliance	Chaparral
Artemisia californica Shrubland Alliance	Coastal Scrub
Baccharis pilularis Shrubland Alliance	Coastal Scrub
Ceanothus thyrsiflorus Shrubland Alliance	Coastal Scrub/Chaparral
Heteromeles arbutifolia Shrubland Alliance	Coastal Scrub/Chaparral
Myrica californica Provisional Alliance	Coastal Scrub
Prunus ilicifolia Shrubland Alliance	Coastal Scrub/Chaparral
Pteridium aquilinum var. *pubescens* Provisional Alliance	Coastal Scrub
Rhamnus californica Shrubland Alliance	Coastal Scrub/Chaparral

Shirokauer et al. (2003) Mapping Unit	Coastal scrub or chaparral (or either)
Ceanothus thyrsiflorus Alliance	Coastal Scrub/Chaparral
Artemisia californica Alliance	Coastal Scrub
Adenostoma fasciculatum Alliance	Chaparral
Rhamnus californica Alliance	Coastal Scrub/Chaparral
Baccharis pilularis Alliance	Coastal Scrub
Chrysolepis chrysophylla Alliance	Chaparral
Prunus ilicifolia Alliance	Coastal Scrub/Chaparral
Toxicodendron diversilobum Alliance	Coastal Scrub/Chaparral

Appendix B: Court Case Parcel Locations

The following map shows the parcel numbers and locations for parcels referenced in the 1907-14 Spring Valley Water Company (SVWC) court case (SVWC vs. San Francisco 1916). Documentation from the court case, including witness testimony and oral arguments, is available through the San Francisco Public Library (http://linkencore.iii.com/iii/encore/record/C__Rb23592065?lang=eng). The testimony provided in these volumes includes descriptions of vegetation cover or other landscape characteristics for individual parcels, many of which are excerpted in this report. Maps showing the locations of individual parcels or "Watershed Properties," created in January 1914 by R.E. Childs and J.N. Hanlon, are available from the California State Archives (F3725:1192(a-t)); these maps were georeferenced in ArcGIS and parcel boundaries were digitized. A set of photographs showing many of the parcels is also available from the California State Archives (F3725:1180).

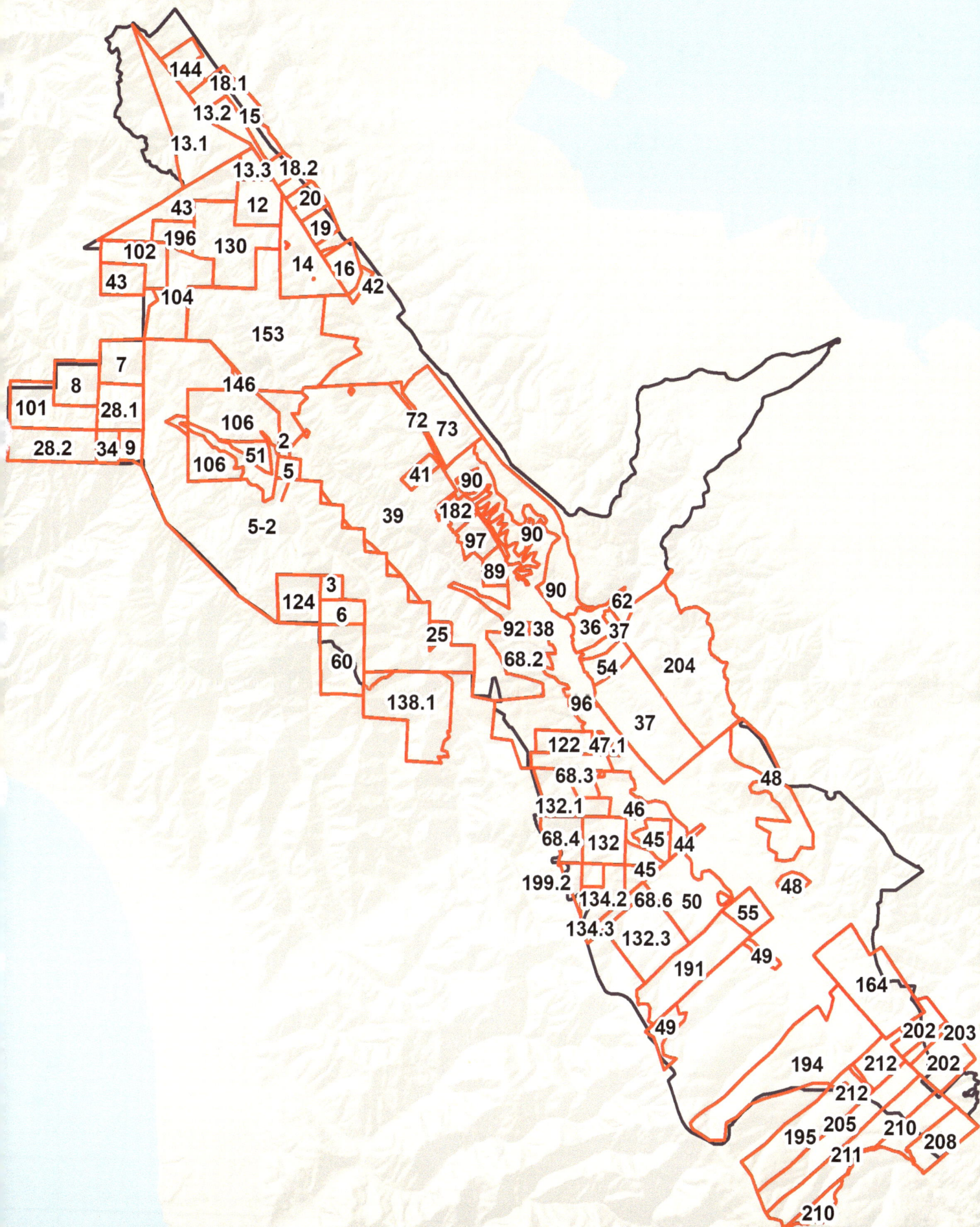

www.ingramcontent.com/pod-product-compliance
Lightning Source LLC
Chambersburg PA
CBHW061148030426
42335CB00003B/151